Classroom Companion: Economics

The Classroom Companion series in Economics includes undergraduate and graduate textbooks alike. It welcomes fundamental textbooks aimed at introducing students to the core concepts, empirical methods, theories and tools of the field, as well as advanced textbooks written for students at the Master and PhD level seeking a deeper understanding of economic theory, mathematical tools and quantitative methods.

Ismail Saglam

Mastering Game Theory

A Comprehensive Introduction to Strategic Decision Making

Springer

Ismail Saglam
TOBB University of Economics
and Technology
Ankara, Türkiye

ISSN 2662-2882 ISSN 2662-2890 (electronic)
Classroom Companion: Economics
ISBN 978-3-031-82093-9 ISBN 978-3-031-82094-6 (eBook)
https://doi.org/10.1007/978-3-031-82094-6

This Springer imprint is published by the registered company Springer Nature Switzerland AG
The registered company address is: Gewerbestrasse 11, 6330 Cham, Switzerland

To my Family

Preface

Game theory is a field that deals with interactive decision making. It is mostly studied by mathematicians, economists, and computer scientists. A game is a situation where a group of players choose their strategic actions simultaneously or sequentially to maximize their payoffs. What makes game theory interesting is that the payoff of each player may depend on the profile of actions chosen by all players. Such a payoff structure makes the optimal actions of the players interdependent and nontrivial to anticipate. The game theory deals with predicting the optimal actions of the players taking this interdependency into account.

The historical roots of game theory, and in particular, the mixed strategies and the ideas of maxmin and minmax, can be dated back to the contributions of the French mathematician Pierre Remond de Montmort in the early eighteenth century (for a discussion see [1]). While the formal inception of game theory as a field had to wait for at least two centuries after his contributions, strategic interactions were modeled in economics even in the nineteenth century, for example, by Cournot [2], Betrand [3], Edgeworth [4], and later by von Stackelberg [5]. A breakthrough in game theory, in its abstraction, came with the pioneering work of Emile Borel [6], analyzing symmetric two-person zero-sum games. His insights were later developed by the seminal work of John von Neumann [7], where the foundations of both noncooperative and cooperative game theory were built successfully. Von Neumann extended this early work of his together with an economist Oscar Morgernstern to a comprehensive book, entitled *Theory of Games and Economic Behavior*. This book contained a detailed analysis on many game theoretical subjects, including cooperative games and stable sets as a cooperative solution concept, extensive form games, utility, zero-sum games, and minmax equilibrium as a noncooperative solution concept.

While the book of von Neumann and Morgenstern [8] marks, according to many game theorists, the beginning of game theory as a field of study, it was undoubtedly the four works of John F. Nash, produced between 1950 and 1953, that carried game theory to the state which we all enjoy today. In these works, John F. Nash made the distinction between cooperative and noncooperative games; introduced a solution concept (equilibrium points, which we call Nash equilibria) for noncooperative games and showed that a solution exists in all finite games;

and introduced a two-person bargaining problem in its abstraction and proposed a solution concept (which we call the Nash bargaining solution).

The contributions of John F. Nash had a swift effect on the game theoretical research in academia starting from the 1950s and the 1960s. Following his works, many noncooperative and cooperative solution concepts were proposed, repeated games and games with incomplete or imperfect information were analyzed, game theory was applied to biology, political science, and engineering among many other fields of study, and games were designed as mechanisms to implement social choice rules. Game theory, and in particular, the idea of Nash equilibrium for noncooperative games, has completely transformed economics, as well. This transformation was enormously huge even in the 1980s, as stated by Aumann [9]:

> "[The Nash] equilibrium is without doubt the single game theoretic solution concept that is most frequently applied in economics. Economic applications include oligopoly, entry and exit, market equilibrium, search, location, bargaining, product quality, auctions, insurance, principal-agent, higher education, discrimination, public goods, what have you. On the political front, applications include voting, arms control and inspection, as well as most international political models (deterrence, etc.). Biological applications all deal with forms of strategic equilibrium; they suggest an interpretation of equilibrium quite different from the usual overt rationalism. We cannot even begin to survey all of this literature here."

This book briefly presents a wide range of topics in both noncooperative and cooperative game theory. With its scope, it differs from many excellent textbooks at the undergraduate or predoctoral level which usually do not (or barely) study any topics in cooperative game theory and some of the advanced-level topics in noncooperative game theory covered by this book. There are a few excellent textbooks at the graduate level, studying both noncooperative and cooperative aspects of game theory. However, they are much lengthier than this book and present the theoretical ideas in a more technical way. This book as a whole can be used as a textbook for a one-semester advanced undergraduate- or graduate-level game theory course. In the last 3 years, I was able to successfully cover all topics in this book in several game theory classes of 42 h.

The intended readership for this book involves students and instructors in game theory; in particular senior undergraduate students majoring in economics or engineering, and Master, MBA, and Ph.D. students majoring in economics, political science, and business administration. However, some chapters, studying topics such as implementation theory, coalitional games with transferable utility, cake cutting and fairness, cooperative bargaining, and matching markets, may also be interesting for readers studying or teaching microeconomics.

Since 1998, I have taught many game theory courses at both undergraduate and graduate levels to students in a number of economics and engineering departments. Also, I have conducted some part of my research in several fields of game theory, including cooperative bargaining, correlated equilibrium, implementation theory and mechanism design, matching markets, and applied game theory. The material and exposition in this book mostly reflect what I have learned so far as a student from my instructors, as a reader of various excellent books, articles, and presentations in game theory, and as an instructor from the questions and comments of many excellent students in my game theory classes. But, definitely, all errors are solely mine.

Structure of the Book

This book consists of five parts, involving 17 chapters in total. A brief description of these parts is as follows:

Static Noncooperative Games with Complete Information. This part includes Chaps. 1–7 of the book. Unlike many undergraduate-level textbooks, these chapters involve perfect equilibrium, evolutionary stable strategies, and correlated equilibrium in addition to several commonly taught topics such as introduction to strategic form games, dominance, rationalizability, and Nash equilibrium.

Static Noncooperative Games with Incomplete Information. This part includes a single chapter (Chap. 8) studying Bayesian Nash equilibrium. Unlike in many undergraduate-level textbooks, this chapter starts with presenting the ex ante version of the Bayesian Nash equilibrium, in addition to the commonly taught the interim version.

Dynamic Noncooperative Games with Perfect or Imperfect Information. This part includes Chaps. 9–12, which involve an introduction to extensive form games, subgame perfect Nash equilibrium, sequential equilibrium, and repeated games.

Implementation Theory. This part includes a single chapter (Chap. 13) studying the implementation of social choice rules. This topic is not covered by a great majority of the undergraduate-level textbooks. This topic was by far the most favorite subject for the majority of all students in my game theory classes. The proof of Proposition 13.2 may be skipped in undergraduate courses, but the rest of the chapter can be taught at all levels of game theory courses.

Cooperative Games. This is the last part of the book. It includes Chaps. 14–17, studying topics such as coalitional games with transferable utility, cake cutting and fairness, cooperative bargaining, and matching markets. Many undergraduate- and graduate-level textbooks do not cover some or any of these topics. The majority of economics, and especially engineering, students in my game theory classes were very pleased to learn the aforementioned four topics in cooperative game theory.

Each chapter starts with theoretical definitions, followed by examples that are presented in detail. Some chapters also contain propositions either dealing with the existence of equilibrium in some abstract games or relating various game theoretical concepts to each other. However, the book does not require the reader to have a strong background in calculus or algebra. The mathematical exposition in each chapter is as simple and self-contained as possible to ensure that the reader can easily apply given theoretical concepts to a set of examples and quickly follow the proofs of propositions. In a few chapters, technical proofs of some propositions are relegated to the appendix at the end of these chapters. The appendix at the end of

Chap. 4 also contains some basic definitions for sets, Weierstrass Extremum Theorem, and Kakutani Fixed Point Theorem with some graphical discussion. This mathematical preliminary may be required for some students to follow the proof of Nash's (1950a) Existence Theorem in Chap. 4. Beside the simplicity of the mathematical exposition, the definitions in the book are almost as formal as one would find in an advanced-level game theory book. Thus, I hope this book will serve to be a bridge, both in content and exposition, between intermediate-level and advanced-level game theory textbooks.

As a final remark, I should indicate that when referring to an individual without any specific gender, I will usually write "she" instead of "he/she" or "he". Also, if there are two players distinctly specified, I will use "she" for player 1 and "he" for player 2.

Ankara, Türkiye
November 2024

Ismail Saglam

References

1. Arrow KJ (2003) Introductory remarks on the history of game theory. Games and Economic Behavior 45:15–18
2. Cournot AA (1897) [Originally published 1838]. Recherches sur les principes math´ematiques de la th´eorie des richesses [Researches into the mathematical principles of the theory of wealth]. Economic classics, translated by Bacon NT.The Macmillan Company, New York
3. Bertrand J (1883) Book review of theorie mathematique de la richesse sociale and of recherches sur les principles mathematiques de la theorie des richesses.Journal de Savants 67:499–508
4. Edgeworth F (1889) The pure theory of monopoly. Reprinted in: Papers relating to political economy (1925) Vol. 1. Macmillan, London
5. von Stackelberg H (1934) Marktform und gleichgewicht. Berlin & Vienna, Springer
6. Borel E (1921) La théorie du jeu et les équations intégrales à noyau symétrique.Comptes Rendus de l'Acad´emie des Sciences 173:1304-308. English translation by Savage LJ (1953) Econometrica 21:97–100
7. von Neumann J (1928) Zur theorie der gesellschaftsspiele. Mathematische Annalen 100:295–320
8. von Neumann J, Morgenstern O (1944) Theory of games and economic behavior.Princeton: Princeton University Press
9. Aumann RJ (1987) Game theory. In: Eatwell J, Milgate M, Newman P (eds)The new palgrave: A dictionary of economics, Vol 2., pp. 460–482. MacMillan, London

Acknowledgements I thank Semih Koray of Bilkent University, who taught me mathematical economics and various topics in game theory and also taught me to think and write more clearly and rigorously. I also thank many excellent game theory instructors at the Massachusetts Institute of Technology, Princeton University, and the University of Michigan, from whom I learned a lot while I was visiting these colleges. I thank my colleagues Asena Caner, Ayca Ozdogan Atabay, Ayse Mutlu Derya, Bugra Caskurlu, Ethem Akyol, Haldun Evrenk, Ozan Eksi, Talha Yalta, and Ufuk Demiroglu for their encouragements for this book. In particular, I thank Ethem Akyol and Haldun Evrenk for detailed comments and suggestions on the earlier drafts of the book. I thank my brother Aziz for providing me with many important resources in game theory. Also, I thank the students in my game theory classes at Bogazici University and TOBB ETU in the past for many excellent comments and questions. In particular, I thank my student Meltem Arman for sharing with me the lecture notes she took in a game theory course of mine, my student Kivanc Tas for proposing the change of a word, and my students Mehmet Ercoskun, Serhat Gokcekli, and Gizem Yilmaz for their encouragements for this book. I thank Gizem Yilmaz also for inspiring me on some graphs and for some comments. Finally, I thank the TOBB University of Economics and Technology for providing me with a peaceful working environment.

Competing Interests This study was not funded by any institution. Also, the author has no conflicts of interest to declare that are relevant to the content of this chapter.

Contents

Acronyms

G	A strategic form game with only pure strategies
G^{Δ}	A strategic form game with mixed strategies
N	A set of players
n	The number of players
$\mathbb{R}$	The set of real numbers
η_i	The number of pure strategies of player i
S_i	The pure strategy space of player i
S_{-i}	The pure strategy space of all players excluding i
S	The space of all pure strategy profiles in game G
s_i	A pure strategy of player i
s_{-i}	A pure strategy profile of all players excluding i
s	A pure strategy profile of all players
u_i	The payoff function of player i in game G
$\Delta(S_i)$	The space of mixed strategies of player i in game G^{Δ}
$\Delta^{N-i}(S_i)$	The space of mixed stragey profiles of all players excluding i in game G^{Δ}
$\Delta^{N}(S)$	The space of mixed strategy profiles of all players in game G^{Δ}
σ_i	A mixed strategy of player i
σ_{-i}	A mixed strategy profile of all players excluding i
σ	A mixed strategy profile of all players
$\tilde{u}_i$	The expected payoff function of player i in game G^{Δ}
SDE	The strongly dominant equilibrium
G^S	The game obtained by eliminating all strongly dominated strategies in G
G^W	The game obtained by eliminating all weakly dominated strategies in G
$S_i^{\infty,sd}$	The set of pure strategies of player i surviving the iterative elimination of strongly dominated strategies in game G
$S_i^{\infty,wd}$	The set of pure strategies of player i surviving the iterative elimination of weakly dominated strategies in game G
IESDS	The iterative elimination of strongly dominated strategies
IEWDS	The iterative elimination of weakly dominated strategies

$co(x, y, z)$	The convex hull of points $x, y,$ and z
BR_i	The best-response correspondence of player i
BR	The product of best-response correspondences of all players
IENBRS	The iterative elimination of never-best-response strategies
NE	The set of Nash equilibria
$B_\epsilon(x)$	The ϵ-neighborhood of x
argmax	The argument that maximizes
$\text{int}(X)$	The interior of X
$\|\|x - y\|\|$	The distance between x and y
ESS	The evolutionary stable strategy
$\Delta(S)$	The set of probability distributions on S
$C\mathcal{E}$	The set of correlated equilibria
G^B	A Bayesian game
A_i	The set of actions of player i in a Bayesian game
A	The set of all action profiles in a Bayesian game
T_i	The set of all types of player i in a Bayesian game
T	The set of all type profiles in a Bayesian game
$p_i(.\|t_i)$	The beliefs of player i with type t_i about the types of other players
$s_i(t_i)$	A pure strategy function of player i with type t_i in a Bayesian game
Γ	An extensive form game
$\mathcal{X}$	The set of all nodes in an extensive form game
$\mathcal{X}^\mathcal{T}$	The set of all terminal nodes in an extensive form game
$\mathcal{X}\backslash\mathcal{X}^\mathcal{T}$	The set of all decision nodes in an extensive form game
x_0	The root node in an extensive form game
$\Gamma[x]$	The subgame of Γ starting at the node x
$\mathcal{A}$	The set of all actions in an extensive form game
$\mathcal{A}(x)$	The set of all actions at the node x in an extensive form game
$\mathcal{H}_i$	The set of all information sets of player i in an extensive form game
$\mathcal{H}$	The set of all information sets in an extensive form game
$b_i(H_i^k)$	A behavior strategy of player i at her information set H_i^k
$i(H)$	The player who can possibly play at the information set H in an extensive form game
SPNE	Subgame Perfect Nash Equilibrium
WPBE	Weak Perfect Bayesian Equilibrium
μ	A system of beliefs in an extensive form game
$\mu(x)$	A belief distribution at a node x in an extensive form game
$G^\delta[T]$	A T-period repeated game with discount factor δ
h^t	A history of actions up to period t in a repeated game
H^t	The set of possible histories of actions up to period t in a repeated game
H^T	The set of possible histories that can occur in the repeated game $G^\delta[T]$
$V_i^{\delta,T}$	The discounted sum of the T-period payoffs of player i in the repeated game $G^\delta[T]$
$s_i(h^t)$	A strategy of player i at a history h^t in a repeated game

$G^{\delta}[\infty]$	The infinitely repeated game with discount factor δ
$V_i^{\delta,[T,\infty)}$	The discounted sum of the payoffs of player i starting from period T onwards in the repeated game $G^{\delta}[\infty]$
$CP(G)$	The convex polytope associated with the stage game G
TU	Transferable utility
$C(N, v)$	The core of a TU game (N, v)
$\varphi(N, v)$	The Shapley value of a TU game (N, v)
(E)	Efficiency in the Cake-Cutting Problem
(EQ)	Equitability in the Cake-Cutting Problem
(EF)	Envy-Freeness in the Cake-Cutting Problem
(P)	Proportionality in the Cake-Cutting Problem
$\sum_0^2$	The set of all admissible two-person bargaining problems with the disagreement point $d = (0, 0)$
$N(S)$	The solution of the Nash rule on the bargaining problem $(S, 0)$
$KS(S)$	The solution of the Kalai-Smorodinsky rule on the problem $(S, 0)$
$E(S)$	The solution of the Egalitarian rule on the problem $(S, 0)$
$a(S)$	The ideal point associated with the bargaining set S
$WPO(S)$	The set of weakly Pareto optimal allocations in S
$PO(S)$	The set of Pareto optimal allocations in S
(WPO)	The axiom of Weak Pareto Optimality
(PO)	The axiom of Pareto Optimality
(S)	The axiom of Symmetry
(CI)	The axiom of Contraction Independence
(SI)	The axiom of Scale Invariance
(SM)	The axiom of Strong Monotonicity
(IM)	The axiom of Individual Monotonicity
(RM)	The axiom of Restricted Monotonicity
DAA-MP	The Deferred Acceptance Algorithm with Men Proposing
DAA-WP	The Deferred Acceptance Algorithm with Women Proposing
C	The set of core matchings
$\mathcal{M}^S$	The set of stable matchings
DAA-SA	The Deferred Acceptance Algorithm with Students Applying
DAA-CP	The Deferred Acceptance Algorithm with Colleges Proposing
TTC	Top Trading Cycles

Part I
Static Noncooperative Games with Complete Information

In this part of the book, we will study games where individuals choose their strategic actions noncooperatively, i.e., without being able to cooperate on, and commit to, any joint action. The games we will study (i) are static in the sense that individuals can act simultaneously and only once, and (ii) involve complete information in the sense that knowledge of each individual about the game, such as the sets of strategies, payoffs, and the rules, is common (available to other individuals in the strategic environment). However, in several chapters in this part, we will reflect on the necessity or redundancy of the common knowledge assumption and discuss how much each player should really know about the other players in order to be able to play a certain strategic form game in the way prescribed by a solution method.

Throughout this book (hence for this part as well), we will assume that (i) the players are *rational*, i.e., they always plan their strategies to maximize their expected payoffs, (ii) the players are *intelligent*; i.e., they know what we know as the writer or readers of this book and they can compute and predict everything that we can. For this part of the book, assumption (ii) implies that players will also know that the game they are playing is noncooperative, static, and it contains complete information.

After an introduction in Chap. 1 to strategic form games, we will see in each subsequent chapter (Chaps. 2–7) a method to solve static noncooperative games with complete information. Chapter 2 will study the concept of dominance, focusing on dominance of strategies from an individual perspective under the title of "strategic dominance" as well as from a social perspective under the title of "Pareto dominance". Chapter 2 will show us that some strategic form games can be solved using concepts related to strategic dominance. Chapter 3 will introduce the concept of best response strategies of players in a strategic form game and show that a certain set of strategies of players can be rationalized using their best response strategies. As these rationalizable sets of strategies will always exist, they will constitute the domain where one should seek a plausible solution to a strategic form game. One such solution is the Nash equilibrium, which is introduced in Chap. 4. After a formal definition, this chapter will state and prove the celebrated theorem of John F. Nash (1950a), ensuring conditions under which a strategic form game will have a Nash equilibrium. This chapter will also define a strict version of Nash equilibrium, relate the Nash equilibrium to equilibrium notions that use strategic dominance, and introduce the concept of risk dominance to select between a pair of Nash equilibria in some simple strategic form games. Chapter 4 will also present alternative interpretations of Nash equilibrium in terms of randomization of players over their pure strategies and in terms of conjectures of players about their opponents' strategies. Chapter 5 will deal with the question concerning whether or when a Nash equilibrium is robust to some small mistakes in the conjectures of players leading to the equilibrium. Chapter 6 will also consider robustness of an equilibrium but from an entirely different perspective. This chapter will be concerned with the question whether or when a large population of players that is randomly matched as pairs to play a symmetric two-person strategic form game is resistant to the strategies of mutants that try to invade the population. The answer will yield the concept of evolutionary stable strategies. Chapter 6 will also present the relation between evolutionary stable strategies and several notions of equilibrium presented in earlier chapters. The last chapter of Part I, Chap. 7, will show how the idea of Nash equilibrium can be extended to a strategic form game where the noncooperative strategies of players can be correlated by a mediator who picks a strategy profile (to be played in the game) randomly according to a commonly known probability distribution (over the set of all strategy profiles) and privately informs each player about the strategy recommended to play. The equilibrium of this game where each player finds it to her best interest to obey the recommendation of the mediator, under the belief that all other players will obey too, is called a correlated equilibrium. Chapter 7 will illustrate that, in some strategic form games, some correlated equilibria lead to expected payoff profiles that players could not achieve by choosing their strategies independently.

Introduction to Strategic Form Games

1

In this chapter, we will present basic structures for strategic form games. These games model situations where individuals choose their strategic actions independently and once. A popular example of these games is the rock-scissors-paper game played by many children all over the world. Section 1.1 will provide the general definition of a strategic form game with pure strategies, followed by various examples; Sect. 1.2 will present some well-known strategic form games that will be used in various chapters of this book; Sects. 1.3 and 1.4 will present two special classes of strategic form games, zero-sum games and symmetric games, respectively; and Sect. 1.5 will present the mixed extension of strategic form games where players are allowed to randomize over their pure strategies.

1.1 Strategic Form Games

A strategic form game where we only consider pure strategies is denoted by a tuple $G = \langle N, (S_i)_{i \in N}, (u_i)_{i \in N} \rangle$. Here, $N = \{1, 2, \ldots, n\}$ denotes the set of players. We assume $n \geq 2$ so that the game is not as trivial as the decision problem in a single player game. S_i denotes the space of pure strategies of player i. We use the word "pure" for a strategy that does not involve any randomness. (In Sect. 1.5, we will introduce unpure strategies that are obtained from pure strategies through randomization). For most of our discussion, we will assume that S_i is a finite space (containing a finite number of strategies). We assume that $|S_i| = \eta_i \geq 2$ for each $i \in N$. Thus, we can write $S_i = \{s_{1,i}, s_{2,i}, \ldots, s_{\eta_i,i}\}$. Let $S = \bigtimes_{i \in N} S_i = S_1 \times \ldots \times S_n$ be the space that involve all *strategy profiles* (i.e., the combination or list of all strategies chosen by all players) in the game G.

For strategic form games that are played only once, strategies can also be called actions (moves or plays). However, in later chapters where we will study Bayesian

I. Saglam, *Mastering Game Theory*, Classroom Companion: Economics,
https://doi.org/10.1007/978-3-031-82094-6_1

Table 1.1 A two-person strategic form game

Player 1 \ Player 2	L	R
U	$(8, 7)$	$(16, 10)$
M	$(8, 6)$	$(-18, 12)$
D	$(-6, 9)$	$(10, 7)$

games, repeated games, or extensive form games, we will need to distinguish a strategy from an action as the former will correspond to a complete plan of actions, prescribing each action a player would take in any possible contingency that might arise.

In our examples, we will usually refer to 2×2 strategic form games that involve two players who both have two pure strategies. Finally, $u_i : S \longrightarrow \mathbb{R}$ is the payoff (utility) function of player i, mapping each strategy profile in S to a real number.

Example 1.1 Consider a strategic form game with $N = \{1, 2\}$, $S_1 = \{U, M, D\}$, $S_2 = \{L, R\}$, and payoff functions that generate the following payoff table.

The number of strategies of players 1 and 2 are $\eta_1 = 3$ and $\eta_2 = 2$, respectively. Table 1.1 involves the payoff of each player at each strategy profile s in $S = S_1 \times S_2 = \{(U, L), (U, R), (M, L), (M, R), (D, L), (D, R)\}$. For example, at the profile $s = (M, R)$, we have $u_1(s) = -18$ and $u_2(s) = 12$. ■

Below, we illustrate a strategic form game with three players.

Example 1.2 Consider the following strategic form game (Table 1.2).

In the game illustrated above, the set of players is $N = \{1, 2, 3\}$ and the strategy spaces of players 1, 2, and 3 are $S_1 = \{U, D\}$, $S_2 = \{L, R\}$, and $S_3 = \{Panel\ a, Panel\ b\}$, respectively. For each $s \in S$, we denote the corresponding payoff profile by a triple of real numbers, $(u_1(s), u_2(s), u_3(s))$. For example, if $s = (U, R, Panel\ b)$, then the payoff profile is $(1, 2, 6)$, representing that $u_1(s) = 1$, $u_2(s) = 2$, and $u_3(s) = 6$. ■

Table 1.2 A three-person strategic form game

Panel a — 1st Strategy of Player 3

Player 1 \ Player 2	L	R
U	$(3, 0, 2)$	$(0, 1, 2)$
D	$(2, 2, 1)$	$(1, 3, 0)$

Panel b — 2nd Strategy of Player 3

Player 1 \ Player 2	L	R
U	$(2, 3, 4)$	$(1, 2, 6)$
D	$(0, 0, 0)$	$(2, 0, 4)$

The strategic form games in the previous examples involve a finite number of pure strategies for each player. Below, we present a well-known strategic form game where each player has uncountably many pure strategies.

Example 1.3 (*Cournot Duopoly Game*) The original version of this game was first considered by Augustine Cournot (1838). The story in our version is as follows. Two firms in a duopolistic industry compete in quantities. The production cost function of firm $i = 1, 2$ is $C_i(q_i) = cq_i$, where the variable q_i denotes the output of firm i and the constant $c \geq 0$ denotes the common marginal cost of the firms. The firms face an inverse demand function $P(Q) = \min\{0, a - Q\}$, where $Q \geq 0$ is the industry output, i.e., $Q = q_1 + q_2$, and $a > c$ is the maximal size of the demand for the good produced by the firms. The profit function of firm $i = 1, 2$, when it produces q_i and its rival (firm $j \neq i$) produces q_j, is given by

$$\pi_i(q_i, q_j) = P(q_i + q_j)q_i - cq_i.$$

Firms 1 and 2 choose their outputs q_1 and q_2 noncooperatively and simultaneously (to maximize their own profits).

We should note that in the duopolistic market game described above, the set of players is $N = \{1, 2\}$ and the strategy spaces of players 1 and 2 are $S_1 = [0, \infty)$ and $S_2 = [0, \infty)$, respectively. Thus, $S = \mathbb{R}^2_+$. Given any $s \in S$, the players' payoffs are $u_1(s) = \pi_1(s_1, s_2)$ and $u_2(s) = \pi_2(s_2, s_1)$. ■

For any $i \in N$, let S_{-i} denote the Cartesian product of the strategy spaces of all players in the game excluding player i; i.e., $S_{-i} = \times_{j \in N-i} S_j$. Thus, for any i, we can write $S = S_i \times S_{-i}$. Note that for the game in Example 1.1, we have $S_{-1} = S_2$ and $S_{-2} = S_1$.

Also, for any i and any $s \in S$, we introduce the notation $s = (s_i, s_{-i})$, where $s_i \in S_i$ and $s_{-i} \in S_{-i}$. For the game in Example 1.2, if $s = (D, L, \textit{Panel b})$, we have $s_1 = D$, $s_2 = L$, and $s_3 = \textit{Panel b}$. Thus, $s_{-1} = (L, \textit{Panel b})$, $s_{-2} = (D, \textit{Panel b})$, and $s_{-3} = (D, L)$.

1.2 Some Well-Known 2 × 2 Games

Below, we will present some well-known 2 × 2 strategic form games.

Prisoner's Dilemma This game was formalized by Alan W. Tucker in 1950. The story of this game is as follows. Two individuals (1 and 2) are strongly suspected to have jointly committed a major crime. There is a small evidence that can sentence each of them to a short imprisonment only. Therefore, the district attorney desires to make the suspects confess their major crime and to make the public relieved. To that aim, he interrogates the suspects in separate rooms and offers to each of them two options: To confess an involvement in the crime (Defection on Silence, D) or to deny any involvement (Cooperation on Silence, C). If both suspects choose to

Table 1.3 Prisoner's dilemma game

		Player 2	
		Deny	Confess
Player 1	Deny	$(-1, -1)$	$(-10, 0)$
	Confess	$(0, -10)$	$(-5, -5)$

Table 1.4 Battle of sexes game

		Player 2	
		O	F
Player 1	O	$(2, 1)$	$(0, 0)$
	F	$(0, 0)$	$(1, 2)$

confess, then each of them is sentenced to 5 years. If both suspects choose to deny, then each of them is sentenced to 1 year. If only one suspect chooses to confess, then that suspect is released, while the other suspect is sentenced to 10 years. The strategic form game corresponding to the described story is presented in Table 1.3.

Battle of Sexes This game, which was introduced in 1957 by R. Duncan Luce and Howard Raiffa, is a simple example of a coordination game. The story of the game is as follows. A woman and a man (or any two individuals, 1 and 2) want to spend leisure time together in one of two locations: An opera house (O) where they can watch a performance and a football arena (F) where they can watch a game. Unfortunately, they cannot communicate and each has to choose a location independently. The woman prefers the possibility that they both go to the opera house to the possibility that they both go to the football arena, while the converse is true for the man. Moreover, both of them prefer the possibility of going to the same location to the possibility of going to different locations. A strategic form game corresponding to the described story is presented in Table 1.4.

Stag Hunt Game This game, which is attributed to the philosopher Jean-Jacques Rousseau (1754), is another example of a coordination game. Two hunters must decide independently whether to hunt a stag (strategy S) or a hare (strategy H). They know that hunting a stag is possible only if they both decide to choose strategy S. If their decisions are aligned to hunt a stag, they become successful and each enjoys a positive payoff of s. In case their decisions are different, the hunter aiming to hunt a stag is unsuccessful and obtains nothing, while the other hunter ends up with hunting a hare yielding a positive payoff h smaller than s. If both hunters decide to hunt a hare, each of them enjoys a payoff of h. The strategic form game corresponding to the described story is presented in Table 1.5.

Game of Chicken This game was first studied by Anatol Rapoport and Albert M. Chammah in 1966. It is a game of anti-coordination (conflict). According to the story

Table 1.5 Stag hunt game

		Player 2	
		S	H
Player 1	S	(s, s)	$(0, h)$
	H	$(h, 0)$	(h, h)

Table 1.6 Game of chicken

		Player 2	
		Go Straight	Swerve
Player 1	Go Straight	(−100, −100)	(5, −10)
	Swerve	(−10, 5)	(0, 0)

Table 1.7 Hawk-dove game

		Player 2	
		H	D
Player 1	H	$\left(\frac{v-c}{2}, \frac{v-c}{2}\right)$	$(v, 0)$
	D	$(0, v)$	$\left(\frac{v}{2}, \frac{v}{2}\right)$

of the game, two drivers drive towards each other on a road with a single lane. Each of them has two options: To swerve or go straight. If none of them swerve, they both die. If only one of them swerves, he is called chicken (coward) and becomes ashamed, while the other driver is called bold and becomes esteemed. If both of them swerve, they feel neither ashamed nor esteemed. Moreover, they both prefer to be a chicken to dying. A strategic form game corresponding to the described story is presented in Table 1.6.

Hawk-Dove Game This game was first presented in 1973 by John Maynard Smith and George R. Price in the biology literature. According to the story of this game, two rival birds of the same species contest for a food resource of value v. Each bird has two options: Acting like a dove (D) and acting like a hawk (H).

A hawk escalates the conflict until either the rival withdraws or it is badly hurt, whereas a dove shares the contested resource if the rival is a dove and withdraws if the rival is a hawk. If both birds choose to act like a hawk, the total damage they incur is c and each of them leaves the contest with a payoff of $(v - c)/2$. On the other hand, if both birds act like a dove, each ends up with a payoff of $v/2$. Finally, if only one bird acts like a dove, it gets nothing from the contest, while the bird acting like a hawk gets the whole resource and gains a payoff of v. The strategic form game corresponding to this story is presented in Table 1.7.

The Hawk-Dove game becomes identical to the Game of Chicken if $c > v$ and it becomes identical to the Prisoner's Dilemma if $c \in (0, v]$.

Matching Pennies A version of this game (played with marbles) is mentioned in the short story "The Purloined Letter" contained by a book of Edgar Allan Poe [1]. According to the conventional story in game theory books, players 1 and 2 both hold a penny hidden in their hands. Players select a face, head (H) or tail (T), of their penny and show to each other simultaneously. If the faces (head or tail) of the pennies match, player 2 pays 1 penny to player 1, and if the faces of the pennies do not match, player 1 pays 1 penny to player 2. This strategic form game is presented in Table 1.8.

Table 1.8 Matching pennies game

		Player 2	
		H	T
Player 1	H	(1, −1)	(−1, 1)
	T	(−1, 1)	(1, −1)

1.3 Zero-Sum Games

A particular class of games puts restrictions on the total payoffs of players.

Definition 1.1 A strategic form game $G = \langle N, (S_i)_{i\in N}, (u_i)_{i\in N}\rangle$ is a zero-sum game if $\sum_{i\in N} u_i(s) = 0$ for all $s \in S$.

If the players' payoffs sum to the same constant at all strategy profiles, the game is called a constant-sum game. The zero-sum games are a special class of constant-sum games. For any zero-sum game, $u_i(s) = -\sum_{j\in N-i} u_j(s)$ for any $s \in S$. Thus, if a player has a positive (negative) payoff at any strategy profile s, then the rest of the players as a whole must have a negative (positive) total payoff at s. So, for any zero-sum game with $n \geq 2$ players, the number of players with positive payoffs (and the number of players with negative payoffs) at any strategy profile $s \in S$ must be less than n.

Many parlor games, sport games, financial games, and war situations are two-player zero-sum games. Zero-sum games are interesting for game theorists since no cooperation can be mutually beneficial in these games. As a fact, each player that maximizes its payoff automatically minimizes the total payoffs of the remaining players. It is easy to observe that in zero-sum games with $n \geq 2$, the number of players whose payoff are both nonzero and maximized at the same strategy profile cannot exceed $n-1$. Thus, in two-player zero-sum games (the case of $n = 2$), any player that maximizes its payoff at some strategy profile minimizes the other player's payoff.

Example 1.4 We first present a zero-sum game with two players (Table 1.9). Apparently, for each $s \in \{U, D\} \times \{L, R\}$, we have $u_1(s) + u_2(s) = 0$. The payoff of player 1 is maximized at the strategy profile (U, R) while the payoff of player 2 is minimized. On the other hand, the payoff of player 2 is maximized at the strategy profile (U, L), while the payoff of player 1 is minimized. Next, we present a zero-sum game with three players (Table 1.10).

When the number of players is 3, the number of players with positive (negative) payoffs can be 0, 1, or 2. For example, when $s = (D, R, \textit{Panel } a)$ in the above game, the number of players with positive payoffs is 2, whereas this number is 1 if $s = (D, L, \textit{Panel } a)$ and 0 if $s = (U, R, \textit{Panel } a)$. Also, the payoff of player 1 is maximized at $(D, L, \textit{Panel } a)$. Thus, the total payoff of players 2 and 3 is minimized at that strategy profile. But, individually their payoffs are not minimized at $(D, L, \textit{Panel } a)$. It is easy to check that the payoff of player 2 is minimized at $(U, R, \textit{Panel } b)$, while the payoff of player 3 is minimized at $(U, L, \textit{Panel } b)$. ■

Table 1.9 A two-person zero-sum strategic form game

		Player 2	
		L	R
Player 1	U	(−10, 10)	(5, −5)
	D	(−2, 2)	(0, 0)

Table 1.10 A three-person zero-sum strategic form game

Panel a — 1st Strategy of Player 3

Player 1 \ Player 2	L	R
U	$(2, 0, -2)$	$(0, 0, 0)$
D	$(3, -1, -2)$	$(1, 3, -4)$

Panel b — 2nd Strategy of Player 3

Player 1 \ Player 2	L	R
U	$(2, 3, -5)$	$(-1, -3, 4)$
D	$(1, 0, -1)$	$(2, -2, 0)$

The noncooperative game theory mainly focused, before the contributions of John F. Nash in the 1950s, on zero-sum games. (See, for example, the book of Von Neumann and Morgenstern, [2]). In this book, we will not be particularly interested in zero-sum games.

1.4 Symmetric Games

There is a class of strategic form games where the payoffs of players are independent of their identities.

Definition 1.2 A strategic form game $G = \langle N, (S_i)_{i \in N}, (u_i)_{i \in N} \rangle$ is symmetric if $S_1 = \ldots = S_n$ and for any permutation $\pi : N \to N$

$$u_i(s_1, \ldots, s_i, \ldots, s_n) = u_{\pi(i)}(s_{\pi(1)}, \ldots, s_{\pi(i)}, \ldots, s_{\pi(n)}) \tag{1.1}$$

for all $s \in S$ and $i \in N$.

For a strategic form game to be symmetric, it is assumed that the strategy spaces of the players must be the same. In addition, the payoff of any player must be independent of its identity. If the set of players and the set of strategy profiles are reshuffled under any given permutation π of player names, then the payoff that any player i enjoys at any strategy profile s in a symmetric game must be the same as the payoff that its new identity (player $\pi(i)$) should enjoy at the permuted strategy profile $s_{\pi(N)}$.

Example 1.5 Consider the strategic form game in Table 1.11. We first observe that $S_1 = S_2 = \{A, B, C\}$. In two-player games, the only permutation that reshuffles players is $\pi(.)$ such that $\pi(1) = 2$ and $\pi(2) = 1$. Under this permutation, (1.1) can be written as

$$u_1(s_1, s_2) = u_2(s_2, s_1) \text{ and } u_2(s_1, s_2) = u_1(s_2, s_1) \text{ for all } s \in S.$$

Table 1.11 A two-person symmetric strategic form game

		Player 2		
		A	B	C
Player 1	A	(1, 1)	(3, 2)	(1, 4)
	B	(2, 3)	(7, 7)	(5, 4)
	C	(4, 1)	(4, 5)	(3, 3)

For example, for $s = (C, B)$, one can check that $u_1(C, B) = 4 = u_2(B, C)$ and $u_2(C, B) = 5 = u_1(B, C)$. Similarly, one can check that the symmetry condition holds at other strategy profiles, as well. ■

The Cournot Duopoly Game, the Prisoner's Dilemma, the Stag-Hunt Game, the Game of Chicken, and the Hawk-Dove Game in Sect. 1.4 are all symmetric. We will use symmetric games in Chap. 6 where we will discuss how stable strategies may arise evolutionarily in a large population of players where the payoffs depend only on the strategies employed, not on the identities of the players.

1.5 Mixed Strategies

Given any strategic form game $G = \langle N, (S_i)_{i\in N}, (u_i)_{i\in N}\rangle$ where players are allowed to use only pure strategies, we can also define the mixed extension $G^{\Delta} = \langle N, (\Delta(S_i))_{i\in N}, (\tilde{u}_i)_{i\in N}\rangle$ where players can randomize over their pure strategies. Here, for each $i \in N$, $\Delta(S_i)$ denotes the set of all probability distributions on S_i. A generic element of this set will be denoted by $\sigma_i = (\sigma_{1,i}, \sigma_{2,i}, \ldots, \sigma_{\eta_i,i})$, where for any $k \in \{1, 2, \ldots, \eta_i\}$ the number $\sigma_{k,i}$ denotes the probability assigned by player i to the pure strategy $s_{k,i}$ in the space $S_i = \{s_{1,i}, s_{2,i}, \ldots, s_{\eta_i,i}\}$.

We can formally define the set $\Delta(S_i)$ as

$$\Delta(S_i) = \left\{ \sigma_i : \sigma_{j,i} \in [0, 1] \text{ for all } j \in \{1, 2, \ldots, \eta_i\} \text{ and } \sum_{j=1}^{\eta_i} \sigma_{j,i} = 1 \right\}.$$

Note that $\Delta(S_i)$ involves uncountable number of elements. It also contains the pure strategies of player i. For example, the mixed strategy σ_i where $\sigma_{j,i} = 1$ if $j = 2$ and $\sigma_{j,i} = 0$ if $j \neq 2$ corresponds to the pure strategy $s_{2,i}$ in $S_i = \{s_{1,i}, s_{2,i}, \ldots, s_{\eta_i,i}\}$. We will call a mixed strategy that is not pure as an unpure (mixed) strategy and call a mixed strategy that puts positive probability on each pure strategy of a player as a totally mixed strategy.

We will denote by $\Delta^N(S)$ the set of all mixed strategy profiles in the game G^{Δ}. This set is the Cartesian product of the mixed strategy spaces of all players; i.e., $\Delta^N(S) = \bigtimes_{i\in N} \Delta(S_i)$. A generic element of $\Delta^N(S)$ is σ. For any $i \in N$, we will sometimes use the notations $\Delta^N(S) = \Delta(S_i) \times \Delta^{N-i}(S_{-i})$ and $\sigma = (\sigma_i, \sigma_{-i}) \in \Delta^N(S)$. Also, for convenience, for any $i \in N$, we will represent each probability distribution $\sigma_i \in \Delta(S_i)$ by a function with the same name $\sigma_i(.)$ such that $\sigma_i(s_k) = \sigma_{k,i}$ for any $k \in \{1, 2, \ldots, \eta_i\}$.

Finally, $\tilde{u}_i$ denotes the expected payoff function of player i derived from its payoff function u in the game G. Given a mixed strategy profile σ, the expected payoff $\tilde{u}_i(\sigma)$ of player i is equal to

$$\tilde{u}_i(\sigma) = \sum_{s \in S} \bigtimes_{i \in N} \sigma_i(s_i)\, u_i(s). \tag{1.2}$$

Since $\sigma = (\sigma_i, \sigma_{-i})$, we can write (1.2) in two additional forms given by

$$\tilde{u}_i(\sigma) = \sum_{s_i \in S_i} \sigma_i(s_i)\, \tilde{u}_i(s_i, \sigma_{-i}) \tag{1.3}$$

and

$$\tilde{u}_i(\sigma) = \sum_{s_{-i} \in S_{-i}} \bigtimes_{j \in N-i} \sigma_j(s_j)\, \tilde{u}_i(\sigma_i, s_{-i}). \tag{1.4}$$

Example 1.6 Consider the following strategic form game (Table 1.12).

Suppose that players 1 and 2 play the mixed strategies $\sigma_1 = (0.3, 0.7)$ and $\sigma_2 = (0.4, 0.6)$, respectively. Using (1.2), we can calculate the expected payoffs of the two players at the mixed strategy profile $\sigma = (\sigma_1, \sigma_2)$ as

$$\begin{aligned}\tilde{u}_1(\sigma) &= (.3)(.4)u_1(U,L) + (.3)(.6)u_1(U,R) + (.7)(.4)u_1(D,L) + (.7)(.6)u_1(D,R)\\ &= (.12)(4) + (.18)(3) + (.28)(2) + (.42)(1)\\ &= 2\end{aligned}$$

and

$$\begin{aligned}\tilde{u}_2(\sigma) &= (.3)(.4)u_2(U,L) + (.3)(.6)u_2(U,R) + (.7)(.4)u_2(D,L) + (.7)(.6)u_2(D,R)\\ &= (.12)(3) + (.18)(2) + (.28)(4) + (.42)(3)\\ &= 3.1,\end{aligned}$$

respectively. Observe that Eq. (1.3) implies

$$\begin{aligned}\tilde{u}_1(\sigma) &= (.3)\,[(.4)u_1(U,L) + (.6)u_1(U,R)] + (.7)\,[(.4)u_1(D,L) + (.6)u_1(D,R)]\\ &= (.3)\tilde{u}_1(U,\sigma_2) + (.7)\tilde{u}_1(D,\sigma_2)\\ &= \sigma_1(U)\tilde{u}_1(U,\sigma_2) + \sigma_1(D)\tilde{u}_1(D,\sigma_2)\end{aligned}$$

and similarly

$$\begin{aligned}\tilde{u}_2(\sigma) &= (.4)\,[(.3)u_2(U,L) + (.7)u_2(U,R)] + (.6)\,[(.3)u_2(D,L) + (.7)u_2(D,R)]\\ &= (.4)\tilde{u}_2(\sigma_1,L) + (.6)\tilde{u}_2(D,\sigma_2)\\ &= \sigma_2(L)\tilde{u}_2(\sigma_1,L) + \sigma_2(R)\tilde{u}_2(\sigma_1,R).\end{aligned}$$

Table 1.12 The strategic form game in Example 1.6

		Player 2	
		L	R
Player 1	U	(4, 3)	(3, 2)
	D	(2, 4)	(1, 3)

On the other hand, Eq. (1.4) implies

$$\begin{aligned}\tilde{u}_1(\sigma) &= (.4)\,[(.3)u_1(U, L) + (.7)u_1(D, L)] + (.6)\,[(.3)u_1(U, R) + (.7)u_1(D, R)] \\ &= (.4)\tilde{u}_1(\sigma_1, L) + (.6)\tilde{u}_1(\sigma_1, R) \\ &= \sigma_2(L)\tilde{u}_1(\sigma_1, L) + \sigma_2(R)\tilde{u}_1(\sigma_1, R)\end{aligned}$$

and similarly

$$\begin{aligned}\tilde{u}_2(\sigma) &= (.3)\,[(.4)u_2(U, L) + (.6)u_2(U, R)] + (.7)\,[(.4)u_2(D, L) + (.6)u_2(D, R)] \\ &= (.3)\tilde{u}_2(U, \sigma_2) + (.7)\tilde{u}_2(D, \sigma_2) \\ &= \sigma_1(U)\tilde{u}_2(U, \sigma_2) + \sigma_1(D)\tilde{u}_2(D, \sigma_2).\end{aligned}$$

■

References

1. Poe EA (1844) The gift: A christmas and new year's present for 1845. Carey and Hart, Philedelphia
2. von Neumann J, Morgenstern O (1944) Theory of games and economic behavior. Princeton University Press, Princeton

Dominance

2

In this chapter, we will study the concept of dominance for strategic form games. In Sect. 2.1, we will introduce strategic dominance. We will first show how a player in a given strategic form game can order her available strategies, whenever possible. Using the dominance relations induced by the individual orderings of players over their strategies, we will next introduce several equilibrium concepts. In Sect. 2.2, we will introduce another dominance concept, called the Pareto dominance. This concept will demonstrate when and how an outside observer who analyzes a strategic form game can order any two pure strategy profiles with respect to the payoff profiles they generate.

2.1 Strategic Dominance

Definition 2.1 Given a strategic form game $G = \langle N, (S_i)_{i \in N}, (u_i)_{i \in N} \rangle$, a strategy $s_i \in S_i$ strongly dominates, for player i, another strategy $s_i' \in S_i$ if

$$u_i(s_i, s_{-i}) > u_i(s_i', s_{-i}) \text{ for all } s_{-i} \in S_{-i}. \tag{2.1}$$

Notice in (2.1) that s_i is a better strategy for player i than the strategy s_i' irrespective of the strategies played by the remaining players in $N - i$. The notion that the domination is strong implies that the inequality sign is strict (>) in the utility comparison appearing in (2.1). When condition (2.1) holds, we can alternatively say that strategy s_i' of player i is strongly dominated by her strategy s_i.

Since the players are assumed to be rational, they do not play strongly dominated strategies. To calculate her strongly dominated strategies, a player must know only her own payoffs in the game; she does not need to know anything about the payoffs or rationality of other players. A player who knows her own payoffs can calculate

I. Saglam, *Mastering Game Theory*, Classroom Companion: Economics,
https://doi.org/10.1007/978-3-031-82094-6_2

(since she is intelligent) and eliminate (since she is rational) any strongly dominated strategy in her strategy space.

Example 2.1 Let us reconsider the game in Example 1.1.

		Player 2	
		L	R
Player 1	U	$(8, 7)$	$(16, 10)$
	M	$(8, 6)$	$(-18, 12)$
	D	$(-6, 9)$	$(10, 7)$

Table 1.1 Reconsidered

Above, the strategy U of player 1 strongly dominates her strategy D. However, U does not strongly dominate M. This is because for player 1 the desirability of U against M is not independent from the strategy of player 2. U is a better strategy for player 1 only if player 2 plays R. When player 2 plays L, playing U or M yield the same payoff (8 utils) to player 1. (Likewise, for player 1 M does not strongly dominate U.) For player 2, no strategy in $S_2 = \{L, R\}$ strongly dominates the remaining strategy. This is because for player 2 the desirability of any strategy is not independent from the strategy of player 1. L would be a better strategy for player 2 than R if (and only if) player 1 were to play D, whereas R would be a better strategy for player 2 than L if (and only if) player 1 were to play either U or M. ■

The dominance relation can be defined in a weaker form, as well.

Definition 2.2 Given a strategic form game $G = \langle N, (S_i)_{i \in N}, (u_i)_{i \in N} \rangle$, a strategy $s_i \in S_i$ weakly dominates, for player i, another strategy $s_i' \in S_i$ if

$$u_i(s_i, s_{-i}) \geq u_i(s_i', s_{-i}) \text{ for all } s_{-i} \in S_{-i} \text{ and}$$
$$u_i(s_i, s_{-i}) > u_i(s_i', s_{-i}) \text{ for some } s_{-i} \in S_{-i}. \tag{2.2}$$

When condition (2.2) holds, we can alternatively say that the strategy s_i' of player i is weakly dominated by her strategy s_i. Comparing (2.2) to (2.1), we should observe that strong domination implies weak domination; i.e., a strategy s_i of player i weakly dominates another strategy s_i' of this player if s_i strongly dominates s_i'. But, the converse is not true. Like in the case of strong dominance, a player that applies weak dominance relation between any two strategies that she can play must know her own payoffs in the game but nothing more. However, unlike in the case of strong dominance, we will not argue that a rational player should not play a weakly dominated strategy. If a strategy s_i' is weakly, but not strongly, dominated for player i by a strategy s_i, then there must exist some strategy profile $s_{-i} \in S_{-i}$ of her opponents that s_i' and s_i yield the same utility to player i when her opponents play s_{-i}. So, if player i cannot rule out the positive likelihood of s_{-i}, then she cannot rule out to play s_i' on the ground of rationality.

Example 2.2 Let us reconsider the game in Table 1.1. The strategy U of player 1 weakly dominates each of her strategies M and D. We already know that the strategy U of player 1 strongly dominates strategy D, hence weak dominance automatically follows. On the other hand, for player 1 the weak dominance of U against M does not follow from any earlier result. We know that the strategy U of player 1 does not strongly dominate strategy M. The dominance of U against M is (really) weak. The strategy U makes player 1 strongly better off than the strategy M if and only if player 2 plays the strategy R. For player 2, no strategy in $S_2 = \{L, R\}$ weakly dominates the remaining strategy. ■

In some games, by checking strong/weak dominance relations, we can calculate a unique undominated strategy for some players.

Definition 2.3 A strategy is the strongly dominant strategy of a player if it strongly dominates every other strategy of this player. Likewise, a strategy is the weakly dominant strategy of a player if it weakly dominates every other strategy of this player.

Clearly, a strategy is the weakly dominant strategy of a player if it is the strongly dominant strategy of that player, but the converse is not true. We should first observe that a player need not have the weakly or strongly dominant strategy. For example, in Example 2.1, player 2 has neither the strongly dominant strategy nor the weakly dominant strategy. On the other hand, player 1 does not have the strongly dominant strategy, whereas she has the weakly dominant strategy (the strategy U that weakly dominates the strategies M and D).

We should also observe that for any player the strongly (or weakly) dominant strategy, whenever exists, must be unique. To see this, suppose there exists a game such that for a player i the distinct strategies s_a and s_b are both strongly (or weakly) dominant. Then, s_a strongly (or weakly) dominates all other strategies in S_i, including s_b. Likewise, s_b strongly (or weakly) dominates all other strategies in S_i, including s_a. But, these two statements cannot be true at the same time! Each player can have at most one strongly (or weakly) dominant strategy. It is also easy to see that for a player if a strategy s_a is the strongly (weakly) dominant strategy, then another strategy s_b cannot be the weakly (strongly) dominant strategy.

If each player in a game has a dominant strategy, we may talk about the dominant equilibrium profile.

Definition 2.4 A strategy profile $s \in S$ is the strongly dominant equilibrium if for each $i \in N$ the strategy s_i is the strongly dominant strategy of player i. Likewise, a strategy profile $s \in S$ is the weakly dominant equilibrium if for each $i \in N$ the strategy s_i is the weakly dominant strategy of player i.

There cannot be an equilibrium concept that is informationally less demanding than the strongly (or weakly) dominant equilibrium where each player needs to know her own payoffs (in addition to the strategy space of the game) but nothing else. Also, if there exists the strongly (or weakly) dominant equilibrium, then it must be unique and it is natural to predict that players will play their strategies in this equilibrium. Unfortunately, a strategic form game need not have the strongly (or weakly) dominant equilibrium. We have seen for the game in Table 1.1 that there exists at least one player (precisely two players) without the strongly dominant strategy (as shown in Example 2.1) and there exists at least one player (player 2) without the weakly dominant strategy (as shown in Example 2.2). Therefore, there exists no strategy profile that is the strongly or weakly dominant equilibrium.

A strategy profile is the weakly dominant equilibrium if it is the strongly dominant equilibrium; but, the converse is not true as illustrated by the following example.

Example 2.3 Consider a strategic form game with $N = \{1, 2\}$, $S_1 = \{U, D\}$, $S_2 = \{L, R\}$, and payoff functions that generate the payoffs presented in Table 2.1. The strategy profile (U, L) is the weakly dominant equilibrium of the above game. But, it is not the strongly dominant equilibrium. This is because for player 1 the strategy U is not strongly dominant; it dominates the strategy D weakly but not strongly. ■

The following example contains a game with no strongly or weakly dominant equilibrium.

Example 2.4 Consider the following two-person strategic form game represented by Table 2.2.

We should note that neither player has a strongly dominant strategy. Since there exists at least one player that has no dominant strategy, there exists no equilibrium in strongly dominant strategies. We should also note that player 1 has a weakly dominant strategy, the strategy U, which weakly dominates both M and D. However, player 2 does not have a weakly dominant strategy. Since there exists at least one player that has no weakly dominant strategy, there exists no equilibrium in weakly dominant strategies. ■

Table 2.1 The strategic form game in Example 2.3

		Player 2	
		L	R
Player 1	U	$(2, 3)$	$(1, 2)$
	D	$(2, 4)$	$(0, 2)$

Table 2.2 The strategic form game in Example 2.4

		Player 2		
		l	m	r
Player 1	U	$(11, 4)$	$(21, 10)$	$(15, 12)$
	M	$(11, 5)$	$(19, 8)$	$(9, 10)$
	D	$(9, 7)$	$(13, 5)$	$(15, 5)$

2.1.1 Iterative Elimination of Strongly Dominated Strategies

Given a strategic form game G, let G^S (G^W) denote the resulting game obtained after the strongly dominated strategies (the weakly dominated strategies) are eliminated. It is usual that the reduced games G^S and G^W contain more than one strategy profile, meaning that the process of eliminating dominated strategies does not yield a unique prediction as to how the players should play the game G. If the games G^S and G^W are identical to G, implying that no strategies in G could be eliminated, then we may stop further inquiries using strong or weak dominance relationships. But if the game G^S (or the game G^W) are smaller in the size of strategy spaces than the original game G, we may ask "why should we stop at this point?". We may apply the process of eliminating strongly (or weakly) dominated strategies on the reduced game G^S (the reduced game G^W) iteratively till no further reductions are possible. Moreover, in this iterative process of elimination, we can use mixed strategies as well. This iterative process would however require the players to know about the rationality of each other as well. Below, we will talk more about this requirement and its consequences.

Now, let us first extend the strong and weak dominance relations in Definitions 2.1 and 2.2 to games allowing mixed strategies.

Definition 2.5 Given a strategic form game with mixed strategies $G^\Delta = \langle N, (\Delta(S_i))_{i\in N}, (\tilde{u}_i)_{i\in N}\rangle$, a strategy $\sigma_i \in \Delta(S_i)$ strongly dominates, for player i, another strategy $\sigma_i' \in \Delta(S_i)$ if

$$\tilde{u}_i(\sigma_i, \sigma_{-i}) > \tilde{u}_i(\sigma_i', \sigma_{-i}) \text{ for all } \sigma_{-i} \in \Delta^{N-i}(S_{-i}). \tag{2.3}$$

Similarly, a strategy $\sigma_i \in \Delta(S_i)$ weakly dominates, for player i, another strategy $\sigma_i' \in \Delta(S_i)$ if

$$\begin{aligned} &\tilde{u}_i(\sigma_i, \sigma_{-i}) \geq \tilde{u}_i(\sigma_i', \sigma_{-i}) \text{ for all } \sigma_{-i} \in \Delta^{N-i}(S_{-i}) \text{ and} \\ &\tilde{u}_i(\sigma_i, \sigma_{-i}) > \tilde{u}_i(\sigma_i', \sigma_{-i}) \text{ for some } \sigma_{-i} \in \Delta^{N-i}(S_{-i}). \end{aligned} \tag{2.4}$$

The above definition requires each player to consider all mixed strategies of her opponents in deciding whether a particular mixed strategy she can play strongly/weakly dominates all other mixed strategies admissible for her. However, recall that equation (1.4) implies that

$$\tilde{u}_i(\sigma_i, \sigma_{-i}) - \tilde{u}_i(\sigma_i', \sigma_{-i}) = \sum_{s_{-i}\in S_{-i}} \bigtimes_{j\in N-i} \sigma_j(s_j)\,[\tilde{u}_i(\sigma_i, s_{-i}) - \tilde{u}_i(\sigma_i', s_{-i})] \quad (2.5)$$

for any $\sigma_i, \sigma_i' \in \Delta(S_i)$ and $\sigma_{-i} \in \Delta^{N-i}(S_{-i})$. So, to check the dominance relations in Definition 2.5, it is sufficient that each player should consider the pure strategies of her opponents. Using this observation, we can restate Definition 2.5 as follows:

Definition 2.6 Given a strategic form game with mixed strategies $G^\Delta = \langle N, (\Delta(S_i))_{i\in N}, (\tilde{u}_i)_{i\in N}\rangle$, a strategy $\sigma_i \in \Delta(S_i)$ strongly dominates, for player i, another strategy $\sigma_i' \in \Delta(S_i)$ if

$$\tilde{u}_i(\sigma_i, s_{-i}) > \tilde{u}_i(\sigma_i', s_{-i}) \text{ for all } s_{-i} \in S_{-i}. \quad (2.6)$$

Similarly, a strategy $\sigma_i \in \Delta(S_i)$ weakly dominates, for player i, another strategy $\sigma_i' \in \Delta(S_i)$ if

$$\begin{aligned} &\tilde{u}_i(\sigma_i, s_{-i}) \geq \tilde{u}_i(\sigma_i', s_{-i}) \text{ for all } s_{-i} \in S_{-i} \text{ and} \\ &\tilde{u}_i(\sigma_i, s_{-i}) > \tilde{u}_i(\sigma_i', s_{-i}) \text{ for some } s_{-i} \in S_{-i}. \end{aligned} \quad (2.7)$$

Now, we are ready to describe the process of iterative elimination of dominated strategies, allowing mixed strategies. Let us first define the set of strategies

$$S_i^{1,sd} = \{s_i \in S_i : s_i \text{ is not strongly dominated given } S_{-i}\},$$

which includes all pure strategies of player i that are not strongly dominated by any mixed strategy in the game $G^\Delta = \langle N, (\Delta(S_i))_{i\in N}, (\tilde{u}_i)_{i\in N}\rangle$. To calculate this set, player i needs to know only her payoffs. Also, the rationality of player i requires her to play a strategy in $S_i^{1,sd}$.

Now, suppose that players have common knowledge about each other's payoffs and rationality, meaning that every player knows that every player knows that ... every player knows that their payoffs are given by $(\tilde{u}_i)_{i\in N}$ and they are rational. Since each player knows $(\tilde{u}_i)_{i\in N}$, each player can calculate the sets $S_1^{1,sd}, S_2^{1,sd}, \ldots, S_n^{1,sd}$. Each player in the game can further calculate the set of strategies

$$S_i^{2,sd} = \left\{s_i \in S_i^{1,sd} : s_i \text{ is not strongly dominated given } S_{-i}^{1,sd}\right\}$$

for each value of $i \in N$, since the aforementioned common knowledge assumption implies that each player knows that other players are rational, too. Note that the set

$S_i^{2,sd}$ is obtained in the second iteration of the process by deleting from $S_i^{1,sd}$ all pure strategies of player i that are strongly dominated given that the space of other players' pure strategies at the beginning of iteration 2 is $S_{-i}^{1,sd}$. We can push this elimination process further to make the following generalization: Set $S_i^{0,sd} = S_i$ and calculate for any integer $k \geq 1$ the set of all pure strategies of player i surviving k iterations of elimination:

$$S_i^{k,sd} = \left\{ s_i \in S_i^{k-1,sd} \ : \ s_i \text{ is not strongly dominated given } S_{-i}^{k-1,sd} \right\}.$$

Note that the common knowledge about the rationality of each player ensures that each player in the game can calculate the set $S_i^{k,sd}$ for any $i \in N$ and for any integer $k \geq 1$. So, consider the limit of this iterative process leading to

$$S_i^{\infty,sd} = \cap_{k=1}^{\infty} S_i^{k,sd} \quad \text{for each } i \in N.$$

If the game G that generates the mixed extension G^{Δ} is a finite strategic form game, then, for each $i \in N$, the set S_i is finite, implying that $S_i^{\infty,sd}$ exists and can be reached after a finite number of iterations; i.e., there exists an integer $k^* \geq 1$ such that $S_i^{k,sd} = S_i^{\infty,sd}$ for all integer $k \geq k^*$. Thus, each player can calculate through the described procedure, after a finite number of iterations, the sets $S_1^{\infty,sd}, S_2^{\infty,sd}, \ldots, S_n^{\infty,sd}$.

Finally, each player can eliminate in the sets $\Delta(S_1^{\infty,sd}), \Delta(S_2^{\infty,sd}), \ldots, \Delta(S_n^{\infty,sd})$ any mixed strategies that are strongly dominated to calculate the final outcome of the elimination process, involving the sets $[\Delta(S_1^{\infty,sd})]^*, [\Delta(S_2^{\infty,sd})]^*, \ldots, [\Delta(S_n^{\infty,sd})]^*$ where

$$[\Delta(S_i^{\infty,sd})]^* = \left\{ \sigma_i \in \Delta(S_i^{\infty,sd}) \ : \ \sigma_i \text{ is not strongly dominated given } \Delta^{N-i}(S_{-i}^{\infty,sd}) \right\}$$

for each $i \in N$. Note that to reach from the sets $(\Delta(S_i^{\infty,sd}))_{i \in N}$ to the sets $([\Delta(S_i^{\infty,sd})]^*)_{i \in N}$ a single step of elimination is sufficient. This is because each player is able to put a zero probability on any pure strategy profile of her opponents while making her calculations and therefore she does not need to know the mixed strategy profiles that her opponents would calculate to be strongly dominated.

Given a strategic form game $G = \langle N, (S_i)_{i \in N}, (u_i)_{i \in N} \rangle$, the iterative elimination procedure described above reduces the strategy space $\times_{i=1}^{n} \Delta(S_i)$ of the mixed extension game G^{Δ} associated with G to the strategy space $\times_{i=1}^{n} [\Delta(S_i^{\infty,sd})]^*$. This reduced strategy space does not have to be a singleton set, containing a unique strategy profile. However, if it is, then we can predict the unique profile in the reduced strategy space to be played by the players in the game, under the assumption that each player's payoffs and rationality are common knowledge among the players.

We should also note that the outcome of the iterative process described above is independent of the order of players in eliminating their strongly dominated strategies. There exist two independent reasons, each of which is sufficient to drive this result.

The first reason is that in a strategic form game with mixed strategies, G^{Δ}, if a player $i \in N$ has a strategy σ_i that is strongly dominated by another strategy σ_i', then this dominance relation will remain to exist in any restriction of the game G^{Δ}, as long as this restriction contains both σ_i and σ_i'. The second reason is that, while constructing the set $S_i^{k,sd}$ for any player $i \in N$ at any iteration $k \geq 1$, we eliminate 'all' strategies that are strongly dominated in $S_i^{k-1,sd}$ given $S_{-i}^{k-1,sd}$. Since the sets $(S_i^{0,sd})_{i\in N}$, where $S_i^{0,sd} = S_i$ for each $i \in N$, are exogenously given for the process of elimination, the sets $(S_i^{1,sd})_{i\in N}$ are not affected by the order of players in eliminating their strongly dominated strategies in iteration 1. By the same argument, the sets $(S_i^{k,sd})_{i\in N}$ are not affected by the order of players in eliminating their strongly dominated strategies at or before iteration k. Hence, the sets $[\Delta(S_i^{\infty,sd})]^*$, where $i \in N$, are not affected by the order of players.

Example 2.5 Consider the following strategic form game, G. Using the mixed extension G^{Δ}, let us calculate the space of strategies surviving the iterative elimination of strongly dominated strategies (Table 2.3).

Note that $S_1 = \{U, M, D\}$ and $S_2 = \{l, m, r\}$. Let $S_1^{0,sd} = S_1$ and $S_2^{0,sd} = S_2$. Since the outcome of the elimination process we presented above is independent of the order of players, we will let player 1 move first at each iteration, without any implication.

Iteration 1: Player 1 has no pure strategy in $S_1^{0,sd}$ that is strongly dominated by any strategy in $\Delta(S_1^{0,sd})$; therefore, $S_1^{1,sd} = S_1^{0,sd}$. Player 2 can only eliminate the pure strategy r in $S_2^{0,sd}$ as it is strongly dominated by m. Thus, $S_2^{1,sd} = \{l, m\}$. At the end of the first iteration, the game G reduces to the game in Table 2.4:

Iteration 2. Player 1 has no pure strategy in $S_1^{1,sd}$ that is strongly dominated by any strategy in $\Delta(S_1^{1,sd})$. The same is true for player 2, as well. He has no pure strategy in $S_2^{1,sd}$ that is strongly dominated by any strategy in $\Delta(S_2^{1,sd})$. Therefore, $S_i^{2,sd} = S_i^{1,sd}$ for each $i = 1, 2$, implying that $S_1^{\infty,sd} = S_1^{1,sd} = \{U, M, D\}$ and $S_2^{\infty,sd} = S_2^{1,sd} = \{l, m\}$.

Table 2.3 The strategic form game in Example 2.5, before IESDS is applied

Player 1 \ Player 2	l	m	r
U	(5, 5)	(3, 3)	(2, 1)
M	(4, 4)	(5, 5)	(0, 4)
D	(2, 2)	(7, 7)	(3, 6)

Table 2.4 The reduced game after Iteration 1 of IESDS

Player 1 \ Player 2	l	m
U	(5, 5)	(3, 3)
M	(4, 4)	(5, 5)
D	(2, 2)	(7, 7)

Now, let us check whether we can eliminate in the sets $\Delta(S_1^{\infty,sd})$ and $\Delta(S_2^{\infty,sd})$ any mixed strategies that are strongly dominated. Note that player 2 has only two strategies in $S_2^{\infty,sd}$. So, we cannot eliminate any mixed strategy of these two strategies using another mixed strategy in $\Delta(S_2^{\infty,sd})$. Thus, $[\Delta(S_2^{\infty,sd})]^* = \Delta(S_2^{\infty,sd})$.

Now, consider player 1. Pick any $\sigma_1 \in \Delta(S_1^{\infty,sd})$. Note that

$$\tilde{u}_1(\sigma_1, l) = \sigma_{11}(5) + \sigma_{21}(4) + \sigma_{31}(2), \quad \text{and}$$

$$\tilde{u}_1(\sigma_1, m) = \sigma_{11}(3) + \sigma_{21}(5) + \sigma_{31}(7).$$

Let $co(A, B, C)$ denote the convex hull (the smallest convex set) connecting the points A, B, and C in $\mathbb{R}^2$. For any $(x, y) \in \mathbb{R}^2$, we have $(x, y) \in co((2, 7), (4, 5), (5, 3))$ if and only if there exists some $\sigma_1 \in \Delta(S_1^{\infty,sd})$ such that $x = \tilde{u}_1(\sigma_1, l)$ and $y = \tilde{u}_1(\sigma_1, m)$.

In Fig. 2.1, we should observe that the red colored area corresponds to $co((2, 7), (4, 5), (5, 3))$, denoting the set of possible expected payoffs of player 1 under strategies in $\Delta(S_1^{\infty,sd})$. Given any two strategies $\sigma_1^a, \sigma_1^b \in \Delta(S_1^{\infty,sd})$, we know that σ_1^a strongly dominates σ_1^b if and only if $\tilde{u}_1(\sigma_1^a, s_2) > \tilde{u}_1(\sigma_1^b, s_2)$ for each $s_2 \in \{l, m\}$. Therefore, the set of all points in $co((2, 7), (4, 5), (5, 3))$ that are not strongly dominated constitute the line segments $[(2, 7), (4, 5)]$ and $[(4, 5), (5, 3)]$, which are highlighted by blue color in Fig. 2.1. That is,

$$[\Delta(S_1^{\infty,sd})]^* = \left\{(\sigma_{11}, \sigma_{21}, \sigma_{31}) \in [0, 1]^3 : \sigma_{11} + \sigma_{21} + \sigma_{31} = 1 \text{ and } \sigma_{11}\sigma_{31} = 0\right\}.$$

The iterative elimination of strongly dominated strategies reduced the strategy space of the game G^Δ, but not to a singleton set. Thus, there is not a unique prediction as to how the players will play in the game. ■

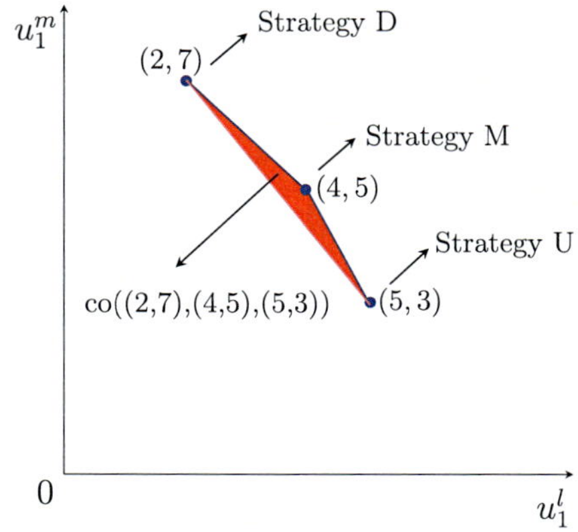

Fig. 2.1 Possible expected payoffs of player 1 under strategies in $\Delta(S_1^{\infty,sd})$

2.1.2 Iterative Elimination of Weakly Dominated Strategies

We can extend the iterative process of eliminating strongly dominated strategies to the case where weakly dominated strategies are eliminated. Suppose that players never use weakly dominated strategies and they will have common knowledge about this prospect as well as about their payoffs. Given a strategic form game with mixed strategies $G^{\Delta} = \langle N, \{\Delta(S_i)\}_{i \in N}, \{\tilde{u}_i\}_{i \in N} \rangle$ (derived from a finite game G), let $S_i^{0,wd} = S_i$. Then, the aforementioned common knowledge assumption ensures that each player in the game can calculate for any integer $k \geq 1$ and for any $i \in N$ the set of strategies

$$S_i^{k,wd} = \left\{ s_i \in S_i^{k-1,wd} \,:\, s_i \text{ is not weakly dominated given } S_{-i}^{k-1,wd} \right\}$$

that survive the elimination process at the end of iteration k. Noting that the limit of this iterative process leads to

$$S_i^{\infty,wd} = \cap_{k=1}^{\infty} S_i^{k,wd} \text{ for each } i \in N,$$

each player can calculate the sets $S_1^{\infty,wd}, S_2^{\infty,wd}, \ldots, S_n^{\infty,wd}$ after a finite number of iterations as S_i is finite for each $i \in N$. Finally, eliminating in the sets $\Delta(S_1^{\infty,wd}), \Delta(S_2^{\infty,wd}), \ldots, \Delta(S_n^{\infty,wd})$ any mixed strategies that are weakly dominated, each player can calculate the outcome of the iterative elimination process, involving the sets $\Delta[(S_1^{\infty,wd})]^*, [\Delta(S_2^{\infty,wd})]^*, \ldots, [\Delta(S_n^{\infty,wd})]^*$, where

$$[\Delta(S_i^{\infty,wd})]^* = \left\{ \sigma_i \in \Delta(S_i^{\infty,wd}) \,:\, \sigma_i \text{ is not weakly dominated given } \Delta^{N-i}(S_{-i}^{\infty,wd}) \right\}$$

for each $i \in N$.

In a strategic form game with mixed strategies, G^{Δ}, if a player $i \in N$ has a strategy σ_i that is weakly dominated by another strategy σ_i', then this dominance relation does not have to remain to exist in every restriction of the game G^{Δ}, since the elimination of some strategies of the opponents may render σ_i and σ_i' equivalent (indistinguishable) on a restricted game. However, the iterative elimination process we presented above is such that the outcome is independent of the order of players even though the players are eliminating their weakly dominated strategies. We have earlier presented the underlying reason while we discussed the elimination process involving strongly dominated strategies. Adapting this reason to the case of weakly dominated strategies, we can say that, while constructing the set $S_i^{k,wd}$ for any player $i \in N$ at any iteration $k \geq 1$, we eliminate 'all' (not some) strategies that are weakly dominated in $S_i^{k-1,wd}$ given $S_{-i}^{k-1,wd}$. Since the sets $(S_i^{0,wd})_{i \in N}$, where $S_i^{0,wd} = S_i$ for each $i \in N$, are exogenously given for the process of elimination, the sets $(S_i^{1,wd})_{i \in N}$ are not affected by the order of players in eliminating their weakly dominated strategies in iteration 1. By the same argument, the sets $(S_i^{k,wd})_{i \in N}$ are not affected by the order of players in eliminating their weakly dominated strategies at or before iteration k. Hence, the sets $[\Delta(S_i^{\infty,wd})]^*$, where $i \in N$, are not affected by the order of players.

However, we should be aware that if the process of elimination were defined in a way (as adopted by most of the textbooks) that would allow players in any iteration k of the process to eliminate any nonempty subset (not necessarily all) of the weakly dominated strategies in the sets $(S_i^{k-1,wd})_{i \in N}$, then the final outcome of the elimination process could depend on the order of players in some strategic form games.

Example 2.6 Reconsider the strategic form game in Table 2.2. Using the mixed extension of this game, let us calculate the space of strategies surviving the iterative elimination of weakly dominated strategies. Note that $S_1 = \{U, M, D\}$ and $S_2 = \{l, m, r\}$. Let $S_1^{0,w} = S_1$ and $S_2^{0,w} = S_2$. Since the outcome of the elimination process we presented above is independent of the order of players, we let player 1 move first at each iteration, without any implication.

		Player 2	
Player 1	l	m	r
U	(11, 4)	(21, 10)	(15, 12)
M	(11, 5)	(19, 8)	(9, 10)
D	(9, 7)	(13, 5)	(15, 5)

Table 2 Reconsidered

Iteration 1: Consider first player 1. Given $S_2^{0,wd}$, the strategies M and D of player 1 are weakly dominated by the strategy U, . Therefore, $S_1^{1,wd} = \{U\}$. Now, consider player 2. Given $S_1^{0,wd}$, player 2 can only eliminate the pure strategy m in $S_2^{0,wd}$, which is weakly dominated by r. Thus, $S_2^{1,wd} = \{l, r\}$. At the end of the first iteration, the game G reduces to the following game (Table 2.5):

Table 2.5 The reduced game after Iteration 1 of IEWDS

		Player 2	
		l	r
Player 1	U	(11, 4)	(15, 12)

Iteration 2: Player 1 has a single strategy in $S_1^{1,wd}$. Therefore, she cannot eliminate any strategy in $S_1^{1,wd}$, implying that $S_1^{2,wd} = S_1^{1,wd}$. Consider player 2. Given $S_1^{1,wd}$, the strategy l of player 2 in $S_2^{1,wd}$ is weakly dominated by the strategy r. Therefore, $S_2^{2,wd} = \{r\}$.

Since $S_i^{2,wd}$ is a singleton set for each $i = 1, 2$, the process of elimination stops at this point. That is, $S_1^{\infty,w} = S_1^{2,s} = \{U\}$ and $S_2^{\infty,w} = S_2^{2,w} = \{r\}$. Clearly, $[\Delta(S_i^{\infty,w})]^* = \Delta(S_i^{\infty,w}) = S_i^{\infty,w}$ for each $i = 1, 2$, implying $[\Delta(S_1^{\infty,w})]^* \times [\Delta(S_2^{\infty,w})]^* = \{(U, r)\}$. So, we have found that there is a unique strategy profile, (U, r), which survives the iterative elimination of the weakly dominated strategies. ∎

2.2 Pareto Dominance

Pareto efficiency (undomination) and Pareto optimality (dominance) are well-known concepts in economics. When we use these concepts in game theory, we should be careful to distinguish between a strategy profile that is strongly or weakly dominant strategy equilibrium and a strategy profile that is Pareto dominant.

Definition 2.7 Given a strategic form game $G = \langle N, (S_i)_{i \in N}, (u_i)_{i \in N} \rangle$, a strategy profile s in S Pareto dominates another profile s' in S if

$$u_i(s) \geq u_i(s') \text{ for all } i \in N \text{ and } u_i(s) > u_i(s') \text{ for some } i \in N. \tag{2.8}$$

If a strategy profile $s \in S$ Pareto dominates a profile $s' \in S$, we can also say that s' is Pareto dominated by s. In Example 2.3, one can check that strategy profile (D, L) Pareto dominates each of the other strategy profiles, (U, R), (D, L), and (D, R).

Definition 2.8 Given a strategic form game $G = \langle N, (S_i)_{i \in N}, (u_i)_{i \in N} \rangle$, a strategy profile s in S is Pareto dominant if it Pareto dominates every other strategy profile in S.

Table 2.6 The strategic form game in Example 2.7

		Player 2	
		L	R
Player 1	U	$(1, -1)$	$(0, 0)$
	D	$(-2, 1)$	$(2, -3)$

We should observe that the Pareto dominance concept and the (strongly or weakly) dominant equilibrium concept are independent from each other. Any of them does not imply the other one. For the strategic form game presented by Example 2.1, one can check that the strategy profile (U, R) is Pareto dominant, but it is not the dominant equilibrium (neither strongly nor weakly). On the other hand, in the strategic form game presented by Example 2.3, the strategy profile (U, L) is the weakly dominant equilibrium, but it is not Pareto dominant. It is Pareto dominated by the strategy profile (D, L), which is Pareto dominant.

Also note that we have defined the strong and weak dominance relations not only on the set of strategy profiles in the game but also on the set of strategies of each player separately. However, the Pareto dominance is a concept that is defined on the set of strategy profiles only.

Here, we can immediately ask whether every strategic form game should have a Pareto dominant strategy profile. The answer is "no", as can be simply shown by using, for example, the Matching Pennies Game in Table 1.8. (In that game, recall that two players simultaneously show the upper face of a penny in their hands. If the faces match, then player 1 takes the penny of player 2. Otherwise, player 2 takes the penny of player 1.) Of course, the game need not be a zero-sum game to ensure the nonexistence of a Pareto dominant strategy profile, as the following example illustrates.

Example 2.7 Consider the strategic form game represented by Table 2.6.
No strategy profile in the above game is Pareto dominant; i.e., there exists no strategy profile that Pareto dominates all other strategy profiles. In fact, no strategy profile Pareto dominates any other profile. Each strategy profile in the game is Pareto undominated. ■

Here, we can ask whether each strategic form game must contain at least one strategy profile that is Pareto undominated. The answer is "yes" if we restrict ourselves to strategic form games where each player has a finite number of strategies and the number of strategy profiles is therefore finite. If at least one player has an infinite number of strategies, then the same answer would not prevail. That is, we can find strategic form games with infinite number of strategy profiles each of which is Pareto dominated, as illustrated by the following example (Table 2.7).

Table 2.7 The strategic form game in Example 2.8

Player 1		Player 2 $s_{1,2}$	$s_{2,2}$
	$s_{1,1}$	$(1, 0)$	$(1, 0)$
	$s_{2,1}$	$(2, 0)$	$(2, 0)$
	$\dots$	$\dots$	$\dots$
	$s_{k,1}$	$(k, 0)$	$(k, 0)$
	$\dots$	$\dots$	$\dots$

Example 2.8 Consider a strategic form game with $N = \{1, 2\}$, $S_1 = \{s_{1,1}, s_{2,1}, s_{3,1}, \dots, \}$, $S_2 = \{s_{1,2}, s_{2,2}\}$, and payoff functions that generate Table 2.7.

Let us consider any strategy profile $(s_1, s_2) \in S_1 \times S_2$. Then $s_1 = s_{k,1}$ for some $k \in \{1, 2, 3, \dots, \infty\}$. We should now observe that strategy profile $(s_{k,1}, s_2)$ is Pareto dominated by strategy profile $(s_{k+1,1}, s_2)$, since $u_1(s_{k+1,1}, s_2) = k + 1 > k = u_1(s_{k,1}, s_2)$, while $u_2(s_{k+1,1}, s_2) = 0 = u_2(s_{k,1}, s_2)$. Thus, each strategy profile in the described (infinite) game is Pareto dominated. ■

You may have noticed that we have defined Pareto dominance relations for any strategic form game G with pure strategies. We can easily extend these relations (Definitions 2.7 and 2.8) to the mixed extension G^{Δ}, as well.

Definition 2.9 Given a strategic form game $G^{\Delta} = \langle N, (\Delta(S_i))_{i \in N}, (\tilde{u}_i)_{i \in N} \rangle$, a strategy profile σ in $\Delta^N(S)$ Pareto dominates another profile σ' in $\Delta^N(S)$ if

$$\tilde{u}_i(\sigma) \geq \tilde{u}_i(\sigma') \text{ for all } i \in N \text{ and } \tilde{u}_i(\sigma) > \tilde{u}_i(\sigma') \text{ for some } i \in N. \quad (2.9)$$

Using the above definition, we can now define Pareto dominance under mixed strategies.

Definition 2.10 Given a strategic form game $G^{\Delta} = \langle N, (\Delta(S_i))_{i \in N}, (\tilde{u}_i)_{i \in N} \rangle$, a strategy profile σ in $\Delta^N(S)$ is Pareto dominant if it Pareto dominates every other strategy profile in $\Delta^N(S)$.

We should note that if there are two pure strategy profiles s' and s'' in a strategic form game G with pure strategies such that s'' Pareto dominates s', then one can find uncountably many unpure strategies in the associated game G^{Δ} with mixed strategies that Pareto dominates s'. To see this, consider the profile $\sigma(x) = xs'' + (1 - x)s'$ for any $x \in [0, 1]$. Note that $\sigma(0) = s'$ and $\sigma(1) = s''$, whereas for any $x \in (0, 1)$ the profile $\sigma(x)$ is an unpure strategy profile in $\Delta^N(S)$ and the number of such profiles is uncountably infinite. Also note that for any $i \in N$ and $x \in (0, 1)$ we have

$$u_i(\sigma(x)) = xu_i(s'') + (1 - x)u_i(s') = u_i(s') + x\,[u_i(s'') - u_i(s')].$$

Since s'' Pareto dominates s' by supposition, we know that $u_i(s'') \geq u_i(s')$ for all $i \in N$ and $u_i(s'') > u_i(s')$ for some $i \in N$. If $i \in N$ is such that $u_i(s'') = u_i(s')$, then the above equation for $u_i(\sigma(x))$ implies that $u_i(\sigma(x)) = u_i(s')$. On the other hand, if $i \in N$ is such that $u_i(s'') > u_i(s')$, then the above equation for $u_i(\sigma(x))$ implies that $u_i(\sigma(x)) > u_i(s')$ since x is positive. Thus, we have established that for any $x \in (0, 1)$ we have $u_i(\sigma(x)) \geq u_i(s')$ for all $i \in N$ and $u_i(\sigma(x)) > u_i(s')$ for some $i \in N$. Therefore, $\sigma(x)$ Pareto dominates s' for any $x \in (0, 1)$ whenever this domination is true for $x = 1$.

However, we should also observe that for any $x, x' \in (0, 1]$ with $x < x'$, $\sigma(x)$ is Pareto dominated by $\sigma(x')$, which puts a higher probability on s''. Thus, in the set of (pure or unpure) mixed strategies $\{\sigma(x)\}_{x \in (0,1]}$ that Pareto dominate $\sigma(0) = s'$, the only Pareto undominated strategy profile is $\sigma(1) = s''$.

Observations above suggest that an unpure strategy profile cannot be Pareto dominant. In other words, if a strategic form game G has a strategy profile that is Pareto dominant, then this profile is Pareto dominant in the mixed extension game G^Δ, as well. The reason is that mixing pure strategies convexifies the payoffs of players. Because of this convexification an unpure mixed strategy profile σ always yields for some player $i \in N$ lower expected payoffs than some mixed strategy profile σ' where every other player continues to play as in σ_{-i}, while player i plays its best pure strategy $s_{k,i}$ in S_i against σ_{-i}, generating the largest expected payoff.

Rationalizability

3

In strategic form games, the strong or weak dominant equilibrium usually does not exist and even the iterative elimination of strongly or weakly dominated strategies cannot always predict a unique strategy profile as an equilibrium. However, as Sect. 3.1 of this chapter will make it clear, in any strategic form game the players can always calculate their strategies that are best response to a strategy profile of their opponents. Using the best-response strategies, we will define in Sect. 3.2 the sets of strategies of each player that can be predicted to be rationally played, or simply "rationalizable".

3.1 Best-Response Strategies

Let us reconsider the strategic form game studied in Example 2.5.

		Player 2		
		l	m	r
Player 1	U	(5, 5)	(3, 3)	(2, 1)
	M	(4, 4)	(5, 5)	(0, 4)
	D	(2, 2)	(7, 7)	(3, 6)

Table 2.3 Reconsidered

The above game does not have the strongly or weakly dominant equilibrium, and in addition the iterated elimination of strongly or weakly dominated strategies does not

I. Saglam, *Mastering Game Theory*, Classroom Companion: Economics,
https://doi.org/10.1007/978-3-031-82094-6_3

Table 3.1 Best-response payoffs highlighted by red and blue colors

		Player 2		
		l	m	r
Player 1	U	(5, 5)	(3, 3)	(2, 1)
	M	(4, 4)	(5, 5)	(0, 4)
	D	(2, 2)	(7, 7)	(3, 6)

lead to a unique strategy profile. However, the players can calculate their strategies that are best response to a strategy profile of their opponents.

Consider player 1 first. Her best-response strategy to the strategy l of player 2 is U which gives a higher payoff than M and D. On the other hand, player 1's best-response strategy to the strategies m and r of player 2 is D. Now, consider player 2. His best-response strategy to the strategy U of player 1 is l which gives a higher payoff than m and r. On the other hand, his best-response strategy to the strategies M and D of player 1 is m.

To simplify the presentation, we will hereafter color the best-response payoffs of player 1 by red color and the best-response payoffs of player 2 by blue color. Thus, we can reillustrate Table 2.3 as in Table 3.1.

Note that in Table 3.1, each player has a unique best response in pure strategies to any pure strategy of the other player. Now, let us change the payoffs in Table 3.1 such that $u_1(D, l)$ becomes 5 as in Table 3.2. Now player 1's best-response strategy to the strategy l of player 2 is not unique: The pure strategies U and D are both best response to l. Moreover, any unpure strategy that mixes U and D with probabilities $p \in (0, 1)$ and $1 - p$, respectively, is also a best response to l.

Thus, we have seen that a mixed (pure or unpure) strategy of some player can be a best response to a pure strategy profile of other players. Moreover, we can also calculate the best-response strategies of each player to unpure strategies of the other players. For example, consider the game in Table 3.2 and the unpure strategy $\sigma_1 = (0.5, 0.5, 0)$ of player 1.

Table 3.2 A strategic form game where best-response payoffs are not unique

		Player 2		
		l	m	r
Player 1	U	(5, 5)	(3, 3)	(2, 1)
	M	(4, 4)	(5, 5)	(0, 4)
	D	(5, 2)	(7, 7)	(3, 6)

Given σ_1, the expected payoff of player 2 is

$$5 \times (0.5) + 4 \times (0.5) = 4.5 \text{ if he plays } l,$$

$$3 \times (0.5) + 5 \times (0.5) = 4 \text{ if he plays } m,$$

$$1 \times (0.5) + 4 \times (0.5) = 2.5 \text{ if he plays } r.$$

Therefore, given $\sigma_1 = (0.5, 0.5, 0)$, the best response of player 2 is l, or equivalently $\sigma_2 = (1, 0, 0)$.

We are now ready to formally define the best-response correspondences.

Definition 3.1 The best-response correspondence of player $i \in N$ is a mapping $BR_i : \times_{j \in N-i} \Delta(S_j) \to \Delta(S_i)$ such that for any $\sigma_{-i} \in \times_{j \in N-i} \Delta(S_j)$ we have

$$BR_i(\sigma_{-i}) = \{\sigma_i \in \Delta(S_i) : \tilde{u}_i(\sigma_i, \sigma_{-i}) \geq \tilde{u}_i(\sigma_i', \sigma_{-i}) \text{ for all } \sigma_i' \in \Delta(S_i)\}.$$

In the following example, we show how to derive the best-response correspondences of players in a 2×2 strategic form game.

Example 3.1 Consider the strategic form game represented by Table 3.3.

Note that, for any $\sigma_2 = (\sigma_{12}, \sigma_{22}) = (\sigma_{12}, 1 - \sigma_{12})$, we have

$$\tilde{u}_1(U, \sigma_2) = (5)\sigma_{12} + (3)(1 - \sigma_{12}) = 3 + 2\sigma_{12}$$

$$\tilde{u}_1(D, \sigma_2) = (4)\sigma_{12} + (5)(1 - \sigma_{12}) = 5 - \sigma_{12}.$$

So, the best-response correspondence of player 1 is given by

$$BR_1(\sigma_2) = \begin{cases} \{(1, 0)\} \text{ if } \sigma_{12} > 2/3 \\ \Delta(S_1) \quad \text{if } \sigma_{12} = 2/3 \\ \{(0, 1)\} \text{ if } \sigma_{12} < 2/3. \end{cases}$$

Table 3.3 The strategic form game in Example 3.1

		Player 2	
		L	R
Player 1	U	(5, 5)	(3, 3)
	D	(4, 4)	(5, 5)

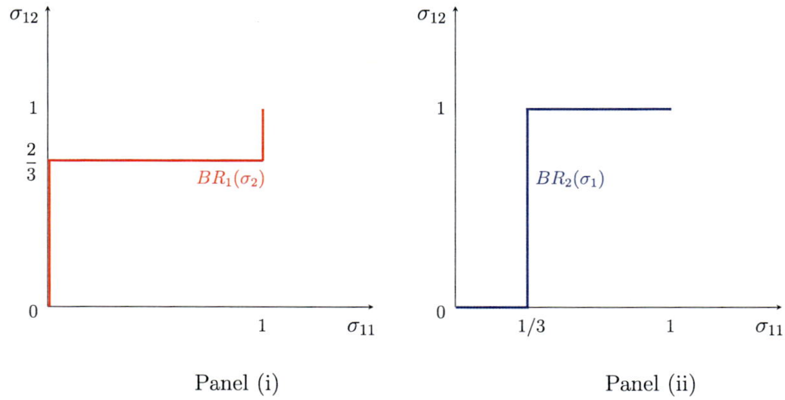

Fig. 3.1 Best-response correspondences for the game in Table 3.3

Also, note that, for any $\sigma_1 = (\sigma_{11}, \sigma_{21}) = (\sigma_{11}, 1 - \sigma_{11})$, we have

$$\tilde{u}_2(\sigma_1, L) = (5)\sigma_{11} + (4)(1 - \sigma_{11}) = 4 + \sigma_{11}$$

$$\tilde{u}_1(\sigma_1, R) = (3)\sigma_{11} + (5)(1 - \sigma_{11}) = 5 - 2\sigma_{11}.$$

So, the best-response correspondence of player 2 is given by

$$BR_2(\sigma_1) = \begin{cases} \{(1,0)\} \text{ if } \sigma_{11} > 1/3 \\ \Delta(S_2) \quad \text{if } \sigma_{11} = 1/3 \\ \{(0,1)\} \text{ if } \sigma_{11} < 1/3. \end{cases}$$

We draw the best-response correspondences in Fig. 3.1.

If player 2 plays the strategy L with a probability higher (lower) than two thirds, then player 1's best response is to play U (play D). If player 2 plays the strategy L with a probability equal to two thirds, then any mixed strategy (pure or unpure) of U and D becomes a best response for player 1. Likewise, if player 1 plays the strategy U with a probability higher (lower) than one third, then player 2's best response is to play L (play R). If player 1 plays the strategy U with a probability equal to one third, then any mixed strategy (pure or unpure) of L and R is a best response for player 2. ■

3.2 Rationalizable Strategies

Recall that given a strategic form game G^Δ where individuals choose their strategies independently, $\Delta(S_i)$ denotes the (mixed) strategy space of player $i \in N$. Consider, for example, player 1. If she believes that each player $j \neq 1$ will choose her strategy from $\Delta(S_j)$, then player 1 should choose a strategy in $\Delta(S_1)$ that is best response to some strategy profile of her opponents in $\times_{j\neq 1} \Delta(S_j)$, since it would not be rational for player 1 to choose a strategy that is never a best response. If player 1 has strategies in $\Delta(S_1)$ that are never best response, then by eliminating them, player 1 will end up with a smaller set of strategies, say Δ_1', each of which can be argued to be rational. But are they really? To answer this, suppose, for simplicity of the argument, $N = \{1, 2\}$, and that the rationality and payoff of each player is common knowledge. Pick a strategy σ_1 from the reduced set Δ_1' we have defined above. Then, σ_1 must be a best response to some strategy $\sigma_2 \in \Delta(S_2)$. Further, suppose that σ_1 is not a best response to any strategy in $\Delta(S_2)$ other than σ_2. Now, the question: What if σ_2 is a strategy that is never best response for player 2 to any strategy in $\Delta(S_1)$? If player 2 will never play σ_2, then player 1 can predict this precisely using her knowledge about player 2 and conclude that she must never play σ_1 because σ_1 is a best response to σ_2 only. But, if σ_1 is a strategy to be eliminated by player 1, then player 2, anticipating this elimination, may find some strategy σ_2' of his to be eliminated if σ_1 is the only strategy in $\Delta(S_1)$ to which σ_2' is a best response for player 2. This may possibly lead player 1 to eliminate some additional strategies, and then may similarly lead player 2 to eliminate some strategies, so on, and so forth, until no player can find any strategy to be eliminated for the reason that it is never best response. This iterative elimination procedure yields for each player a set of strategies, called the rationalizable set. To formally define this set, let us first introduce the following.

Consider any sets of strategies $(R_i)_{i\in N}$, with $R_i \subseteq \Delta(S_i)$ for each $i \in N$. Let $BR_i(R_{-i})$ for any $i \in N$ denote the set of strategies in $\Delta(S_i)$ each of which is a best response to some strategy profile in R_{-i}. We will be interested in the sets of strategies $(R_i)_{i\in N}$ that satisfy the following best-response condition:

$$R_i \subseteq BR_i(R_{-i}) \;\; \text{for each } i \in N \tag{3.1}$$

Definition 3.2 Given a strategic form game $G^\Delta = \langle N, (\Delta(S_i))_{i\in N}, (\tilde{u}_i)_{i\in N}\rangle$, the sets of strategies $(R_i^*)_{i\in N}$, with $R_i^* \subseteq \Delta(S_i)$ for each $i \in N$, are called the rationalizable sets of the players in N if (i) they satisfy the best-response condition (3.1) and (ii) there exist no other sets of strategies $(R_i)_{i\in N}$, with $R_i \subseteq \Delta(S_i)$ for each $i \in N$, that satisfy the best-response condition (3.1) and the relation $R_i \supsetneq R_i^*$ for some $i \in N$.

Condition (ii) in the above definition requires that the rationalizable sets of strategies for the society N must be the collection of "largest" sets $(R_i)_{i\in N}$, with $R_i \subseteq \Delta(S_i)$ for each $i \in N$, that satisfy the best-response condition (3.1).

We know by the works of Bernheim [1] and Pearce [2] that for every strategic form game the rationalizable sets of strategies exist. To find these sets, we can consider the iterative elimination of never best-response strategies in the same way as the iterative elimination of strongly dominated strategies. We can easily see that the order of players cannot affect the final outcome of the elimination procedure because if a strategy of any player is never best response at the beginning of the procedure, then it remains to be never best response irrespective of which strategies of her opponents are eliminated during the procedure and when they are eliminated.

Here, we can ask whether we can benefit from the procedure of the iterative elimination of strongly dominated strategies to calculate the sets of rationalizable strategies. This question is appropriate since in any strategic form game a strategy is never best response if it is strongly dominated. Thus eliminating strategies that are strongly dominated, we get closer to characterizing the sets of rationalizable strategies. The remaining question is whether a strategy is never best response only if it is strongly dominated. The answer to this question is always yes if the strategic form game involves only two players ($n = 2$). However, if the number of players is at least three ($n \geq 3$), then the answer is yes in general if and only if the opponents of each player have correlated strategies (i.e., they can coordinate their strategies through randomization). Note that in describing the game G^{Δ}, we have assumed that the mixed strategies of players are independent; i.e., for any strategy profile σ, we have $\sigma_i \in \Delta(S_i)$ for each $i \in N$. Under the restrictions that $n \geq 3$ and the strategies of players are independent, one can find strategic form games where some strategies of some players can be never best response even if they are not strongly dominated.

In Example 2.5, we considered a two-person strategic form game where each player has three pure strategies, and using the mixed extension of that game we calculated for each player the set of strategies surviving the iterative elimination of strongly dominated strategies. Below, we will directly show that for each player this set is equal to the rationalizable set of strategies.

Example 3.2 Reconsider the strategic form game in Example 2.5. We will iteratively eliminate the pure strategies that are never best response.

		Player 2	
Player 1	l	m	r
U	(5, 5)	(3, 3)	(2, 1)
M	(4, 4)	(5, 5)	(0, 4)
D	(2, 2)	(7, 7)	(3, 6)

Table 2.3 Reconsidered

Since the order of players in elimination is not consequential, let us start with player 1. We can check that $U \in BR_1(l)$, $D \in BR_1(m) \cap BR_1(r)$, while M is never best response. So, we eliminate M from the set of strategies of player 1 and obtain the set $\{U, D\}$. For player 2, we can check that $l \in BR_2(U)$ and $m \in BR_2(D)$, while r is never best response. So, we eliminate r from the set of strategies of player 2 and obtain the set $\{l, m\}$. Let $Z_1^* = \{U, D\}$ and $Z_2^* = \{l, m\}$. Note that $Z_1^* = BR_1(Z_2^*)$

and $Z_2^* = BR_2(Z_1^*)$. Hence, the sets of strategies $(Z_i^*)_{i \in N}$ satisfy the best-response condition (3.1) and there exist no other sets of strategies $(Z_i)_{i \in N}$, with $Z_i \subseteq S_i$ for each $i \in N$, that satisfy the best-response condition (3.1) and the relation $Z_i \supsetneq Z_i^*$ for some $i \in N$. Therefore, the sets $(Z_i^*)_{i \in N}$ are the rationalizable sets of strategies when mixed strategies are not allowed.

Now, we suppose that mixed strategies are allowed. Now, we can consider the mixed extension of the game in Table 2.3. For this extension, we will first find the set of strategy profiles that survive the Iterative Elimination of Never Best Response Strategies (IENBRS) and check whether this set is equal to the set of strategy profiles that survive the Iterative Elimination of Strictly Dominated Strategies (IESDS). As the order of players in elimination is inconsequential, we can start with player 2. Strategy r is never best response for player 2 and we can eliminate it to obtain the reduced game in Table 3.4.

Now, we turn to player 1. Looking at the best responses, we observe $U \in BR_1(l)$ and $D \in BR_1(m)$. On the other hand, M is not best response to any pure strategy of player 2. To check whether it is best response to any unpure strategy of player 2, consider the strategy $\sigma_2 = (\sigma_{12}, \sigma_{22})$ of player 2. Then,

$$\begin{aligned} \tilde{u}_1(M, \sigma_2) &= 4\sigma_{12} + 5(1 - \sigma_{12}) = 5 - \sigma_{12}, \\ \tilde{u}_1(U, \sigma_2) &= 5\sigma_{12} + 3(1 - \sigma_{12}) = 3 + 2\sigma_{12}, \\ \tilde{u}_1(D, \sigma_2) &= 2\sigma_{12} + 7(1 - \sigma_{12}) = 7 - 5\sigma_{12}. \end{aligned}$$

In order M to be a best response to σ_2, we must have

$$\tilde{u}_1(M, \sigma_2) - \tilde{u}_1(U, \sigma_2) = 2 - 3\sigma_{12} \geq 0 \quad \text{and}$$

$$\tilde{u}_1(M, \sigma_2) - \tilde{u}_1(D, \sigma_2) = -2 + 4\sigma_{12} \geq 0.$$

These inequalities both hold if and only if $1/2 \leq \sigma_{12} \leq 2/3$. Note that we do not need to check that M fares weakly better than the mixed strategies $(\sigma_{11}, 0, 1 - \sigma_{11})$, thanks to the linearity of expected utilities with respect to σ_{ji} for $i \in \{1, 2\}$ and $j \in \{1, \ldots, |S_i|\}$. Thus, each pure strategy of player 1 is a best-response.

Let $T_1 = S_1$ and $T_2 = \{l, m\}$. So, we have established that $T_1 = BR_1(T_2)$ and $T_2 = BR_2(T_1)$. After this point, no elimination of pure strategies is possible. Thus, the set of pure strategy profiles that survive IENBRS is T_1 and T_2 for player 1 and player 2, respectively. Recall that this is the set of pure strategy profiles that survive IESDS, as we showed in Example 2.5. When we allowed mixed strategies in that

Table 3.4 The reduced game (from Table 2.3) where each pure strategy is a best-response

		Player 2	
		l	m
Player 1	U	$(5, 5)$	$(3, 3)$
	M	$(4, 4)$	$(5, 5)$
	D	$(2, 2)$	$(7, 7)$

example, we also showed that for player 1 the set of all strategy profiles that survive IESDS is

$$[\Delta(S_1^{\infty,sd})]^* = \left\{(\sigma_{11}, \sigma_{21}, \sigma_{31}) \in [0,1]^3 : \sigma_{11} + \sigma_{21} + \sigma_{31} = 1 \text{ and } \sigma_{11}\sigma_{31} = 0\right\},$$

whereas, for player 2, the set of all strategy profiles that survive IESDS is

$$[\Delta(S_2^{\infty,sd})]^* = \left\{(\sigma_{12}, \sigma_{22}) \in [0,1]^2 : \sigma_{12} + \sigma_{22} = 1\right\}.$$

Since a strategy profile that is strongly dominated cannot be a best response, the set of mixed strategy profiles in T_i that survive IENBRS (which we may denote by $[\Delta(T_i)]^*$ for any $i = 1, 2$) cannot be larger than $[\Delta(S_i^{\infty,sd})]^*$; i.e., it is true that

$$[\Delta(T_i)]^* \subseteq [\Delta(S_i^{\infty,sd})]^* \text{ for each } i = 1, 2.$$

We will show that the above inclusions hold with equality; i.e.,

$$[\Delta(T_i)]^* = [\Delta(S_i^{\infty,sd})]^* \text{ for each } i = 1, 2.$$

To show this, we will prove that

$$[\Delta(S_i^{\infty,sd})]^* \subseteq BR_i([\Delta(S_{-i}^{\infty,sd})]^*) \text{ for each } i = 1, 2.$$

The proof of the statement $[\Delta(S_1^{\infty,sd})]^* \subseteq BR_1([\Delta(S_2^{\infty,sd})]^*)$:

Let us pick any $\sigma_1 = (\sigma_{11}, 1 - \sigma_{11}, 0)$ where $\sigma_{11} \in [0, 1]$. Recall that $\sigma_1 \in [\Delta(S_1^{\infty,sd})]^*$. Let $\sigma_2 = (\sigma_{12}, 1 - \sigma_{12})$ for some $\sigma_{12} \in [0, 1]$. Note that $\sigma_2 \in [\Delta(S_2^{\infty,sd})]^*$. Then,

$$\tilde{u}_1(\sigma_1, \sigma_2) = 5\sigma_{11}\sigma_{12} + 3\sigma_{11}(1 - \sigma_{12}) + 4(1 - \sigma_{11})\sigma_{12} + 5(1 - \sigma_{11})(1 - \sigma_{12}).$$

It follows that

$$\frac{\partial \tilde{u}_1(\sigma_1, \sigma_2)}{\partial \sigma_{12}} = 5\sigma_{12} + 3(1 - \sigma_{12}) - 4\sigma_{12} - 5(1 - \sigma_{12}) = 3\sigma_{12} - 2.$$

So, if $\sigma_{12} = 2/3$ (implying $\sigma_2 = (2/3, 1/3)$), then the expected utility of player 1 becomes equal to $13/3$ for any strategy $(\sigma_{11}, 1 - \sigma_{11}, 0)$ in $[\Delta(S_1^{\infty,sd})]^*$. On the other hand, for any strategy $(0, \sigma'_{11}, 1 - \sigma'_{11})$ in $\Delta(S_1^{\infty,sd})]^*$, her expected utility would always between $13/3$ (implied by the strategy $(0, 1, 0)$) and $11/3$ (implied by the strategy $(0, 0, 1)$). Therefore, $\sigma_1 \in BR_1((2/3, 1/3))$.

Now, consider any $\sigma_1 = (0, \sigma_{11}, 1 - \sigma_{11})$ where $\sigma_{11} \in [0, 1]$. Recall that $\sigma_1 \in [\Delta(S_1^{\infty,sd})]^*$. Note that the payoffs induced by strategy $(0, 1, 0)$ of player 1 sum up to 9 and this is also true for strategy $(0, 0, 1)$. So, if $\sigma_2 = (1/2, 1/2)$, the average expected utility induced by each of these two strategies becomes

$9/2$, and therefore $\tilde{u}_1(\sigma_1, (1/2, 1/2)) = 9/2$. On the other hand, for any $\sigma_1' = (\sigma_{11}', 1 - \sigma_{11}', 0) \in [\Delta(S_1^{\infty,s})]^*$, we have $\tilde{u}_1(\sigma_1', (1/2, 1/2)) \leq 9/2$. This is because $\tilde{u}_1((1, 0, 0), (1/2, 1/2)) = 4$ and $\tilde{u}_1((0, 1, 0), (1/2, 1/2)) = 9/2$. Therefore, $\sigma_1 \in BR_1((1/2, 1/2))$.

Thus, we have established that $[\Delta(S_1^{\infty,sd})]^* \subseteq BR_1([\Delta(S_2^{\infty,sd})]^*)$.

The proof of the statement $[\Delta(S_2^{\infty,sd})]^* \subseteq BR_2([\Delta(S_1^{\infty,sd})]^*)$:

Let us pick any $\sigma_2 = (\sigma_{12}, 1 - \sigma_{12})$ with $\sigma_{12} \in [0, 1]$. Recall that $\sigma_2 \in [\Delta(S_2^{\infty,sd})]^*$. Let $\sigma_1 = (\sigma_{11}, 1 - \sigma_{11}, 0)$ with $\sigma_{11} \in [0, 1]$. Note that $\sigma_1 \in [\Delta(S_1^{\infty,sd})]^*$. Then,

$$\tilde{u}_2(\sigma_1, \sigma_2) = 5\sigma_{11}\sigma_{12} + 3\sigma_{11}(1 - \sigma_{12}) + 4(1 - \sigma_{11})\sigma_{12} + 5(1 - \sigma_{11})(1 - \sigma_{12}).$$

If we differentiate this expected utility with respect to σ_{12}, we obtain

$$\frac{\partial \tilde{u}_2(\sigma_1, \sigma_2)}{\partial \sigma_{12}} = 5\sigma_{11} - 3\sigma_{11} + 4(1 - \sigma_{11}) - 5(1 - \sigma_{11}) = 3\sigma_{11} - 1.$$

So, if $\sigma_{11} = 1/3$ (implying $\sigma_1 = (1/3, 2/3, 0)$), then the expected utility of player 2 is maximized at any $\sigma_2 \in [\Delta(S_2^{\infty,sd})]^*$. Hence, $\sigma_2 \in BR_2((1/3, 2/3, 0))$ if $\sigma_2 \in [\Delta(S_2^{\infty,sd})]^*$. Thus, we have established that $[\Delta(S_2^{\infty,sd})]^* \subseteq BR_2([\Delta(S_1^{\infty,sd})]^*)$. ■

References

1. Bernheim D (1984) Rationalizable strategic behavior. Econometrica 52:1007–1028
2. Pearce D (1984) Rationalizable strategic behavior and the problem of perfection. Econometrica 52:1029–1050

Nash Equilibrium

4

In this chapter, we will introduce the Nash equilibrium, one of the most basic (and famous) concepts in game theory. This concept was formulated, for strategic form games, by John F. Nash [1], who also showed that the mixed extension of every finite strategic form game has a Nash equilibrium. Section 4.1 will give definitions for the Nash equilibrium concept, followed by several examples where the Nash equilibria of the given strategic form games will be calculated using best-response correspondences. This section will also state Nash's [1] existence theorem (which will be proven in the appendix of this chapter after some mathematical preliminaries). Next, Sect. 4.2 will calculate the Nash equilibria of some well-known strategic form games that were presented in Chap. 1. A short section (Sect. 4.3) will define a strict version of the Nash equilibrium, followed by Sect. 4.4 that will relate the Nash equilibrium of a strategic form game to equilibrium notions that uses dominance. Section 4.5 will introduce the risk dominance concept for two-person strategic form games and Sect. 4.6 will discuss epistemic conditions leading to Nash equilibrium. Finally, Sect. 4.7 will present some mathematical preliminaries and the proof of Nash's [1] Existence Theorem.

4.1 Definitions, Examples, and Existence Theorem

We will first define the Nash equilibrium concept for strategic form games that do not allow randomization.

Definition 4.1 Given a strategic form game $G = \langle N, (S_i)_{i \in N}, (u_i)_{i \in N} \rangle$, a (pure) strategy profile $s^* \in S$ is a Nash equilibrium if

$$u_i(s_i^*, s_{-i}^*) \geq \tilde{u}_i(s_i, s_{-i}^*) \quad \text{for all } i \in N \text{ and } s_i \in S_i.$$

I. Saglam, *Mastering Game Theory*, Classroom Companion: Economics,
https://doi.org/10.1007/978-3-031-82094-6_4

Below, we extend the above definition of the Nash equilibrium concept to strategic form games that allow for randomization.

Definition 4.2 Given a strategic form game $G^{\Delta} = \langle N, (\Delta(S_i))_{i\in N}, (\tilde{u}_i)_{i\in N}\rangle$, a strategy profile $\sigma^* \in \Delta^N(S)$ is a Nash equilibrium if

$$\tilde{u}_i(\sigma_i^*, \sigma_{-i}^*) \geq \tilde{u}_i(\sigma_i, \sigma_{-i}^*) \quad \text{for all } i \in N \text{ and } \sigma_i \in \Delta(S_i).$$

Recall that, in Sect. 3.1, we defined for player $i \in N$ the best-response correspondence $BR_i : \bigtimes_{j\in N-i} \Delta(S_j) \to \Delta(S_i)$ such that for any $\sigma_{-i} \in \bigtimes_{j\in N-i} \Delta(S_j)$ we have

$$BR_i(\sigma_{-i}) = \{\sigma_i \in \Delta(S_i) : \tilde{u}_i(\sigma_i, \sigma_{-i}) \geq \tilde{u}_i(\sigma_i', \sigma_{-i}) \text{ for all } \sigma_i' \in \Delta(S_i)\}.$$

Using the best-response correspondences, Definition 4.2 can be equivalently written as follows:

Definition 4.3 Given a strategic form game $G^{\Delta} = \langle N, (\Delta(S_i))_{i\in N}, (\tilde{u}_i)_{i\in N}\rangle$, a strategy profile $\sigma^* \in \Delta^N(S)$ is a Nash equilibrium if $\sigma_i^* \in BR_i(\sigma_{-i}^*)$ for all $i \in N$.

Some strategic form games, like in the following example, have no Nash equilibrium in pure strategies. However, a celebrated result by John F. Nash [1], which we will formally state in this section, ensures that every finite strategic form game has a mixed (pure or unpure) strategy Nash equilibrium. Therefore, a strategic form game that has no Nash equilibrium in pure strategies must have a Nash equilibrium in unpure strategies.

Example 4.1 Consider the Matching Pennies Game in the following table (Table 4.1).

We have emphasized above the best-response payoffs of player 1 and player 2 by red and blue colors, respectively. Whenever both payoffs in any cell are colored, the strategy profile associated with that cell is a pure-strategy Nash equilibrium. One can easily see that there exists no Nash equilibrium in pure strategies. According to

Table 4.1 The matching pennies game in Example 4.1

Player 1 \ Player 2	H	T
H	$(1, -1)$	$(-1, 1)$
T	$(-1, 1)$	$(1, -1)$

Nash's [1] existence result that we will state in Proposition 4.1, this game must have an equilibrium. So, this equilibrium must be in unpure strategies. To validate this, let us formally calculate the best-response correspondence of each player.

Note that, for any $\sigma_2 = (\sigma_{12}, \sigma_{22}) = (\sigma_{12}, 1 - \sigma_{12})$, we have

$$u_1(H, \sigma_2) = (1)\sigma_{12} + (-1)(1 - \sigma_{12}) = 2\sigma_{12} - 1$$
$$u_1(T, \sigma_2) = (-1)\sigma_{12} + (1)(1 - \sigma_{12}) = 1 - 2\sigma_{12}.$$

So, the best-response correspondence of player 1 is given by

$$BR_1(\sigma_2) = \begin{cases} \{(1, 0)\} \text{ if } \sigma_{12} > 0.5 \\ \Delta(S_1) \quad \text{if } \sigma_{12} = 0.5 \\ \{(0, 1)\} \text{ if } \sigma_{12} < 0.5. \end{cases}$$

Also, note that, for any $\sigma_1 = (\sigma_{11}, \sigma_{21}) = (\sigma_{11}, 1 - \sigma_{11})$, we have

$$u_2(\sigma_1, H) = (-1)\sigma_{11} + (1)(1 - \sigma_{11}) = 1 - 2\sigma_{11}$$
$$u_1(\sigma_1, T) = (1)\sigma_{11} + (-1)(1 - \sigma_{11}) = 2\sigma_{11} - 1.$$

So, the best-response correspondence of player 2 is given by

$$BR_2(\sigma_1) = \begin{cases} \{(0, 1)\} \text{ if } \sigma_{11} > 0.5 \\ \Delta(S_2) \quad \text{if } \sigma_{11} = 0.5 \\ \{(1, 0)\} \text{ if } \sigma_{11} < 0.5. \end{cases}$$

We draw the best-response correspondences in Fig. 4.1.

Remember that $\sigma^* \in \Delta^N(S)$ is a Nash equilibrium if and only if $\sigma_1^* \in BR_1(\sigma_2^*)$ and $\sigma_2^* \in BR_2(\sigma_1^*)$. Thus, we intersect the best-response correspondences $BR_1(.)$ and $BR_2(.)$ and obtain the unique Nash equilibrium $\sigma^* = ((0.5, 0.5), (0.5, 0.5))$, where each player randomizes between H and T with probability 0.5. ■

A strategic-form game may have equilibria in both pure and unpure strategies. Moreover, the equilibria in unpure strategies may be uncountably many, as the next example illustrates.

Example 4.2 Consider the strategic form game represented by Table 4.2.

We can easily see from the colored payoff profiles that (U, L) and (U, R) are pure-strategy Nash equilibria of the given game. To check whether this game has any unpure-strategy equilibrium as well, let us formally calculate the best-response correspondence of each player.

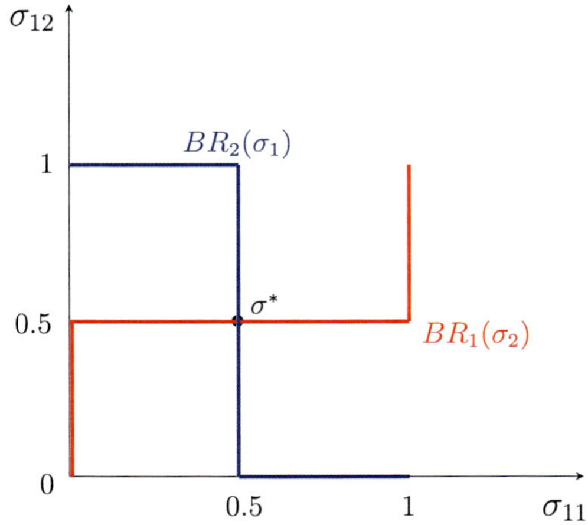

Fig. 4.1 Best-response correspondences in matching pennies game

Table 4.2 The strategic form game in Example 4.2

Player 1 \ Player 2	L	R
U	(2, 1)	(2, 1)
D	(0, 0)	(1, 2)

Note that, for any $\sigma_2 = (\sigma_{12}, \sigma_{22}) = (\sigma_{12}, 1 - \sigma_{12})$, we have

$$u_1(U, \sigma_2) = (2)\sigma_{12} + (2)(1 - \sigma_{12}) = 2$$
$$u_1(D, \sigma_2) = (0)\sigma_{12} + (1)(1 - \sigma_{12}) = 1 - \sigma_{12} \in [0, 1].$$

Therefore, playing U (the mixed strategy $(1, 0)$) is the unique best response of player 1 to any mixed strategy of player 2; i.e.,

$$BR_1(\sigma_2) = \{(1, 0)\} \text{ for any } \sigma_2 \in \Delta(S_2).$$

Also note that, for any $\sigma_1 = (\sigma_{11}, \sigma_{21}) = (\sigma_{11}, 1 - \sigma_{11})$, we have

$$u_2(\sigma_1, L) = (1)\sigma_{11} + (0)(1 - \sigma_{11}) = \sigma_{11} \in [0, 1]$$
$$u_2(\sigma_1, R) = (1)\sigma_{11} + (2)(1 - \sigma_{11}) = 2 - \sigma_{11} \in [1, 2].$$

It follows that playing R (the mixed strategy $(0, 1)$) is the unique best response of player 2 unless player 1 plays U with probability one, i.e., $\sigma_{11} = 1$. If player 1 plays

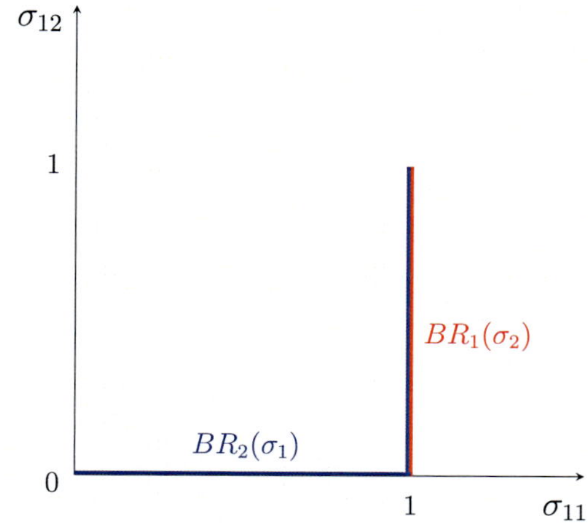

Fig. 4.2 Best-response correspondences for the game in Table 4.2

U with probability one, then any mixture of L and R is a best response for player 2. Thus,

$$BR_2(\sigma_1) = \begin{cases} \{(0,1)\} & \text{if } \sigma_{11} \in [0,1), \\ \Delta(S_2) & \text{if } \sigma_{11} = 1. \end{cases}$$

We draw the best-response correspondences in Fig. 4.2.

Remember that $\sigma^* \in \Delta^N(S)$ is a Nash equilibrium if and only if $\sigma_1^* \in BR_1(\sigma_2^*)$ and $\sigma_2^* \in BR_2(\sigma_1^*)$. Thus, we intersect the best-response correspondences $BR_1(.)$ and $BR_2(.)$ and obtain the set of Nash equilibria

$$NE = \{(\sigma_1^*, \sigma_2^*) : \sigma_1^* = (1,0) \text{ and } \sigma_2^* \in \Delta(S_2)\},$$

which contains uncountably many unpure equilibrium profiles in addition to the pure equilibrium profiles $((1,0),(1,0))$ and $((1,0),(0,1))$. ■

Above, we have considered examples involving 2×2 games. When one of the players have more than two strategies, finding a Nash equilibrium may be more tedious as illustrated by the next example.

Example 4.3 Consider the strategic form game in Table 4.3. Apparently, this game has no Nash equilibrium in pure strategies. Below, we will calculate the Nash equilibria in unpure strategies.

Consider any strategy profile $\sigma \in \Delta^N(S)$. Recall that $\sigma_1 = (\sigma_{11}, \sigma_{21})$ and $\sigma_2 = (\sigma_{12}, \sigma_{22}, \sigma_{32})$. We can calculate the expected payoffs of player 1 and player 2 as

$$u_1(U, \sigma_2) = 0\sigma_{12} + 2\sigma_{22} + 4\sigma_{32} = 2\sigma_{22} + 4(1 - \sigma_{12} - \sigma_{22}) = 4 - 4\sigma_{12} - 2\sigma_{22}$$

$$u_1(D, \sigma_2) = 4\sigma_{12} + 2\sigma_{22} + 0\sigma_{32} = 4\sigma_{12} + 2\sigma_{22}$$

Table 4.3 The strategic form game in Example 4.3

Player 1 \ Player 2	L	M	R
U	(0, 8)	(2, 4)	(4, −1)
D	(4, 0)	(2, 4)	(0, 7)

and

$$\begin{aligned} u_2(\sigma_1, L) &= 8\sigma_{11} \\ u_2(\sigma_1, M) &= 4\sigma_{11} + 4\sigma_{21} = 4 \\ u_2(\sigma_1, R) &= (-1)\sigma_{11} + 7\sigma_{21} = 7 - 8\sigma_{11}, \end{aligned}$$

respectively. To find which σ profile can be a Nash equilibrium, we have to consider the following four possible cases:

Case 1: $\sigma_{12}\sigma_{22}\sigma_{32} > 0$.

If σ is a Nash equilibrium, then player 2 has to play each of its three strategies L, M, and R with positive probabilities. This implies that the expected payoffs of its three strategies must be equal. Note that $u_2(\sigma_1, L) = u_2(\sigma_1, M)$ implies $\sigma_{11} = 1/2$, whereas $u_2(\sigma_1, M) = u_2(\sigma_1, R)$ implies $\sigma_{11} = 3/8$, a contradiction. Therefore, there exists no Nash equilibrium in Case 1.

Case 2: $\sigma_{12} = 0$ and $\sigma_{22}\sigma_{32} > 0$.

If σ is a Nash equilibrium, then we must have $u_2(\sigma_1, L) < u_2(\sigma_1, M) = u_2(\sigma_1, R)$. We have $u_2(\sigma_1, M) = u_2(\sigma_1, R)$ only if $\sigma_{11} = 3/8$. In that case, $u_2(\sigma_1, L) = 3 < 4 = u_2(\sigma_1, M) = u_2(\sigma_1, R)$. Now, let us calculate the expected payoffs of player 1. Since $\sigma_{12} = 0$, we have $u_1(U, \sigma_2) = 4 - 2\sigma_{22}$ and $u_1(D, \sigma_2) = 2\sigma_{22}$, implying that

$$BR_1(\sigma_2) = \begin{cases} \{(1, 0)\} & \text{if } \sigma_{22} \in [0, 1) \\ \Delta(S_1) & \text{if } \sigma_{22} = 1. \end{cases}$$

Since we also found that $\sigma_1 = (3/8, 5/8)$, we have $\sigma_1 \in BR_1(\sigma_2)$ only if $\sigma_{22} = 1$. But, in that case $\sigma_{32} = 0$, contradicting $\sigma_{22}\sigma_{32} > 0$. Therefore, there exists no Nash equilibrium in Case 2.

Case 3: $\sigma_{22} = 0$ and $\sigma_{12}\sigma_{32} > 0$.

If σ is a Nash equilibrium, then we must have $u_2(\sigma_1, M) < u_2(\sigma_1, L) = u_2(\sigma_1, R)$. We have $u_2(\sigma_1, L) = u_2(\sigma_1, R)$ only if $\sigma_{11} = 7/16$. In that case, $u_2(\sigma_1, M) =$

$4 > 7/2 = u_2(\sigma_1, L) = u_2(\sigma_1, R)$. Therefore, $BR_2(\sigma_1) = \{(0, 1, 0)\}$, i.e., the best response of player 2 to σ_1 is to play M, contradicting $\sigma_{12}\sigma_{32} > 0$. Therefore, there exists no Nash equilibrium in Case 3.

Case 4: $\sigma_{32} = 0$ and $\sigma_{12}\sigma_{22} > 0$.

If σ is a Nash equilibrium, then we must have $u_2(\sigma_1, R) < u_2(\sigma_1, L) = u_2(\sigma_1, M)$. We have $u_2(\sigma_1, L) = u_2(\sigma_1, M)$ only if $\sigma_{11} = 1/2$. In that case, $u_2(\sigma_1, R) = 3 < 4 = u_2(\sigma_1, L) = u_2(\sigma_1, M)$. Now, let us calculate the expected payoffs of player 1. Since $u_1(U, \sigma_2) = 2\sigma_{22} = 2 - 2\sigma_{12}$ and $u_1(D, \sigma_2) = 4\sigma_{12} + 2\sigma_{22} = 2 + 2\sigma_{12}$, we have

$$BR_1(\sigma_2) = \begin{cases} \Delta(S_1) & \text{if } \sigma_{12} = 0 \\ \{(0, 1)\} & \text{if } \sigma_{12} \in (0, 1]. \end{cases}$$

Since we also found that $\sigma_1 = (1/2, 1/2)$, we have $\sigma_1 \in BR_1(\sigma_2)$ only if $\sigma_{12} = 0$. But, this contradicts $\sigma_{12}\sigma_{22} > 0$. Therefore, there exists no Nash equilibrium in Case 4.

Case 5: $\sigma_2 = (1, 0, 0)$.

If σ is a Nash equilibrium, then $u_2(\sigma_1, L) > \max\{u_2(\sigma_1, M),\ u_2(\sigma_1, R)\}$. We have $u_2(\sigma_1, L) > u_2(\sigma_1, M)$ only if $\sigma_{11} > 1/2$ and $u_2(\sigma_1, L) > u_2(\sigma_1, R)$ only if $\sigma_{11} > 7/16$. Thus, σ is a Nash equilibrium only if $\sigma_{11} > 7/16$. Also, note that $u_1(U, \sigma_2) = 0$ and $u_1(D, \sigma_2) = 4$, implying $BR_1(\sigma_2) = \{(0, 1)\}$. Since we also found that $\sigma_{11} > 7/16$, we have $\sigma_1 \notin BR_1(\sigma_2)$. Therefore, there exists no Nash equilibrium in Case 5.

Case 6: $\sigma_2 = (0, 1, 0)$.

If σ is a Nash equilibrium, then $u_2(\sigma_1, M) > \max\{u_2(\sigma_1, L),\ u_2(\sigma_1, R)\}$. We have $u_2(\sigma_1, M) > u_2(\sigma_1, L)$ only if $\sigma_{11} < 1/2$ and $u_2(\sigma_1, M) > u_2(\sigma_1, R)$ only if $\sigma_{11} > 3/8$. Thus, σ is a Nash equilibrium only if $\sigma_{11} \in (3/8, 1/2)$. Also, note that $u_1(U, \sigma_2) = 2$ and $u_1(D, \sigma_2) = 2$, implying $BR_1(\sigma_2) = \Delta(S_1)$. Clearly, $\sigma_1 \in BR_1(\sigma_2)$ if $\sigma_{11} \in (3/8, 1/2)$. Thus, we have found that the set of Nash equilibria in Case 6 is

$$NE = \{\sigma^* \in \Delta^N(S) : \ \sigma^*_{11} \in (3/8, 1/2) \text{ and } \sigma^*_2 = (0, 1, 0)\}.$$

Case 7: $\sigma_2 = (0, 0, 1)$.

If σ is a Nash equilibrium, then $u_2(\sigma_1, R) > \max\{u_2(\sigma_1, L),\ u_2(\sigma_1, M)\}$. We have $u_2(\sigma_1, R) > u_2(\sigma_1, L)$ only if $\sigma_{11} < 7/16$ and $u_2(\sigma_1, R) > u_2(\sigma_1, M)$ only if $\sigma_{11} < 3/8$. Thus, σ is a Nash equilibrium only if $\sigma_{11} < 3/8$. Also, note that $u_1(U, \sigma_2) = 4$ and $u_1(D, \sigma_2) = 0$, implying $BR_1(\sigma_2) = \{(1, 0)\}$. Clearly, if $\sigma_{11} < 3/8$, then $\sigma_1 \notin BR_1(\sigma_2)$. Therefore, there exists no Nash equilibrium in Case 7.

Since we have covered all possible strategy profiles above, the set of Nash equilibria of the strategic form game in this example is equal to the set of Nash equilibria that we found in Case 6 above. ■

Before we conclude this part, we should note that if a strategic form game has more than one totally mixed strategy equilibrium, then it must have infinitely many. To prove this, suppose the game has more than one totally mixed strategy equilibrium. Pick any two of these equilibria and call them σ^a and σ^b. Since both equilibria are totally mixed, it is true that for any $i \in N$ and any $s_i', s_i'' \in S_i$, and any $k \in \{a, b\}$, we have $\tilde{u}_i(s_i', \sigma_{-i}^k) = \tilde{u}_i(s_i'', \sigma_{-i}^k)$. Now, consider the totally mixed strategy $\sigma^p = p\sigma^a + (1-p)\sigma^a$ for any $p \in (0, 1)$. Then, for any i and any $s_i', s_i'' \in S_i$, we have $\tilde{u}_i(s_i', \sigma_{-i}^p) = \tilde{u}_i(s_i'', \sigma_{-i}^p)$. Therefore, σ_i^p is a best response to σ_{-i}^p, completing the proof.

Another interesting feature of Nash equilibria is that the number of Nash equilibria is odd in almost all finite strategic form games, as proven by Wilson [2]. There are exceptional games where the number of equilibria is even. We illustrate one example below.

Example 4.4 Consider the strategic form game represented by Table 4.4, where $a > b$.

Let us find the set of all Nash equilibria of the mixed extension of the above game. Note that, for any $\sigma_2 = (\sigma_{12}, \sigma_{22}) = (\sigma_{12}, 1 - \sigma_{12})$, we have

$$u_1(U, \sigma_2) = (a)\sigma_{12} + (b)(1 - \sigma_{12}) = b + (a - b)\sigma_{12}$$
$$u_1(D, \sigma_2) = (b)\sigma_{12} + (b)(1 - \sigma_{12}) = b.$$

So, the best-response correspondence of player 1 is given by

$$BR_1(\sigma_2) = \begin{cases} \{(1, 0)\} \text{ if } \sigma_{12} > 0 \\ \Delta(S_1) \quad \text{if } \sigma_{12} = 0. \end{cases}$$

Also, note that, for any $\sigma_1 = (\sigma_{11}, \sigma_{21}) = (\sigma_{11}, 1 - \sigma_{11})$, we have

$$u_2(\sigma_1, L) = (a)\sigma_{11} + (b)(1 - \sigma_{11}) = b + (a - b)\sigma_{11}$$
$$u_1(\sigma_1, R) = (b)\sigma_{11} + (b)(1 - \sigma_{11}) = b.$$

So, the best-response correspondence of player 2 is given by

$$BR_2(\sigma_1) = \begin{cases} \{(1, 0)\} \text{ if } \sigma_{11} > 0 \\ \Delta(S_2) \quad \text{if } \sigma_{11} = 0. \end{cases}$$

We draw the best-response correspondences in Fig. 4.3.

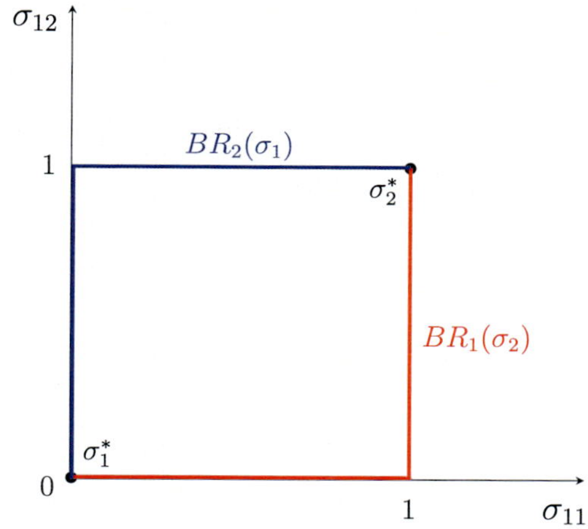

Fig. 4.3 Best-response correspondences for the game in Table 4.4

Remember that $\sigma^* \in \Delta^N(S)$ is a Nash equilibrium if $\sigma_1^* \in BR_1(\sigma_2^*)$ and $\sigma_2^* \in BR_2(\sigma_1^*)$. So, we intersect the best-response correspondences $BR_1(.)$ and $BR_2(.)$ and obtain two Nash equilibria, involving $\sigma_1^* = ((0, 1), (0, 1))$ corresponding to (D, R) and $\sigma_2^* = ((1, 0), (1, 0))$ corresponding to (U, L). Thus, we have established that the number of Nash equilibria is even for this (exceptional) game. ■

We should simply note that a Nash equilibrium strategy profile is always contained by the rationalizable sets of strategies, since each player's strategy in a Nash equilibrium is a best-response strategy by definition. However, every rationalizable strategy profile is not a Nash equilibrium. For example, one can easily see from the best-response correspondences calculated in Example 4.4 that the rationalizable sets of strategies for the game in Table 4.4 is $\Delta^N(S) = [0, 1] \times [0, 1]$, whereas the set of Nash equilibrium is $\{((0, 1), (0, 1)), ((1, 0), (1, 0))\}$. What distinguishes the Nash equilibrium in the sets of rationalizable strategies is the consistency of beliefs of players with the actual play. To see this, let us consider the strategy profile (U, R) or equivalently $((1, 0), (0, 1))$ in Example 4.4 and let us restrict ourselves to pure strategies, for simplicity. Note that, for player 1, the strategy U is a best response to R, whereas, for player 2, the strategy R is a best response to D. Here, one can argue that player 1 will play U since she believes that player 2 will play R since player 2 believes that player 1 will play D. But, since this chain of justification must be available to all players (as each of them is assumed to be intelligent to make the necessary

Table 4.4 The strategic form game in Example 4.4

Player 1 \ Player 2	L	R
U	(a, a)	(b, b)
D	(b, b)	(b, b)

calculations underlying this reasoning), how can player 2 be expected to play R in an equilibrium of the game knowing that player 1, as a rational player, would play U, but not D, as a best-response strategy, contradicting the beliefs of player 2? In fact, given a (rationalizable) strategy profile that is not a Nash equilibrium, we can always find at least two players i and j in N such that what player i is prescribed to play in the given strategy profile differs from the strategy player j believes that player i will play. On the other hand, in a Nash equilibrium profile (like (U, L) and (D, R) in the above example), we always see that the strategy to be played by any player is consistent with the beliefs of other players' about her strategy.

In the previous examples, we always found a Nash equilibrium in mixed (pure or unpure) strategies. In fact, a celebrated result by John F. Nash [1] shows that the mixed extension of every finite strategic form game has a Nash equilibrium (in pure or unpure strategies).

Proposition 4.1 (Nash's Existence Theorem, 1950a) *The mixed extension of every strategic form game with a finite number of players and pure strategies has a Nash equilibrium.*

Proof See the appendix at the end of this chapter. ∎

The above theorem is only a sufficiency result. There are infinite strategic form games that have a Nash equilibrium, as we will illustrate through Example 4.10 in the next section.

4.2 Nash Equilibria of Well-Known Games

Below, we will present some well-known 2×2 strategic form games. We will emphasize by red color each payoff of player 1 produced by a pure strategy that is a best response to a pure strategy of player 2, and likewise emphasize by blue color each payoff of player 2 produced by a pure strategy that is a best response to a pure strategy of player 1. Whenever both payoffs in any cell of the payoff table are colored, the strategy profile associated with that cell is a pure-strategy Nash equilibrium.

Table 4.5 Prisoner's dilemma game in Example 4.5

		Player 2	
		Deny	Confess
Player 1	Deny	$(-1, -1)$	$(-10, 0)$
	Confess	$(0, -10)$	$(-5, -5)$

Example 4.5 (*Prisoner's Dilemma*)

As the colored best-response payoffs imply, the prisoner's dilemma game in Table 4.5 has a Nash equilibrium in pure strategies, (Confess,Confess), which can also be denoted as $((0, 1), (0, 1))$. This game has no other Nash equilibrium ensured by the fact that Confess is the strongly dominant strategy for each player in this game. Formally, we can calculate for any player $i \in N$, the best-response correspondence $BR_i(.)$ such that $BR_i(\sigma_{-i}) = \{(0, 1)\}$ for any $\sigma_{-i} \in S_{-i}$. A strategy profile $\sigma^* \in \Delta^N(S)$ is then a Nash equilibrium only if $\sigma_i^* \in BR_i(\sigma_{-i}^*) = \{(0, 1)\}$ for each $i \in N$, implying that $\sigma^* = ((0, 1), (0, 1))$. ■

Example 4.6 (*Battle of Sexes*)

The battle of sexes game in Table 4.6 has two pure-strategy Nash equilibria (O, O) and (F, F), as implied by the colored response payoffs in the payoff table. To check whether the game has any unpure strategy equilibrium, let us calculate the best-response correspondences.

Note that $u_1(O, \sigma_2) = 2\sigma_{12}$ and $u_1(F, \sigma_2) = \sigma_{22} = 1 - \sigma_{12}$. Then, we can calculate the best-response correspondence of player 1 as

$$BR_1(\sigma_2) = \begin{cases} \{(0, 1)\} & \text{if } \sigma_{12} < 1/3 \\ \Delta(S_1) & \text{if } \sigma_{12} = 1/3 \\ \{(1, 0)\} & \text{if } \sigma_{12} > 1/3. \end{cases}$$

Also, note that $u_2(\sigma_1, O) = \sigma_{11}$ and $u_2(\sigma_1, F) = 2\sigma_{21} = 2 - 2\sigma_{11}$. Then, we can calculate the best-response correspondence of player 2 as

$$BR_2(\sigma_1) = \begin{cases} \{(0, 1)\} & \text{if } \sigma_{11} < 2/3 \\ \Delta(S_2) & \text{if } \sigma_{11} = 2/3 \\ \{(1, 0)\} & \text{if } \sigma_{11} > 2/3. \end{cases}$$

A profile σ^* is a Nash equilibrium if and only if $\sigma_1^* \in BR_1(\sigma_2^*)$ and $\sigma_2^* \in BR_2(\sigma_1^*)$. Therefore, the set of Nash equilibria is

$$NE = \{((1, 0), (1, 0)), ((0, 1), (0, 1)), ((2/3, 1/3), (1/3, 2/3))\}.$$

Table 4.6 Battle of sexes game in Example 4.6

		Player 2	
		O	F
Player 1	O	(2, 1)	(0, 0)
	F	(0, 0)	(1, 2)

Table 4.7 Stag hunt game in Example 4.7

		Player 2	
		S	H
Player 1	S	(s, s)	$(0, h)$
	H	$(h, 0)$	(h, h)

The first two (pure) equilibria can be written as (O, O) and (F, F) while the last (unpure) equilibrium can be written as $((2/3)O + (1/3)F, (1/3)O + (2/3)F)$. ■

Example 4.7 (*Stag Hunt Game*)

In Table 4.7, it is assumed that $s > h > 0$. Under this assumption, the stug hunt game has two pure-strategy Nash equilibria (S, S) and (H, H), as implied by the colored response payoffs in the payoff table. To check whether the game has any unpure mixed strategy equilibrium, let us calculate the best-response correspondences.

Note that $u_1(S, \sigma_2) = s\sigma_{12}$ and $u_1(H, \sigma_2) = h$. Then, we can calculate the best-response correspondence of player 1 as

$$BR_1(\sigma_2) = \begin{cases} \{(0, 1)\} & \text{if } \sigma_{12} < h/s \\ \Delta(S_1) & \text{if } \sigma_{12} = h/s \\ \{(1, 0)\} & \text{if } \sigma_{12} > h/s. \end{cases}$$

Also, note that $u_2(\sigma_1, S) = s\sigma_{11}$ and $u_2(\sigma_1, H) = h$. Then, we can calculate the best-response correspondence of player 2 as

$$BR_2(\sigma_1) = \begin{cases} \{(0, 1)\} & \text{if } \sigma_{11} < h/s \\ \Delta(S_2) & \text{if } \sigma_{11} = h/s \\ \{(1, 0)\} & \text{if } \sigma_{11} > h/s. \end{cases}$$

A profile σ^* is a Nash equilibrium if and only if $\sigma_1^* \in BR_1(\sigma_2^*)$ and $\sigma_2^* \in BR_2(\sigma_1^*)$. Therefore, the set of Nash equilibria is

$$NE = \{((1, 0), (1, 0)), ((0, 1), (0, 1)), ((h/s, 1 - (h/s)), (h/s, 1 - (h/s)))\}.$$

The first two (pure) equilibria can be written as (S, S) and (H, H), whereas the last (unpure) equilibrium can be written as $((h/s)S + (1 - (h/s))H, (h/s)S + (1 - (h/s))H)$. ■

Table 4.8 Game of chicken in Example 4.8

		Player 2	
		Go Straight	Swerve
Player 1	Go Straight	$(-100, -100)$	$(5, -10)$
	Swerve	$(-10, 5)$	$(0, 0)$

Example 4.8 (*Game of Chicken*)

As the colored best-response payoffs in Table 4.8 imply, the above game has two pure-strategy Nash equilibria (G, S) and (S, G), where $G =$ Go Straight and $S =$ Swerve. To check whether the game has any unpure mixed strategy equilibrium, let us calculate the best-response correspondences.

Note that $u_1(G, \sigma_2) = -100\sigma_{12} + 5(1 - \sigma_{12})$ and $u_1(S, \sigma_2) = -10\sigma_{12}$. We have $u_1(G, \sigma_2) > u_1(S, \sigma_2)$ if and only if $\sigma_{12} < 1/19$. Then, we can calculate the best-response correspondence of player 1 as

$$BR_1(\sigma_2) = \begin{cases} \{(1,0)\} \text{ if } \sigma_{12} < 1/19 \\ \Delta(S_1) \quad \text{if } \sigma_{12} = 1/19 \\ \{(0,1)\} \text{ if } \sigma_{12} > 1/19. \end{cases}$$

Also, note that $u_2(\sigma_1, G) = -100\sigma_{11} + 5(1 - \sigma_{11})$ and $u_1(\sigma_1, D) = -10\sigma_{11}$. We have $u_2(\sigma_1, G) > u_2(\sigma_1, S)$ if and only if $\sigma_{11} < 1/19$. Then, we can calculate the best-response correspondence of player 2 as

$$BR_2(\sigma_1) = \begin{cases} \{(1,0)\} \text{ if } \sigma_{11} < 1/19 \\ \Delta(S_2) \quad \text{if } \sigma_{11} = 1/19 \\ \{(0,1)\} \text{ if } \sigma_{11} > 1/19. \end{cases}$$

A profile σ^* is a Nash equilibrium if and only if $\sigma_1^* \in BR_1(\sigma_2^*)$ and $\sigma_2^* \in BR_2(\sigma_1^*)$. Therefore, the set of Nash equilibria is

$$NE = \{((1,0),(0,1)), ((0,1),(1,0)), ((1/19, 18/19), (1/19, 18/19))\}.$$

The first two (pure) equilibria can be written as (H, D) and (D, H), whereas the last (unpure) equilibrium can be written as $((1/19)G + (18/19)S,\ (1/19)G + (18/19)S)$. ■

Example 4.9 (*Hawk-Dove Game*)

In Table 4.9, it is assumed that $c > v > 0$. Then, the associated strategic form game has two pure-strategy Nash equilibria (D, H) and (H, D), as implied by the colored best-response payoffs. To check whether the game has any unpure mixed strategy equilibrium, let us calculate the best-response correspondences.

Table 4.9 Hawk-dove game in Example 4.9

		Player 2 H	Player 2 D
Player 1	H	$(\frac{v-c}{2}, \frac{v-c}{2})$	$(v, 0)$
	D	$(0, v)$	$(\frac{v}{2}, \frac{v}{2})$

Note that $u_1(H, \sigma_2) = \sigma_{12}(v - c)/2 + (1 - \sigma_{12})v$ and $u_1(D, \sigma_2) = (1 - \sigma_{12})v/2$. We have $u_1(H, \sigma_2) > u_1(D, \sigma_2)$ if and only if $(1 - \sigma_{12})v/2 > \sigma_{12}(c - v)/2$ or $\sigma_{12} < v/c$. Then, we can calculate the best-response correspondence of player 1 as

$$BR_1(\sigma_2) = \begin{cases} \{(1,0)\} \text{ if } \sigma_{12} < v/c \\ \Delta(S_1) \quad \text{if } \sigma_{12} = v/c \\ \{(0,1)\} \text{ if } \sigma_{12} > v/c. \end{cases}$$

Also, note that $u_2(\sigma_1, H) = \sigma_{11}(v - c)/2 + (1 - \sigma_{11})v$ and $u_1(\sigma_1, D) = (1 - \sigma_{11})v/2$. We have $u_2(\sigma_1, H) > u_2(\sigma_1, D)$ if and only if $(1 - \sigma_{11})v/2 > \sigma_{11}(c - v)/2$ or $\sigma_{11} < v/c$. Then, we can calculate the best-response correspondence of player 2 as

$$BR_2(\sigma_1) = \begin{cases} \{(1,0)\} \text{ if } \sigma_{11} < v/c \\ \Delta(S_2) \quad \text{if } \sigma_{11} = v/c \\ \{(0,1)\} \text{ if } \sigma_{11} > v/c. \end{cases}$$

A profile σ^* is a Nash equilibrium if and only if $\sigma_1^* \in BR_1(\sigma_2^*)$ and $\sigma_2^* \in BR_2(\sigma_1^*)$. Therefore, the set of Nash equilibria is

$$NE = \{((1,0), (0,1)), ((0,1), (1,0)), ((v/c, 1 - (v/c)), (v/c, 1 - (v/c)))\}.$$

The first two (pure) equilibria can be written as (H, D) and (D, H), whereas the last (unpure) equilibrium can be written as $((v/c)H + (1 - (v/c))D, (v/c)H + (1 - (v/c))D)$. ■

Example 4.10 (*Cournot Duopoly Game*) Recall from Example 1.3 that two firms in a duopolistic industry compete in quantities. Thus, the set of players is $N = \{1, 2\}$. The strategy spaces of players 1 and 2 involve all possible nonnegative quantities, i.e., $S_1 = [0, \infty)$ and $S_2 = [0, \infty)$. Thus, $S = \mathbb{R}_+^2$. For any strategy (quantity) profile $q \in S$, the payoff (operational profit) of firm $i \in N$ is

$$u_i(q) = \pi_i(q_i, q_{-i}) = P(q_i + q_{-i})q_i - cq_i = \min\{0, (a - (q_i + q_{-i}))\}q_i - cq_i$$

where $a > c > 0$. The best-response correspondence of player i to any $q_{-i} \in S_{-i}$ is defined as

$$BR_i(q_{-i}) = \{q_i : u_i(q_i, q_{-i}) \geq u_i(q_i', q_{-i}) \text{ for all } q_i' \in S_i\}.$$

To find this correspondence, we have to find the maximizer(s) of $u_i(q_i, q_{-i})$. If $q_{-i} \geq a$, then $P(q_i + q_{-i}) = 0$ for any $q_i \geq 0$. In that case, $u_i(q_i, q_{-i}) = -cq_i$. Thus, $q_i = 0$ maximizes $u_i(q_i, q_{-i})$ if $q_{-i} \geq a$. Now, suppose $q_{-i} < a$ and also suppose $P(q_i + q_{-i}) > 0$. Then, the first-order partial derivative of $u_i(q_i, q_{-i})$ with respect to q_i becomes

$$\frac{\partial}{\partial q_i} u_i(q) = \frac{\partial}{\partial q_i} \pi_i(q_i, q_{-i}) = a - c - q_{-i} - 2q_i.$$

If $q_{-i} \geq a - c$, then $\frac{\partial}{\partial q_i} u_i(q) < 0$ for each $q_i > 0$. Therefore, $q_i = 0$ maximizes $u_i(q_i, q_{-i})$ if $q_{-i} \geq a - c$. If $q_{-i} < a - c$, then the first-order condition $\frac{\partial}{\partial q_i} u_i(q) = 0$ (for an interior maximum) is satisfied at $q_i = (a - c - q_{-i})/2$, which is positive and not greater that $(a - c)/2$. We can also check that $a - q_i - q_{-i} = (a + c - q_{-i})/2 > 0$, implying that the price $P(q_i + q_{-i})$ is positive, as earlier supposed.

Also note that the second-order sufficient condition for a maximum with respect to q_i is satisfied since

$$\frac{\partial^2}{\partial (q_i)^2} u_i(q) = \frac{\partial^2}{\partial (q_i)^2} \pi_i(q_i, q_{-i}) = -2 < 0.$$

Thus,

$$BR_i(q_{-i}) = \begin{cases} (a - c - q_{-i})/2 & \text{if } q_{-i} < a - c \\ 0 & \text{if } q_{-i} \geq a - c. \end{cases}$$

Note that, for each q_{-i}, the best-response correspondence $BR_i(q_{-i})$ is single-valued, i.e., it is a function which we denote by $\hat{q}_i(q_{-i})$. Moreover this function is negatively sloped over the range where $q_{-i} < a - c$. We draw the best-response correspondences $BR_1(q_2)$ and $BR_2(q_1)$ in Fig. 4.4.

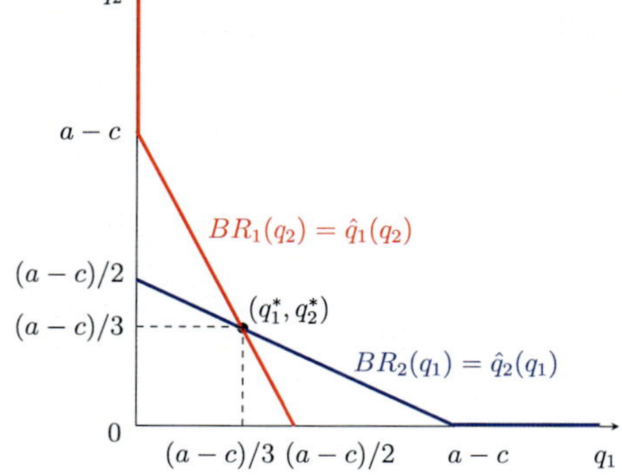

Fig. 4.4 Best-response correspondences for the Cournot duopoly game

Recall that a strategy profile q^* is a (Cournot) Nash equilibrium if and only if $q_i^* \in BR_i(q_{-i}^*)$ for each $i \in N$; i.e., $q_1^* = \hat{q}_1(q_2^*) = (a - c - q_2^*)/2$ and $q_2^* = \hat{q}_2(q_1^*) = (a - c - q_1^*)/2$. Solving these two equations together we obtain $q_1^* = q_2^* = (a - c)/3$. Thus, the (Cournot) Nash equilibrium of the described symmetric Cournot duopoly is $q^* = ((a - c)/3, (a - c)/3))$. ■

4.3 Strict Nash Equilibrium

We should recall from Definition 4.1 that a strategy profile is a Nash equilibrium of a strategic form game if no player has a strict incentive to unilaterally deviate from that profile. According to this definition, a weak incentive for unilateral deviation does not prevent a strategy profile from being an equilibrium. This definition can be strengthened to obtain the following refinement of Nash equilibrium.

Definition 4.4 Given a strategic form game $G^{\Delta} = \langle N, (\Delta(S_i))_{i\in N}, (\tilde{u}_i)_{i\in N}\rangle$, a strategy profile $\sigma^* \in \Delta^N(S)$ is a strict Nash equilibrium if

$$\tilde{u}_i(\sigma_i^*, \sigma_{-i}^*) > \tilde{u}_i(\sigma_i', \sigma_{-i}^*) \quad \text{for all } i \in N \text{ and } \sigma_i' \in \Delta(S_i).$$

Notice that a strategy profile is a Nash equilibrium if it is a strict Nash equilibrium, but the converse is not true. Also notice that a strict Nash equilibrium cannot be an unpure-strategy profile. In other words, if a strategy profile $\sigma^* \in \Delta^N(S)$ is a strict Nash equilibrium, then $\sigma^* \in S$. To see this suppose towards a contradiction that σ^* is unpure. Then there exists $i \in N$ such that $\sigma_i^* \notin S_i$, implying that there exists $s, s' \in S_i$ such that $\tilde{u}_i(\sigma_i^*, \sigma_{-i}^*) = \tilde{u}_i(s, \sigma_{-i}^*) = \tilde{u}_i(s', \sigma_{-i}^*)$, contradicting that σ^* is a Nash equilibrium.

Since an unpure-strategy profile cannot be a strict Nash equilibrium, Nash's [1] existence result (Proposition 4.1) does not apply for strict Nash equilibrium. Every strategic form game has a Nash equilibrium, but not necessarily a strict Nash equilibrium, as well. In Example 4.1, we saw that the unique Nash equilibrium of the Matching Pennies Game is totally unpure; therefore this game has no strict Nash equilibrium. In Example 4.4, we found that the strategic form game (in Table 4.4) has two pure-strategy Nash equilibria (U, L) and $D, R)$. But, only one of them, (U, L), is a strict Nash equilibrium.

4.4 Nash Equilibrium and Strategic Dominance

In Example 4.5, we showed that Prisoner's Dilemma Game has a unique (and strict) Nash equilibrium profile, which is (Confess, Confess). This profile is also the strongly dominant equilibrium of the game. Moreover, this profile is the unique profile in the game that is Pareto dominated (inefficient). Hence, we have learned that a Nash equilibrium does not have to be Pareto undominated (efficient). Moreover, this conclusion is independent of whether the equilibrium profile of strategies is pure or unpure. The Battle of Sexes Game that we analyzed in Example 4.6 has an unpure-strategy Nash equilibrium $((2/3)O + (1/3)F, (1/3)O + (2/3)F)$ that yields the expected payoff profile $(2/3, 2/3)$. This unpure equilibrium is Pareto dominated by the pure-strategy equilibria (O, O) and (F, F) which, respectively, yield the expected payoffs (2, 1) and $(1, 2)$. In addition, our equilibrium calculations for the Matching Pennies Game in Example 4.1 showed that a strategic form game may not have a Nash equilibrium in pure strategies even if each pure-strategy profile of this game is Pareto efficient. Overall, we have so far seen that there exists no causal relationship between the concepts of Nash equilibrium and Pareto dominance. However, a relationship may exist between Nash equilibrium and some concepts of strategic dominance.

Proposition 4.2 *Given a strategic form game $G = \langle N, (S_i)_{i\in N}, (u_i)_{i\in N}\rangle$, if a pure-strategy profile $s^* \in S$ is the weakly dominant equilibrium, then it is a Nash equilibrium.*

Proof Consider any strategic form game $G = \langle N, (S_i)_{i\in N}, (u_i)_{i\in N}\rangle$. Suppose a strategy profile $s^* \in S$ is the weakly dominant equilibrium. Then,

$$u_i(s_i^*, s_{-i}) \geq u_i(s_i', s_{-i}) \text{ for all } i \in N, \ s_i' \in S_i, \text{ and } s_{-i} \in S_{-i}.$$

Since $s_{-i}^* \in S_{-i}$, the above inequality must be true for s_{-i}^* as well. That is,

$$u_i(s_i^*, s_{-i}^*) \geq u_i(s_i', s_{-i}^*) \text{ for all } i \in N \text{ and } s_i' \in S_i,$$

implying that s^* is a Nash equilibrium. ■

In Prisoner's Dilemma Game, the strategy profile (Confess, Confess) is the weakly (and strongly) dominant equilibrium, and automatically a Nash equilibrium by Proposition 4.2. We also know from our previous observations that the converse of Proposition 4.2 is not true. A Nash equilibrium does not have to be the weakly dominant equilibrium. For example, The Battle of Sexes Game in Example 4.6 has three Nash equilibrium profiles, none of which is the weakly dominant equilibrium.

Proposition 4.3 *Given a strategic form game $G = \langle N, (S_i)_{i\in N}, (u_i)_{i\in N}\rangle$, if $s^* \in S$ is the unique strategy profile that survives IESDS, then it is a Nash equilibrium.*

Proof Consider any strategic form game $G = \langle N, (S_i)_{i\in N}, (u_i)_{i\in N}\rangle$. Suppose that $s^* \in S$ is the unique strategy profile that survives IESDS in G and assume towards a contradiction that s^* is not a Nash equilibrium. Then, there exists some player k such that s_k^* is not a best response to s_{-k}^*. Let $V = \{t \in S_k : u_k(t, s_{-k}^*) > u_k(s_k^*, s_{-k}^*)\}$. We know that $V \neq \emptyset$ since s_k^* is not a best response to s_{-k}^*. However, we also know that $s^* \in S$ is the unique strategy profile that survives IESDS. Hence, all of the strategies in V must be eliminated by IESDS. Consider the stage where the last strategy, say l, of player k in the set V is eliminated via IESDS. For the strategy l to be eliminated, there must exist a strategy $d \in S_k$ that strongly dominates l. This would imply

$$u_k(d, s^*_{-k}) > u_k(l, s^*_{-k}) > u_k(s^*_k, s^*_{-k}),$$

where the second inequality follows from the fact that $l \in V$. Then, we must have $d \in V$. Moreover, since $s^* \in S$ is the unique strategy profile that survives IESDS, the strategy d of player k must be eliminated, too. But, this contradicts the supposition that l was her last eliminated strategy in set V. Therefore, s^* must be a Nash equilibrium. ∎

The converse of the above proposition is not true. A Nash equilibrium does not have to be the unique strategy profile that survives IESDS. For example, in the Battle of Sexes Game analyzed in Example 4.6, one can check that both of the pure-strategy Nash equilibrium profiles (O, O) and (F, F) survive IESDS. Regarding the converse of Proposition 4.3, we have the following weaker result.

Proposition 4.4 *Given a strategic form game $G = \langle N, (S_i)_{i \in N}, (u_i)_{i \in N} \rangle$, if a strategy profile $s^* \in S$ is a Nash equilibrium, then it is not eliminated by IESDS.*

Proof Consider any strategic form game $G = \langle N, (S_i)_{i \in N}, (u_i)_{i \in N} \rangle$. Suppose towards a contradiction that a Nash equilibrium $s^* \in S$ is eliminated during IESDS. Then, there exist some (iteration) stage T and player k such that (i) before stage T no strategy in s^* was eliminated for any player and (ii) at stage T strategy s^*_k of player is eliminated. So, at stage T there must exist a strategy $s'_k \in S_k$ that strongly dominates s^*_k. This implies that

$$u_k(s'_k, s^*_{-k}) > u_k(s^*_k, s^*_{-k}),$$

further implying that strategy s^*_k is not a best response, for player k, to s^*_{-k}. This contradicts that $(s^*_k, s^*_{-k}) = s^*$ is a Nash equilibrium. Therefore, the proposition is true. ∎

4.5 Nash Equilibrium and Risk Dominance

Consider the strategic form game in Table 4.10. One can easily check that there are two pure-strategy Nash equilibria, (U, L) and (D, R). Player 1 prefers to be in the first equilibrium while player 2 prefers to be in the second one. Thus, this game is a coordination game.

None of the players is certain about which equilibrium will arise. Each player is expected to choose her strategy depending on her beliefs about the strategy of her opponent. Player 1 is expected to play U if she believes Player 2 will play L with a probability higher than 0.10 and to play D if she believes Player 2 will play R with a probability higher than 0.90. Thus, for player 1, strategy U is less risky than D.

Table 4.10 A two-person coordination game

		Player 2	
		L	R
Player 1	U	$(9, 4)$	$(0, 0)$
	D	$(0, 0)$	$(1, 6)$

Player 2 plays L if he believes Player 1 will play U with a probability higher than 0.60 and plays R if he believes Player 1 will play D with a probability higher than 0.40. Thus, for player 2, R is less risky than L.

(U,L) and (D,R) are Nash equilibria. In terms of risk, player 1 prefers (U,L) and player 2 prefers (D,R). But, the relative riskiness of D for player 1 with respect to U and the relative riskiness of L for player 2 with respect to R are not equal. To compromise between the players' risk preferences, Harsanyi and Selten [3] define a measure for risk dominance, simply calculating the product of deviation losses incurred when one of the players decides to change its strategy. In any 2×2 game, an equilibrium s^* strictly risk dominates another equilibrium s if

$$[u_1(s_1^*, s_2^*) - u_1(s_1, s_2^*)][u_2(s_1^*, s_2^*) - u_2(s_1^*, s_2)]$$
$$> [u_1(s_1, s_2) - u_1(s_1^*, s_2)][u_2(s_1, s_2) - u_2(s_1, s_2^*)]. \quad (4.1)$$

The above inequality means that it is more costly for the players to mistakenly think that they are playing s than to mistakenly think that they are playing s^*.

Note that given the strategic form game in Table 4.10, the strategy profile (U, L) strictly risk dominates (D, R) since

$$[u_1(U, L) - u_1(D, L)][u_2(U, L) - u_2(U, R)] = (9 - 0)(4 - 0) = 36 >$$
$$6 = (1 - 0)(6 - 0) = [u_1(D, R) - u_1(U, R)][u_2(D, R) - u_2(D, L)]. \quad (4.2)$$

The risk dominance condition in (4.1) becomes even simpler when the game is a symmetric coordination game. Note that, for a 2×2 game, the symmetry condition implies $S_1 = S_2 = \{s_a, s_b\}$, $u_1(s_a, s_a) = u_2(s_a, s_a)$, $u_1(s_a, s_b) = u_2(s_b, s_a)$, $u_2(s_a, s_b) = u_1(s_b, s_a)$, and $u_1(s_b, s_b) = u_2(s_b, s_b)$. Moreover, if the game is a coordination game, then (s_a, s_a) and (s_b, s_b) become pure-strategy Nash equilibria. Using these observations and (4.1), we can say that the equilibrium (s_a, s_a) strictly risk dominates the equilibrium (s_b, s_b) if

$$[u_1(s_a, s_a) - u_1(s_b, s_a)][u_2(s_a, s_a) - u_2(s_a, s_b)] >$$
$$[u_1(s_b, s_b) - u_1(s_a, s_b)][u_2(s_b, s_b) - u_2(s_b, s_a)] \quad (4.3)$$

or equivalently if

$$u_1(s_a, s_a) - u_1(s_b, s_a) > u_1(s_b, s_b) - u_1(s_a, s_b). \quad (4.4)$$

The last condition can be rewritten as

$$u_1(s_a, s_a) + u_1(s_a, s_b) > u_1(s_b, s_a) + u_1(s_b, s_b). \quad (4.5)$$

Multiplying each side of the last inequality by 1/2 and rearranging it yields

$$u_1(s_a, 0.5s_a + 0.5s_b) > u_1(s_b, 0.5s_a + 0.5s_b). \tag{4.6}$$

Note that by symmetry, we also have

$$u_2(0.5s_a + 0.5s_b, s_a) > u_2(0.5s_a + 0.5s_b, s_b). \tag{4.7}$$

The last two conditions simply say that the equilibrium (s_a, s_a) strictly risk dominates the equilibrium (s_b, s_b) if, for each player $i \in N$, the strategy s_a yields a higher expected payoff than the strategy s_b when this player believes that his opponent $j \neq i$ plays the strategies s_a and s_b with equal probabilities.

Example 4.11 Reconsider the Stag Hunt Game in Example 4.7.

		Player 2	
		S	H
Player 1	S	(s, s)	$(0, h)$
	H	$(h, 0)$	(h, h)

Table 4.7 Reconsidered

It is assumed that $s > h > 0$. Note that the above game is not only symmetric but also a coordination game. (The pure-strategy Nash equilibria are (S, S) and (H, H).) Thus, we can use the condition in (4.6) to check which of the equilibrium is risk-dominant. Note that (S, S) strictly risk dominates (H, H) if and only if

$$u_1\,(S, (0.5)S + (0.5)H) > u_1\,(H, (0.5)S + (0.5)H)$$

implying

$$(0.5)u_1(S, S) + (0.5)u_1(S, H) = \frac{s}{2} > h = (0.5)u_1(H, S) + (0.5)u_1(H, H)$$

or equivalently if and only if $s > 2h$. One can similarly show that (H, H) strongly risk dominates (S, S) if and only if $s < 2h$.

We should observe that Pareto dominance and strict risk dominance become aligned in the above game and choose the same equilibrium profile, (S, S), if and only if $s > 2h$. ■

4.6 Informational Assumptions

In this section we will discuss what epistemic conditions lead to Nash equilibrium. Until the 1990s, the dominant interpretation of a Nash equilibrium in unpure strategies was the act of randomization. A player who is playing an unpure strategy as part of the Nash equilibrium profile was viewed to be randomizing between several pure strategies in her strategy space. Under this view, if each player is rational and knows its own payoff function and the strategies chosen by the other players, then the strategy choices of the players may constitute a Nash equilibrium. This is so because each rational player, knowing the strategy choices of the other players, is expected to choose her best-response strategy. Under this interpretation of a Nash equilibrium, no common knowledge assumption is ever needed. The players need to have only mutual knowledge about the equilibrium strategy profile. Given this knowledge, no player needs to know whether the other players are rational or what payoff functions the other players are endowed with.

An alternative interpretation of a Nash equilibrium, made by Harsanyi [4] and others, suggested that players use mixed strategies not to randomize. An unpurely mixed strategy used by a player just reflects the other players' uncertainty as to which available pure strategy will be chosen by that particular player. Thus, a Nash equilibrium is just a profile of equilibrium conjectures made by the players. Epistemic conditions under this interpretation of the Nash equilibrium were provided by Aumann and Brandeburger [5]. For strategic form games with two players, they showed that if the rationality, payoffs, and conjectures of all players (about their strategies) are mutual knowledge, then the players' conjectures constitute a Nash equilibrium. For strategic form games with at least three players, their result is much more demanding since the common knowledge assumption is also needed. To see this, pick any player i, where $i \in N = \{1, 2, \ldots, n\}$ with $n \geq 3$. Players in $N - i$ may have different conjectures about the strategy of player i. Since the strategy of player i must be some unique element of her strategy space $\Delta(S_i)$, its representation as the conjecture of other players must also be unique.

To ensure that all other players can agree on a particular conjecture for the strategy of player i, Aumann and Brandeburger [5] had to assume that the players have a common prior about the strategy profile to be played in the game and their individual conjectures are due only to differences in their information before the game is played. Using this assumption they were able to show that in any n-player strategic form game with $n \geq 3$, if the players have a common prior, their rationality and payoff functions are mutually known, and their conjectures are commonly known, then for any $i \in N$ all players in $N - i$ can agree on the same conjecture about the strategy of player i and the resulting profiles of conjectures made by the players constitute a Nash equilibrium.

In Table 4.11, we list the informational assumptions about the Nash equilibrium under the views of randomization and conjectures, in comparison to the assumptions about some other solution concepts, including the strongly dominant equilibrium (SDE), the iterated elimination of strongly dominated strategies (IESDS), and the rationalizable sets.

4.7 Chapter Appendix

4.7.1 Mathematical Preliminaries

We say that the set $\mathbb{R}^m$ and the distance function $d : \mathbb{R}^m \times \mathbb{R}^m \to \mathbb{R}$ with

$$d(x, y) = \left[\sum_{i=1}^{m} (x_i - y_i)^2 \right]^{\frac{1}{2}} \quad \text{for any } x, y \in \mathbb{R}^m$$

defines an m-dimensional Euclidean (metric) space. Let for any $x \in \mathbb{R}^m$, the set $B_\epsilon(x)$ denote the ϵ-neighborhood of x; i.e.,

$$B_\epsilon(x) = \{y \in \mathbb{R}^m : d(x, y) < \epsilon\}.$$

A set $S \in \mathbb{R}^m$ is called *open* in $(\mathbb{R}^m, d)$ If, for each $x \in S$, there exists an $\epsilon > 0$ such that $B_\epsilon(x) \subseteq S$. A set $S \in \mathbb{R}^m$ is called *closed* in $(\mathbb{R}^m, d)$ if $\mathbb{R}^m \setminus S$ is open in $(\mathbb{R}^m, d)$. A set $S \in \mathbb{R}^m$ is called *bounded* if there exists $\epsilon > 0$ such that $S \subseteq B_\epsilon(x)$ for some $x \in S$.

For any $x, y \in \mathbb{R}^m$, we denote by $[x, y]$ the line segment between x and y; i.e.,

$$[x, y] = \{x + t(y - x) : t \in [0, 1]\}.$$

A set $S \in \mathbb{R}^m$ is called *convex* if for any $x, y \in S$, the line segment $[x, y]$ is a subset of S; i.e., $[x, y] \subseteq S$.

Given a nonempty set A in $\mathbb{R}^m$, a function $f : A \to \mathbb{R}$ is called *continuous* if for any $a \in A$, $\lim_{x \to a} f(x) = f(a)$.

Weierstrass Theorem. *Let A be a nonempty, closed, and bounded set in a multidimensional Euclidean space and $f : A \to \mathbb{R}$ is a continuous function. Then f has a maximum and a minimum on A.*

We will use Weierstrass Theorem in the proof of Proposition 4.1 (Nash's Existence Theorem, 1950a), while we show that given the mixed extension of a finite strategic form game the best-response correspondence of each individual is nonempty.

Kakutani Fixed Point Theorem. *Let A be a nonempty, closed, bounded, and convex subset of $\mathbb{R}^m$ and $f : A \to A$ is a correspondence, with $f(x) \subseteq A$ for all $x \in A$, that satisfies the following assumptions:*

Table 4.11 Informational assumptions for several solution concepts

Informational assumption	Solution concept					
	SDE	IESDS	Rationalizable sets	Nash equilibrium (randomization) $n \geq 2$	Nash equilibrium (conjectures) $n = 2$	Nash equilibrium (conjectures) $n \geq 3$
Each player is rational	✓	✓	✓	✓	✓	✓
Common prior						✓
Common knowledge about payoffs		✓	✓			
Common knowledge about rationality		✓	✓			
Common knowledge about conjectures						✓
Mutual knowledge about payoffs		Implied	Implied		✓	✓
Mutual knowledge about rationality		Implied	Implied		✓	✓
Mutual knowledge about conjectures					✓	Implied
Mutual knowledge about randomization				✓		
Each player knows its payoffs	✓	Implied	Implied	✓	Implied	Implied

(i) *$f(x)$ is nonempty for all $x \in A$,*

(ii) *$f(x)$ is convex for all $x \in A$, and*

(iii) *the graph of $f(x)$ is closed, i.e., for any sequence $(x^n, y^n) \to (x, y)$ such that $y^n \in f(x^n)$, it is true that $y \in f(x)$.*

Then, there exists $x \in A$ such that $x \in f(x)$.

We will use Kakutani Fixed Point Theorem in the proof of Proposition 4.1 (Nash's Existence Theorem, 1950a), while we show that given the mixed extension of a finite strategic form game the profile of best-response correspondences has a fixed point. The following example illustrates that none of the assumptions in the above fixed point theorem is redundant.

Example 4.12 Let A be a nonempty subset of a multidimensional Euclidean space and $f : A \to A$ be a correspondence, with $\emptyset \neq f(x) \subseteq A$ for all $x \in A$, satisfying the following assumptions:

Assumption-1: A is closed,

Assumption-2: A is bounded,

Assumption-3: A is convex,

Assumption-4: $f(x)$ is convex for all $x \in A$,

Assumption-5: The graph of $f(x)$ is closed for all $x \in A$.

This example has five parts. For part $i \in \{1, 2, \ldots, 5\}$, we will find a pair (A, f) that satisfies all assumptions except for Assumption-i and is such that f has no fixed point, i.e., there exists no $x \in A$ such $x \in f(x)$.

Part (i): We will write a pair (A, f) that only violates Assumption 1. Let $A = [0, 1)$ and $f : A \to A$ be a correspondence such that $f(x) = \{0.5 + 0.5x\}$.

The set A is not closed since it does not contain all of its boundaries. (Note that the boundary $\{1\}$ is not contained by A.) On the other hand, A is bounded. (Note that it can be enclosed by a finite set $B = [-1, 2]$, for example.) Also, A is convex. (For any two points $a_1, a_2 \in [0, 1) = A$, we have $[a_1, a_2] \subseteq A$.)

The correspondence $f(x)$ is single-valued at each $x \in A$, and therefore convex. Also the graph of $f(x)$ is closed at each $x \in A$. Take any sequence (x^n, y^n), with $x^n \in A$ and $y^n \in f(x^n)$, that is converging to some point (x, y). We have $y = \lim_{n\to\infty} y^n = \lim_{n\to\infty} 0.5 + 0.5x^n = 0.5 + 0.5x$. Therefore, $y \in f(x)$.

We should note that $x \in f(x)$ cannot be true whenever $x \neq 1$. Since $\{1\}$ is not in A, f does not have a fixed point in A.

Part (ii): We will write a pair (A, f) that only violates Assumption 2. Let $A = \cup_{i=1}^{\infty} A_i$ where $A_i = [i-1, i]$ for each $i = 1, 2, \ldots, \infty$. Thus, $A = [0, 1] \cup [1, 2] \cup [2, 3] \ldots$, which is the nonnegative real line. Let $f : A \to A$ be a correspondence such that $f(x) = \{1 + x\}$.

Clearly, the set A_i is closed for each $i = 1, 2, \ldots, \infty$. Since the union of any number of closed sets will also be closed, the set A is closed. Also, A is convex. (For any two points $a_1, a_2 \in A$, we have $[a_1, a_2] \subseteq A$.) However, A is not bounded, since $\lim_{i\to\infty} \max\{A_i\} = \lim_{i\to\infty} i = \infty$, implying that we cannot contain A by any finite set.

The correspondence $f(x)$ is single-valued at each $x \in A$, and therefore convex. Also the graph of $f(x)$ is closed at each $x \in A$. Take any sequence (x^n, y^n), with $x^n \in A$ and $y^n \in f(x^n)$, that is converging to some point (x, y). We have $y = \lim_{n\to\infty} y^n = \lim_{n\to\infty} 1 + x^n = 1 + x$. Therefore, $y \in f(x)$.

We should note that there exists no $x \in A$ such that $x \in \{1 + x\} = f(x)$. Therefore, f does not have a fixed point in A.

Part (iii): We will write a pair (A, f) that only violates Assumption 3. Let $A = [0, 0.5] \cup [1.5, 2]$. Let $f : A \to A$ be a correspondence such that $f(x) = \{2\}$ if $x \in [0, 0.5]$ and $f(x) = \{0.5\}$ if $x \in [1.5, 2]$. Note that $f(x) \subset A$ for all $x \in A$. Clearly, the set A is closed because it is the union of two closed sets in $\mathbb{R}$. Also, A is bounded. (Note that it can be enclosed by a finite set, e.g., $B = [-1, 3]$.) However, A is not convex. (The points $a_1 = 0.5$ and $a_2 = 1.5$ are in A, while the line segment $[0.5, 1.5]$ is not contained by A.)

The correspondence $f(x)$ is single-valued at each $x \in A$, and therefore convex. Also the graph of $f(x)$ is closed at each $x \in A$. Take any sequence (x^n, y^n), with $x^n \in A$ and $y^n \in f(x^n)$, that is converging to some point (x, y). If $x \in [0, 0.5]$, then $y = \lim_{n\to\infty} y^n = 2$, implying $y \in f(x)$. Else if $x \in [1.5, 2]$, then $y = \lim_{n\to\infty} y^n = 0.5$, again implying $y \in f(x)$.

We should note that $x \in f(x)$ can never be true on A, since $f(x)$ is above 0.5 for any $x \leq 0.5$ and $f(x)$ is below 1.5 for any $x \geq 1.5$. Therefore, f does not have a fixed point in A.

Part (iv): We will write a pair (A, f) that only violates Assumption 4. Let $A = [0, 2]$. Let $f : A \to A$ be a correspondence such that

$$f(x) = \begin{cases} \{1.5\} & \text{if } x \in [0, 1) \\ [0.5, 0.75] \cup [1.25, 1.5] & \text{if } x = 1 \\ \{0.5\} & \text{if } x \in (1, 2]. \end{cases}$$

Clearly, the set A is closed because it contains its boundaries. Also, A is bounded. (Note that it can be enclosed by a finite set, e.g., $B = [-1, 3]$.) Moreover, A is convex. (For any two points $a_1, a_2 \in [0, 2] = A$, we have $[a_1, a_2] \subseteq [0, 2]$.)

The graph of $f(x)$ is closed at each $x \in A$. Take any sequence (x^n, y^n), with $x^n \in A$ and $y^n \in f(x^n)$, that is converging to (x, y). If $x < 1$, then $y = \lim_{n\to\infty} y^n = 1.5$ and $y \in f(x)$. If $x > 1$, then $y = \lim_{n\to\infty} y^n = 0.5$ and $y \in f(x)$. If $x = 1$, then $y = \lim_{n\to\infty} y^n \in \{0.5, 1.5\}$ and $y \in f(x)$.

However, the correspondence $f(x)$ is not convex-valued at $x = 1$. Note that $\{0.75\} \in f(1)$ and $\{1.25\} \in f(1)$, whereas the line segment $[0.75, 1.25]$ is not contained by $f(1)$.

We should note that $x \in f(x)$ does not hold at any $x \in A$. Therefore, f does not have a fixed point in A.

Part (v): We will write a pair (A, f) that only violates Assumption 5. Let $A = [0, 2]$. Let $f : A \to A$ be a correspondence such that

$$f(x) = \begin{cases} \{1\} & \text{if} x \in [0, 1) \\ [0.5, 1) & \text{if } x = 1 \\ \{0.5\} & \text{if } x \in (1, 2]. \end{cases}$$

Clearly, the set A is closed because it contains its boundaries. Also, A is bounded. (Note that it can be enclosed by a finite set $B = [-1, 3]$, for example.) Moreover, A is convex. (For any two points $a_1, a_2 \in [0, 2] = A$, we have $[a_1, a_2] \subseteq [0, 2]$.)

Also note that $f(x)$ is convex-valued at all points in $A \setminus \{1\}$ since $f(x)$ is single-valued if $x \neq 1$. Moreover, $f(1)$ is convex-valued since for any $y_1, y_2 \in f(1) = [0.5, 1)$ we have $[y_1, y_2] \subseteq f(1)$.

However, it is not true that the graph of $f(x)$ is closed at each $x \in A$. Consider the point $x = 1$. Take any sequence (x^n, y^n), with $x^n \in [0, 1)$ and $y^n \in f(x^n)$, that is converging to (x, y) where $x = 1$. Note that $y = \lim_{n\to\infty} y^n = 1$. But, $1 = y \notin f(x) = f(1) = [0.5, 1)$.

We should note that $x \in f(x)$ cannot be true on A. If $x < 1$, then $f(x) = \{1\}$; so $x \notin f(x)$. If $x > 1$, then $f(x) = \{0.5\}$; so $x \notin f(x)$. Finally, if $x = 1$, then $f(x) = [0.5, 1)$; so $x \notin f(x)$. Therefore, f does not have a fixed point in A.
The panels (a)–(e) of Fig. 4.5 portray the set A and the correspondence f in parts (i)–(v), respectively.

4.7.2 Proof of Nash's Existence Theorem

Proof of Proposition 4.1 (Nash's Existence Theorem, 1950a). Consider any finite strategic form game $G = \langle N, (S_i)_{i\in N}, (u_i)_{i\in N}\rangle$ and its mixed extension G^{Δ}. Let $BR_i : \times_{j\neq i} \Delta(S_j) \to \Delta(S_i)$ denote the best-response correspondence of player $i \in N$. Define $BR = \times_{i\in N} BR_i$. Note that BR is a correspondence from $\Delta^N(S) \equiv \times_{i=1}^{n} \Delta(S_i)$ to $\Delta^N(S)$. Using the Kakutani Fixed Point Theorem, we will prove that the best-response correspondence BR has a fixed point in $\Delta^N(S)$.

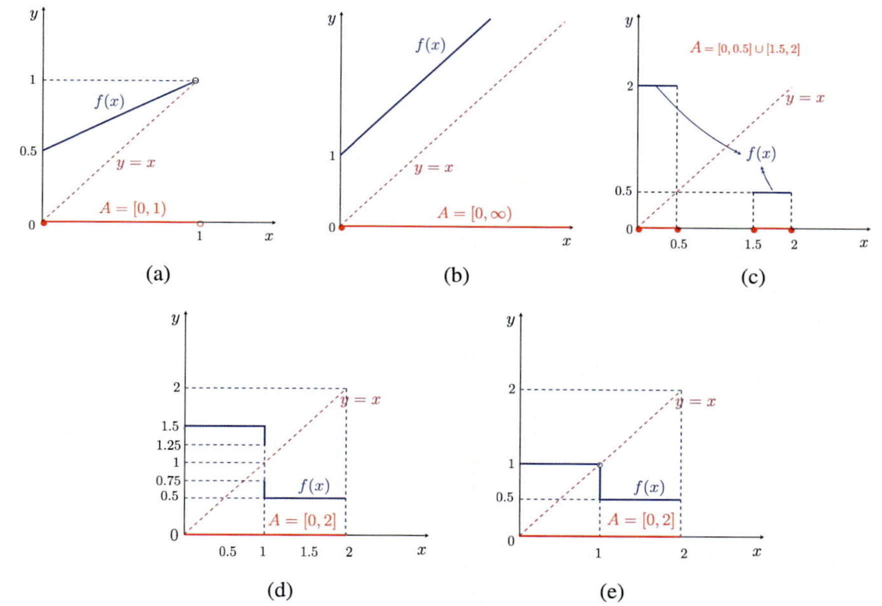

Fig. 4.5 Examples violating the assumptions of Kakutani fixed point theorem

First, we have to show that condition (i) in the Kakutani Fixed Point Theorem holds; i.e., $BR(\sigma)$ is nonempty for all $\sigma \in \Delta^N(S)$. This condition holds if and only if, for each $i \in N$, BR_i is nonempty for all $\sigma_{-i} \in \bigtimes_{j \neq i} \Delta(S_j)$. Recall that, for any $\sigma_{-i} \in \bigtimes_{j \neq i} \Delta(S_j)$, we have

$$BR_i(\sigma_{-i}) = \{\sigma_i \in \Delta(S_i) : \tilde{u}_i(\sigma^i, \sigma_{-i}) \geq \tilde{u}_i(\sigma_i', \sigma_{-i}) \text{ for all } \sigma_i' \in \Delta(S_i)\}.$$

Thus, if $\sigma_i^* \in BR_i(\sigma_{-i})$, we must have $\sigma_i^* = \text{argmax}_{\sigma_i \in \Delta(S_i)} \tilde{u}_i(\sigma_i, \sigma_{-i})$. Also recall that

$$\Delta(S_i) = \{\sigma_i : \sigma_{j,i} \in [0, 1] \text{ for all } j \in \{1, 2, \dots, \eta_i\} \text{ and } \sum_{j=1}^{\eta_i} \sigma_{j,i} = 1\},$$

where $\eta_i = |S_i|$. Clearly, for each $i \in N$, the set $\Delta(S_i)$ is nonempty, closed, bounded, and convex. Therefore, $\Delta^N(S)$ is nonempty, closed, bounded, and convex. We also know that $\tilde{u}_i : \Delta^N(S) \to \mathbb{R}$ is a continuous function by assumption. Therefore, by Weierstrass Theorem $\tilde{u}$ has a maximum on $\Delta^N(S)$, implying that $BR_i(\sigma_{-i})$ is nonempty for all $\sigma_{-i} \in \bigtimes_{j \neq i} \Delta(S_j)$. This further implies that $BR(\sigma)$ is nonempty for all $\sigma \in \Delta^N(S)$.

Now, we will show that condition (ii) in the Kakutani Fixed Point Theorem holds, i.e., $BR(\sigma)$ is convex for all $\sigma \in \Delta^N(S)$. So, pick any $i \in N$, any $\sigma_{-i} \in \bigtimes_{j \neq i} \Delta(S_j)$, and any $\sigma^1, \sigma^2 \in BR_i(\sigma_{-i})$. By the definition of BR_i, we must have

$$\tilde{u}_i(\sigma_i^1, \sigma_{-i}) \geq \tilde{u}_i(\sigma_i', \sigma_{-i}) \text{ for all } \sigma_i' \in \Delta(S_i),$$

$$\tilde{u}_i(\sigma_i^2, \sigma_{-i}) \geq \tilde{u}_i(\sigma_i', \sigma_{-i}) \text{ for all } \sigma_i' \in \Delta(S_i).$$

It follows that, for any $\lambda \in [0, 1]$,

$$\lambda\tilde{u}_i(\sigma_i^1, \sigma_{-i}) + (1 - \lambda)\tilde{u}_i(\sigma_i^2, \sigma_{-i}) \geq \tilde{u}_i(\sigma_i', \sigma_{-i}) \text{ for all } \sigma_i' \in \Delta(S_i),$$

or equivalently

$$\tilde{u}_i(\lambda\sigma_i^1 + (1 - \lambda)\sigma_i^2, \sigma_{-i}) \geq \tilde{u}_i(\sigma_i', \sigma_{-i}) \text{ for all } \sigma_i' \in \Delta(S_i),$$

implying that $\lambda\sigma_i^1 + (1 - \lambda)\sigma_i^2 \in BR_i(\sigma_{-i})$. Since λ was picked arbitrarily, $[\sigma^1, \sigma^2] \subseteq BR_i(\sigma_{-i})$. Hence, $BR_i(\sigma_{-i})$ is convex for all $\sigma_{-i} \in \bigtimes_{j \neq i} \Delta(S_j)$. Therefore, $BR(\sigma)$ is convex for all $\sigma \in \Delta^N(S)$.

Finally, let us prove that condition (iii) of Kakutani Fixed Point Theorem holds; i.e., the graph of $BR(\sigma)$ is closed for all $\sigma \in \Delta^N(S)$. Let us pick any sequence $(\sigma^n, \hat{\sigma}^n) \to (\sigma, \hat{\sigma})$ such that $\hat{\sigma}^n \in BR(\sigma^n)$ for all n. We will show that $\hat{\sigma} \in BR(\sigma)$. Suppose towards a contradiction that $\hat{\sigma} \notin BR(\sigma)$. Then, there exists $i \in N$ such that $\hat{\sigma}_i \notin BR(\sigma_{-i})$. Also, there exists some $\sigma_i' \in \Delta(S_i)$ and $\epsilon > 0$ such that

$$\tilde{u}_i(\sigma_i', \sigma_{-i}) > \tilde{u}_i(\hat{\sigma}_i, \sigma_{-i}) + 2\epsilon.$$

Moreover, using the fact that $\sigma_{-i}^n \to \sigma_{-i}$ and the continuity of $\tilde{u}_i$, there exists an integer n_1 such that

$$\tilde{u}_i(\sigma_i', \sigma_{-i}^n) \geq \tilde{u}_i(\sigma_i', \sigma_{-i}) - \epsilon$$

for any integer $n \geq n_1$. From the preceding two inequalities it follows that

$$\tilde{u}_i(\sigma_i', \sigma_{-i}^n) > \tilde{u}_i(\hat{\sigma}_i, \sigma_{-i}) + \epsilon$$

for any integer $n \geq n_1$. Finally, by the continuity of $\tilde{u}_i$, there exists an integer n_2 such that

$$\tilde{u}_i(\hat{\sigma}_i, \sigma_{-i}) \geq \tilde{u}_i(\hat{\sigma}_i^n, \sigma_{-i}^n) - \epsilon$$

for any integer $n \geq n_2$. From the last two inequalities, it then follows that

$$\tilde{u}_i(\sigma_i', \sigma_{-i}^n) > \tilde{u}_i(\hat{\sigma}_i^n, \sigma_{-i}^n)$$

for any integer $n \geq \max\{n_1, n_2\}$. This contradicts the supposition that $\hat{\sigma}_i^n \in BR_i(\sigma_{-i}^n)$ for all n. Therefore, there exists no $i \in N$ such that $\hat{\sigma}_i \notin BR_i(\sigma_{-i})$, implying that $\hat{\sigma} \in BR(\sigma)$. Thus, we have established that the graph of $BR(\sigma)$ is closed for all $\sigma \in \Delta^N(S)$. Since all conditions of the Kakutani Fixed Point Theorem holds, it follows that there exists $\sigma \in \Delta^N(S)$ such that $\sigma \in BR(\sigma)$. ■

References

1. Nash JF (1950) Equilibrium points in n-person games. PNAS 36(1):48–49
2. Wilson R (1971) Computing equilibria of n-person games. SIAM J Appl Math 21:80–87
3. Harsanyi JC, Selten R (1988) A general theory of equilibrium selection in games. MIT Press, Cambridge, Massachusetts
4. Harsanyi JC (1973) Games with randomly disturbed payoffs: A new rationale of mixed strategy equilibrium points. Int J Game Theory 2:1–23
5. Aumann R, Brandenburger A (1995) Epistemic conditions for Nash equilibrium. Econometrica 63(5):1161–1180

5 Perfect Equilibrium

In the previous chapter, we mentioned that the Nash equilibrium has an interpretation in which each profile of equilibrium strategies reflects the conjectures of players about the other players' strategies. In this chapters we will deal with the question concerning whether or when a Nash equilibrium is robust (resistant) to some small mistakes (trembles) in these conjectures. Section 5.1 will define such a robust equilibrium, called the perfect equilibrium, and show how it can be calculated. We will refer to the perfect equilibrium as a refinement of the Nash equilibrium, as such an equilibrium will be more plausible than a Nash equilibrium that is not perfect. Section 5.2 will present an existence result for the perfect equilibrium in strategic form games, whereas Sects. 5.3 and 5.4 will, respectively, relate the perfect equilibrium to the Nash equilibrium and to the strategy profiles that are weakly undominated.

5.1 Definitions and Examples

The concept of (trembling hand) perfect equilibrium was introduced by Reinhard Selten [1]. According to this concept, each player—when calculating her best-response conjectures— must take into account the possibility that the other players may make some small mistakes in forming their conjectures. There are various definitions of a perfect equilibrium, which were shown to be equivalent. In this section, we will present two of them.

Definition 5.1 Consider any strategic form game with mixed strategies $G^{\Delta} = \langle N, (\Delta(S_i))_{i \in N}, (\tilde{u}_i)_{i \in N} \rangle$. A strategy profile σ^* in $\Delta^N(S)$ is a (trembling-hand) perfect equilibrium of G^{Δ} if there exists a totally mixed strategy profile sequence σ^k in $\Delta^N(S)$ that converges to σ^* and is such that $\sigma_i^* \in BR_i(\sigma_{-i}^k)$ for each $i \in N$ and $k = 1, 2, \ldots, \infty$.

I. Saglam, *Mastering Game Theory*, Classroom Companion: Economics,
https://doi.org/10.1007/978-3-031-82094-6_5

Table 5.1 The strategic form game in Example 5.1

Player 1 \ Player 2	L	R
U	$(3, 3)$	$(-8, 6)$
D	$(4, 0)$	$(0, 0)$

Below, we will apply this definition to a strategic form game.

Example 5.1 Consider the strategic form game represented by Table 5.1.

Let G^{Δ} be the mixed extension of the above game. Using Definition 5.1, let us check whether the strategy profile (D, R), which we can write as $\sigma^* = ((0, 1), (0, 1))$, is a perfect equilibrium of G^{Δ}. Consider a sequence $\sigma^k = ((\delta^k, 1 - \delta^k), (\delta^k, 1 - \delta^k))$ where $\delta \in (0, 1)$ and $k = 1, 2, \ldots, \infty$. Note that σ^k converges to σ. Moreover, $u_1(U, \sigma_2^k) = 3\delta^k - 8(1 - \delta^k) = 11\delta^k - 8$ and $u_1(D, \sigma_2^k) = 4\delta^k$. We have $\sigma_1^* \in BR_1(\sigma_2^k)$ if $7\delta^k < 8$ or $\delta^k < 8/7$, which holds for each $\delta \in (0, 1)$ and $k = 1, 2, \ldots, \infty$. On the other hand, $u_2(L, \sigma_1^k) = 3\delta^k$ and $u_2(R, \sigma_1^k) = 6\delta^k$. We have $\sigma_2^* \in BR_2(\sigma_1^k)$ for each $\delta \in (0, 1)$ and $k = 1, 2, \ldots, \infty$. Thus, we have established that $\sigma_i^* \in BR_i(\sigma_{-i}^k)$ for each $\delta \in (0, 1)$ and $k = 1, 2, \ldots, \infty$. This proves that there exists a totally mixed strategy profile sequence σ^k in $\Delta^N(S)$ that converges to σ^* and is such that $\sigma_i^* \in BR_i(\sigma_{-i}^k)$ for each $i \in N$ and $k = 1, 2, \ldots, \infty$. So, σ^* is a perfect equilibrium of G^{Δ}. ■

For any $\epsilon > 0$ and $\sigma \in \Delta^N(S)$, we define the ϵ-neighborhood of σ as the set $\{\tau \in \mathbb{R}^n : ||\tau - \sigma|| \leq \epsilon\}$. Below, we present a result by Selten [1] showing that there is an alternative way to check whether a strategy profile is a perfect equilibrium.

Proposition 5.1 (Selten, [1]) *A strategy profile σ^* in $\Delta^N(S)$ is a perfect equilibrium if and only if for each $\epsilon > 0$ the ϵ-neighborhood of σ contains some totally mixed strategy profile $\sigma \in int(\Delta^N(S))$ such that $\sigma^* \in BR(\sigma)$.*

The above proposition implies that a mixed strategy profile σ is not a perfect equilibrium if there is some $\epsilon > 0$ such that for every completely mixed strategy profile $\tau \in \text{int}(\Delta^N(S))$ that lies in the ϵ-neighborhood of σ, we can find some player $i \in N$ such that $\sigma_i \notin BR_i(\tau_{-i})$.

In the following example, we will illustrate how to apply the alternative definition of perfect equilibrium in Proposition 5.1.

Example 5.2 Let us reconsider the game G^{Δ} in Example 5.1. Using Proposition 5.1, let us check whether the strategy profile (D, L), or $\sigma^* = ((0, 1), (1, 0))$, is a perfect equilibrium of G^{Δ}. If it is not, we will find some $\epsilon > 0$ such that, for every totally mixed strategy τ in Δ^N with $||\tau - \sigma^*|| < \epsilon$, there will be some player $i \in N$ such that σ_i^* is not a best response to τ_{-i}. So, consider $\tau = ((\delta_1, 1 - \delta_1), (1 - \delta_2, \delta_2))$ for some $\delta_1, \delta_2 \in (0, 1)$. Note that $||\tau - \sigma^*|| = \sqrt{2\delta_1^2 + 2\delta_2^2}$. We have

$u_1(U, \tau_2) = 3(1 - \delta_2) - 8\delta_2 = 3 - 11\delta_2$ and $u_1(D, \tau_2) = 4 - 4\delta_2$. So, $\sigma_1^* \in BR_1(\tau_2)$ if $1 > -7\delta_2$, which holds for any $\delta_2 \in (0, 1)$. On the other hand, $u_2(L, \tau_1) = 3\delta_1$ and $u_2(R, \tau_1) = 6\delta_1$. We have $\sigma_2^* \notin BR_2(\tau_1)$ for any $\delta_1 \in (0, 1)$. Note that δ_1 and δ_2 cannot exceed 1. Therefore, we can set $\epsilon = 2$ to ensure that the above conclusions will be valid for any $\delta_1, \delta_2 \in (0, 1)$. Thus, for $\epsilon = 2$, we can show that for every totally mixed strategy profile $\tau \in \Delta^N(S)$ that lies in the ϵ-neighborhood of σ^*, the strategy σ_2^* of player 2 is not a best response to τ_{-2}. Therefore, (D, L) is not a perfect equilibrium of G^Δ. ■

5.2 Existence

Below, we show that a perfect equilibrium always exists in the mixed extension of any finite strategic form game.

Proposition 5.2 The mixed extension of any finite strategic form game has a perfect equilibrium.

Proof Consider a finite strategic form game $G = \langle N, (S_i)_{i \in N}, (u_i)_{i \in N} \rangle$. Pick any profile of totally mixed strategies $\bar{\sigma} \in \Delta^N(S)$. For each $i \in N$, pick a sequence of $(\delta_i^k)_k$ converging to zero and define the payoff $u_i^k(s) = u_i((\delta_i^k \bar{\sigma}_i + (1 - \delta_i^k) s_i)$ for all $s \in S$. Let G^k be the strategic form game obtained from G by replacing u_i with u_i^k for each i. We know from Nash's [2] existence theorem that the mixed extension of G^k, which we denote by $G^{k,\Delta}$, has a Nash equilibrium, say $\sigma^k \in \Delta^N(S)$, for each $k = 1, 2, \ldots, \infty$. There exists a convergent subsequence of $(\sigma^k)_k$, since $\Delta^N(S)$ is a closed and bounded set. Let $(\sigma^{*,k})_k$ be such a subsequence and let σ^* be its limit. We know that, for any s_i with $\sigma^*(s_i) > 0$, there is an integer $z(s_i)$ such that $\sigma^{*,k}(s_i) > 0$ for all $k \geq z(s_i)$. Let $z = \max\{z(s_i) : s_i \in S_i \text{ and } i \in N\}$. It follows that, for any s_i with $\sigma^{*,k}(s_i) > 0$, we have $s_i \in BR_i(\sigma_{-i}^{*,k})$ for all $k > z$. Therefore, $\sigma_i^* \in BR_i(\sigma_{-i}^{*,k})$ for all $k > z$. Finally, by eliminating the first z elements of the sequence $(\sigma^{*,k})_k$ and renumbering the remaining elements starting from 1, we obtain $\sigma_i^* \in BR_i(\sigma_{-i}^{*,k})$ for all k and i, implying that σ^* is a perfect equilibrium. ■

5.3 Relation with Nash Equilibrium

The following result relates the perfect and Nash equilibrium concepts.

Proposition 5.3 Every perfect equilibrium in the mixed extension of a finite strategic form game is a Nash equilibrium.

Proof Let $G = \langle N, (S_i)_{i \in N}, (u_i)_{i \in N} \rangle$ be any finite strategic form game and G^Δ be its mixed extension. By Proposition 5.2, the set of perfect equilibrium profiles in G^Δ

is nonempty. So, let $\sigma^* \in \Delta^N(S)$ be a perfect equilibrium. Pick any player $i \in N$ and any strategy $\sigma_i' \in \Delta(S_i) - \sigma_i^*$. By Definition 5.1, there exists a sequence $(\sigma^k)_k$ of strategy profiles in $\Delta^N(S)$ converging to the perfect equilibrium σ^* such that

$$\tilde{u}_i(\sigma_i^*, \sigma_{-i}^k) \geq \tilde{u}_i(\sigma_i', \sigma_{-i}^k)$$

for all $k = 1, 2, \ldots, \infty$. The continuity of u_i implies

$$\tilde{u}_i(\sigma_i^*, \sigma_{-i}^*) \geq \tilde{u}_i(\sigma_i', \sigma_{-i}^*),$$

further implying $\sigma_i^* \in BR_i(\sigma_{-i}^*)$. Since i was arbitrarily picked, the above result is true for all i. Therefore, σ^* is a Nash equilibrium. ■

Propositions 5.2 and 5.3 together show that the perfectness notion is a strict refinement of the Nash equilibrium concept, since we know that every Nash equilibrium is not a perfect equilibrium. However, the set of Nash equilibria may include certain subsets where each member is a perfect equilibrium, as illustrated by the next two results.

Proposition 5.4 If a strategy profile in the mixed extension of a finite strategic form game is a strict Nash equilibrium, then this strategy profile is a perfect equilibrium.

Proof Let $G = \langle N, (S_i)_{i \in N}, (u_i)_{i \in N} \rangle$ be any finite strategic form game and G^Δ be its mixed extension. Suppose $s \in S$ is a strict Nash equilibrium of G^Δ. (Recall that an unpure strategy cannot be strict.) Then, for all $i \in N$, $u_i(s_i, s_{-i}) > u_i(s_i', s_{-i})$ for all $s_i' \in S$ with $s_i' \neq s_i$. Now, consider any $\sigma \in \text{int}(\Delta^N(S))$ such that $\sigma_i(s_i)$ is arbitrarily close to 1 for each $i \in N$. The continuity of u_i with respect to its arguments implies that $s_i \in BR_i(\sigma_{-i})$ implying $s \in BR(\sigma)$. Then, by Proposition 5.1, the strategy profile s is a perfect equilibrium. ■

Proposition 5.5 If a totally (unpure) mixed strategy profile in the mixed extension of a finite strategic form game is a Nash equilibrium, then this profile is a perfect equilibrium.

Proof Let $G = \langle N, (S_i)_{i \in N}, (u_i)_{i \in N} \rangle$ be any finite strategic form game and G^Δ be its mixed extension. Suppose $\sigma^* \in \text{int}(\Delta^N(S))$ is a totally unpure Nash equilibrium profile in G^Δ. To show that σ^* is a perfect equilibrium, we have to prove that for any $\epsilon > 0$ the ϵ-neighborhood of σ^* contains some $\tau \in \text{int}(\Delta^N(S))$ such that $\sigma^* \in BR(\tau)$. Note that this condition is trivially satisfied for $\tau = \sigma^*$ since $\sigma^* \in \text{int}(\Delta^N(S))$. Therefore, σ^* is a perfect equilibrium. ■

The mixed extension of a finite strategic form game may not have a totally unpure Nash equilibrium profile. Therefore, a totally unpure perfect equilibrium may not exist. On the other hand, Propositions 5.2 and 5.3 ensure that a perfect equilibrium (either in pure or unpure strategies) always exists and it is a Nash equilibrium.

5.4 Relation with Weak Dominance

Below, we show a one-sided relation between perfectness and dominance.

Proposition 5.6 Every perfect equilibrium in the mixed extension of a finite strategic form game is weakly undominated.

Proof Let $G = \langle N, (S_i)_{i \in N}, (u_i)_{i \in N} \rangle$ be any finite strategic form game and G^{Δ} be its mixed extension. Also, let σ be a perfect equilibrium in G^{Δ}. Suppose towards a contradiction that σ is weakly dominated, implying that there exists $i \in N$ such that σ_i is weakly dominated for player i. Then, for any totally unpure strategy profile $x \in \text{int}(\Delta^N(S))$, it is true that $\sigma_i \notin BR_i(x_{-i})$. So, by Proposition 5.1, σ cannot be a perfect equilibrium, a contradiction. Therefore, σ is weakly undominated. ■

To better understand why a weakly undominated strategy profile cannot be a perfect equilibrium, let us see the next example.

Example 5.3 Consider the game in Table 5.2 borrowed from Myerson [3].

This game has two Nash equilibria, (α_1, β_1) and (α_2, β_2). The equilibrium (α_1, β_1) is a strict Nash equilibrium; therefore, by Proposition 5.4, we know that it is a perfect equilibrium. Let us check whether the equilibrium (α_2, β_2) is also perfect. First, note that the profile (α_2, β_2) is equivalent to $\sigma = ((0, 1), (0, 1))$. To use Proposition 5.1, we will consider $\tau = ((\delta_1, 1 - \delta_1), (\delta_2, 1 - \delta_2))$ for some $\delta_1, \delta_2 \in (0, 1)$. Note that $||\tau - \sigma|| = \sqrt{2}\sqrt{\delta_1^2 + \delta_2^2}$. We have $u_1(\alpha_1, \tau_2) = \delta_2$ and $u_1(\alpha_2, \tau_2) = 0$. It follows that $u_1(\alpha_2, \tau_2) < u_1(\alpha_1, \tau_2)$ for any $\delta_2 \in (0, 1)$. Let us choose ϵ to be $\sqrt{2}$. This implies that $||\tau - \sigma|| < \epsilon$ only if $\delta_2 < 1$. However, α_2 cannot be a best response for player 1 to τ_2 at any $\delta_2 < 1$. Therefore, the Nash equilibrium (α_2, β_2) is not a perfect equilibrium. The reason is that it is weakly dominated. ■

In the above example, the strategy α_2 is weakly dominated by α_1 for player 1 and the strategy β_2 is weakly dominated by β_1 for player 2. Consequently, (α_1, β_1) Pareto dominates (α_2, β_2). However, this does not prevent (α_2, β_2) from being a Nash equilibrium, since by definition each player believes that his opponent is unlikely to play a strategy not mentioned in the equilibrium profile. However, in a perfect equilibrium, as beautifully stated by Myerson [3, p. 76].

Table 5.2 The strategic form game in Example 5.3

		Player 2	
		β_1	β_2
Player 1	α_1	(1, 1)	(0, 0)
	α_2	(0, 0)	(0, 0)

> ... a player cannot ignore any point in $S_1 \times \ldots \times S_n$ as "impossible" when he computes his best responses, since every one of his opponents' pure strategies has a positive probability; and yet no player wants to make any substantial change in his strategy since each player is already putting almost all his probability weight on his best responses.

Propositions 5.2, 5.3, and 5.6 together imply that the mixed extension of every finite strategic form game has at least one Nash equilibrium in which no player uses a weakly dominated strategy. Van Damme [4, pp. 49–50] shows that there exists a partial converse for this result: If the mixed extension of a strategic form game with two players and a finite number of strategies has an undominated Nash equilibrium, then this equilibrium is perfect. Van Damme [4, p. 29] further shows that when the number of players is at least three, this result does not hold. The following example of his verifies this.

Example 5.4 Consider the following three-player strategic form game represented by Table 5.3. Clearly, $(D, L, Panel\, a)$ is a weakly undominated equilibrium of the above game. In terms of probabilities, this strategy profile is represented by $\sigma = ((0, 1), (1, 0), (1, 0))$.

Using Proposition 5.1, let us check whether σ is a perfect equilibrium. If it is not, we will find some $\epsilon > 0$ such that for every totally mixed strategy τ in Δ^N with $||\tau - \sigma|| < \epsilon$ there will be some player $i \in N$ such that σ_i is not a best response to τ_{-i}. So, consider $\tau = ((\delta_1, 1 - \delta_1), (1 - \delta_2, \delta_2), (1 - \delta_3, \delta_3))$ for some $\delta_1, \delta_2, \delta_3 \in (0, 1)$. Note that $||\tau - \sigma|| = \sqrt{2}\sqrt{\delta_1^2 + \delta_2^2 + \delta_3^2}$. We have

$$\begin{aligned} u_1(U, \tau_{-1}) &= 1(1 - \delta_2)(1 - \delta_3) + 1(\delta_2)(1 - \delta_3) + 1(1 - \delta_2)(\delta_3) + 0(\delta_2)(\delta_3) \\ &= (1 - \delta_3) + (1 - \delta_2)(\delta_3) \end{aligned}$$

and

$$\begin{aligned} u_1(D, \tau_{-1}) &= 1(1 - \delta_2)(1 - \delta_3) + 0(\delta_2)(1 - \delta_3) + 0(1 - \delta_2)(\delta_3) + 1(\delta_2)(\delta_3) \\ &= (1 - \delta_2)(1 - \delta_3) + (\delta_2)(\delta_3). \end{aligned}$$

Table 5.3 The strategic form game in Example 5.4

Panel a — 1st Strategy of Player 3 (rows: Player 1; columns: Player 2)

	L	R
U	(1, 1, 1)	(1, 0, 1)
D	(1, 1, 1)	(0, 0, 1)

Panel b — 2nd Strategy of Player 3 (rows: Player 1; columns: Player 2)

	L	R
U	(1, 1, 0)	(0, 0, 0)
D	(0, 1, 0)	(1, 0, 0)

It follows that $u_1(D, \tau_{-1}) < u_1(U, \tau_{-1})$ if $\delta_2 \in (0, 0.5)$. Let us choose ϵ to be $\sqrt{2}(0.01)$. This implies that $||\tau - \sigma|| < \epsilon$ only if $\delta_2 < 0.1$. However, in that case D cannot be a best response for player 1 to τ_{-1}. Therefore, $(D, L, Panel\ a)$ is not a perfect equilibrium. ■

A major criticism to the perfect equilibrium concept is that a perfect equilibrium needs to be robust only to a particular sequence of trembles (mistakes), not to all kind of trembles. As a remedy, one may require the perfect equilibrium to be strict in the sense that it is robust to all possible trembles. However, this stronger equilibrium, called the strictly perfect equilibrium, has a downside, too: it may not exist.

A second criticism, implicitly remarked by Myerson [3], is that there are games where the presence of a perfect equilibrium may be sensitive to the addition/deletion of weakly dominated strategies. Below, we illustrate this annoying fact using the games he analyzed.

Example 5.5 Reconsider the strategic form game in Example 5.3. Recall that (α_2, β_2) is a Nash equilibrium which is not a perfect equilibrium since α_2 is weakly dominated for player 1 (and β_2 is weakly dominated for player 2).

Suppose now that both players add a weakly undominated strategy (α_3 and β_3, respectively) to their strategy spaces, leading to the following modified game (Table 5.4).

We will show that the strategy profile (α_2, β_2) is now a perfect equilibrium. We can write this profile as $\sigma^* = ((0, 1, 0), (0, 1, 0))$. So, consider a sequence $\sigma^k = ((\delta^k, 1 - 2\delta^k, \delta^k), (\delta^k, 1 - 2\delta^k, \delta^k))$ where $\delta \in (0, 1)$ and $k = 1, 2, \ldots, \infty$. Note that σ^k converges to σ. Moreover, $u_1(\alpha_1, \sigma_2^k) = -8\delta^k$, $u_1(\alpha_2, \sigma_2^k) = -7\delta^k$, and $u_1(\alpha_3, \sigma_2^k) = -7 - 2\delta^k$. For any $\delta \in (0, 1)$, we find $\alpha_2 \in BR_1(\sigma_2^k)$ for any $k = 1, 2, \ldots, \infty$. On the other hand, $u_2(\beta_2, \sigma_1^k) = -8\delta^k$, $u_2(\beta_2, \sigma_1^k) = -7\delta^k$, and $u_2(\beta_2, \sigma_1^k) = -7 - 2\delta^k$. For any $\delta \in (0, 1)$, we find $\beta_2 \in BR_2(\sigma_1^k)$ for any $k = 1, 2, \ldots, \infty$. Thus, we have established that there exists a totally unpure strategy profile sequence σ^k in $\Delta^N(S)$ that converges to σ^* and is such that $\sigma_i^* \in BR_i(\sigma_{-i}^k)$ for each $i \in N$ and $k = 1, 2, \ldots, \infty$. So, σ^* is a perfect equilibrium of the modified game. ■

Table 5.4 The modified strategic form game in Example 5.5

Player 1 \ Player 2	β_1	β_2	β_3
α_1	$(1, 1)$	$(0, 0)$	$(-9, -9)$
α_2	$(0, 0)$	$(0, 0)$	$(-7, -7)$
α_3	$(-9, -9)$	$(-7, -7)$	$(-7, -7)$

References

1. Selten R (1975) Reexamination of the perfectness concept for equilibrium points in extensive games. Int J Game Theory 4:25–55
2. Nash JF (1950) Equilibrium points in n-person games. PNAS 36(1):48–49
3. Myerson RB (1978) Refinements of the Nash equilibrium concept. Int J Game Theory 7(2):73–80
4. Van Damme E (1987) Stability and perfection of Nash equlibria. Springer-Verlag, Berlin Heidelberg

Evolutionary Stable Strategies

6

In this chapter, we will consider robustness of an equilibrium from an evolutionary perspective. Briefly, we will be dealing with the question whether or when a large population of players that is randomly matched as pairs to play a symmetric two-person strategic form game is resistant against the strategies of mutants that try to invade the population. The answer will yield a concept called evolutionary stable strategies. In this chapter, we will also present the relation between this concept and several notions of equilibrium presented in earlier chapters.

Two researchers in biology, John Maynard Smith and George R. Price asked an interesting question in the early 1970s: Given a large population of animals, can we identify the set of strategies in this population that are immune or resistant against the strategies of mutants that try to infect/invade the population? In other words, when or how can we ensure that equilibrium actions in such a population will be evolutionary stable?

To convey the question and answer of Maynard Smith and Price [1,2] formally, suppose that a large population of individuals (or animals) is randomly matched as pairs to play a symmetric two-person strategic form game allowing mixed strategies. Also suppose that initially the population is monomorphic; i.e., every individual is playing the same mixed strategy, say x. When an individual is randomly matched with another individual from this population, its expected payoff becomes $u(x, x)$. Suppose now, a small group of individuals change their strategies to another mixed strategy y, which we call a mutant strategy. Let the fraction of these individuals in the whole population be $\epsilon \in (0, 1)$. We can call ϵ as the invasion probability of y to a monomorphic population or the infection probability of this population by y. After the invasion of y at the probability of ϵ, only $(1 - \epsilon)$ percent of the population will play x. So, the average strategy in the post-entry population will be $(1 - \epsilon)x + \epsilon y$. Consequently, when an individual playing the incumbent strategy x is randomly matched with an individual from the post-entry population, its expected payoff will

I. Saglam, *Mastering Game Theory*, Classroom Companion: Economics,
https://doi.org/10.1007/978-3-031-82094-6_6

be $u(x, (1-\epsilon)x + \epsilon y)$. On the other hand, if this individual changes its strategy to the mutant strategy y, its expected payoff will be $u(y, (1-\epsilon)x + \epsilon y)$. Comparing these two payoffs, we can say that an individual that plays the incumbent strategy x in the pre-entry population would not switch to the mutant strategy y with an invasion probability ϵ if and only if

$$u(x, (1-\epsilon)x + \epsilon y) > u(y, (1-\epsilon)x + \epsilon y).$$

If the above condition holds, we can say that the incumbent strategy x is immune (resistant) against the mutant strategy y invading a percentage ϵ of the population. Moreover, whenever a strategy x becomes immune to any mutant strategy $y \neq x$ invading the population at a sufficiently small probability, we will say that x is evolutionary stable. We will formally define this idea below.

6.1 Definitions and Results

Let $G = \langle N, (S_i)_{i\in N}, (u_i)_{i\in N}\rangle$ be a finite and symmetric two-person strategic form game and G^{Δ} be its mixed extension. Let $S_1 = S_2 = \bar{S}$ and $u_1 = u_2 = u$.

Definition 6.1 A mixed (pure or unpure) strategy x in $\Delta(\bar{S})$ is an evolutionary stable strategy (ESS) if for any $y \in \Delta(\bar{S}) - x$ there exists $\bar{\epsilon}(y)$ such that

$$u(x, (1-\epsilon)x + \epsilon y) > u(y, (1-\epsilon)x + \epsilon y) \text{ for all } \epsilon \in (0, \bar{\epsilon}(y)). \tag{6.1}$$

We can rewrite the inequality condition in (6.1) after we define a relative payoff function $R(x, y, \epsilon)$ where

$$\begin{aligned} R(x, y, \epsilon) &= u(x, (1-\epsilon)x + \epsilon y) - u(y, (1-\epsilon)x + \epsilon y) \\ &= (1-\epsilon)u(x, x) + \epsilon u(x, y) - [(1-\epsilon)u(y, x) + \epsilon u(y, y)] \\ &= (1-\epsilon)[u(x, x) - u(y, x)] + \epsilon[u(x, y) - u(y, y)]. \end{aligned}$$

We can restate Definition 6.1 as follows.

Definition 6.2 A mixed (pure or unpure) strategy x in $\Delta(\bar{S})$ is an evolutionary stable strategy (ESS) if for any $y \in \Delta(\bar{S}) - x$ there exists $\bar{\epsilon}(y)$ such that

$$R(x, y, \epsilon) > 0 \text{ for all } \epsilon \in (0, \bar{\epsilon}(y)). \tag{6.2}$$

The formulation in Definition 6.2 will allow us to write two simple conditions to identify whether any given strategy is an ESS or not.

Proposition 6.1 *A mixed (pure or unpure) strategy x in* $\Delta(\bar{S})$ *is an ESS if and only if the following two conditions hold:*

$$(i)\ u(x,x) \geq u(y,x) \ \text{for all}\ \ y \in \Delta(\bar{S}), \tag{6.3}$$

$$(ii)\ u(x,x) = u(y,x) \ \text{only if}\ y = x \ \ \text{or}\ \ u(x,y) > u(y,y). \tag{6.4}$$

Proof Let us first prove the "if part" of the proposition. Given the game G^{Δ}, pick two distinct mixed strategies x and y in $\Delta(\bar{S})$ and any $\epsilon \in (0, 1)$. The conditions (i) and (ii) in (6.3) and (6.4) imply that $R(x, y, \epsilon) > 0$. Set $\bar{\epsilon}(y) = 1$. Since $\epsilon \in (0, 1)$, we have just proved that $R(x, y, \epsilon) > 0$ for all $\epsilon \in (0, \bar{\epsilon}(y))$. So, the condition in (6.2) is satisfied.

Now, let us prove the "only if part" of the proposition. Note that

$$\lim_{\epsilon \to 0} R(x, y, \epsilon) = u(x, x) - u(y, x).$$

If $u(x, x) - u(y, x) < 0$, then by the continuity of $R(x, y, \epsilon)$ in ϵ we can say that $R(x, y, \epsilon) > 0$ cannot be true for some ϵ sufficiently close to zero. Thus, condition (i) in (6.3) must hold. Now, suppose this condition holds with equality. Then, we have

$$R(x, y, \epsilon) = \epsilon[u(x, y) - u(y, y)].$$

Since $x \neq y$, the inequality $R(x, y, \epsilon) > 0$ can hold only if $u(x, y) > u(y, y)$, as stated by condition (ii) in (6.4). ■

Maynard Smith and Price [1] defined ESS in terms of conditions (i) and (ii) in the above proposition. Condition (i) says that a mixed strategy x is an ESS only if x is a best response to itself. On the other hand, condition (ii) says that if there exists another mixed strategy, y, that is a best response to x, then x must be a better response to y than y itself.

Unlike the set of Nash equilibria, the set of ESS may be empty in some strategic games. (We will illustrate this fact in Example 6.1 in the next section.) Whenever in a strategic form game we find a strategy that is an ESS, the following result may be helpful to quickly find out whether there can be other strategies that are also ESS.

Proposition 6.2 *If* $x \in \Delta(\bar{S})$ *is a totally mixed (unpure) strategy and ESS, then it is the unique ESS.*

Proof Suppose x is a totally mixed strategy and it is an ESS. Then, (x, x) is a Nash equilibrium as ensured by condition (i) in Proposition 6.1. Since x is a totally mixed strategy, each strategy in $\bar{S}$ is a best reply to x (for otherwise any player would assign zero probability to some strategies in $\bar{S}$ to increase its expected utility). Hence, for any $y \in \Delta(\bar{S}) - x$, we have $u(y, x) = u(x, x)$. Since x is an ESS, condition (ii) of ESS in Proposition 6.1 implies that $u(x, y) > u(y, y)$ for any $y \in \Delta(\bar{S}) - x$, further implying that y is not a best reply to itself, i.e., (y, y) is not a Nash equilibrium, hence y is not an ESS. ■

The above proposition implies that if two pure strategy profiles in a symmetric strategic game are ESS, then a strategy profile mixing the strategies of each player at these two equilibria is not an ESS.

Using condition (i) in Proposition 6.1, we can also state the following relationship between the ESS and Nash equilibrium concepts in strategic form games.

Proposition 6.3 *A mixed strategy x in $\Delta(\bar{S})$ is an ESS only if (x, x) is a Nash equilibrium.*

Proof Suppose x is an ESS. Condition (i) for ESS says that $u(x, x) \geq u(y, x)$ for any $y \in \Delta(\bar{S}) - x$, implying that (x, x) is a Nash equilibrium. ■

The above result indicates that evolution leads to Nash equilibrium play in symmetric games. Moreover, if the Nash equilibrium concept is strengthened, we have the following sufficiency result.

Proposition 6.4 *A mixed strategy x in $\Delta(\bar{S})$ is an ESS if (x, x) is a strict Nash equilibrium.*

Proof Suppose $x \in \Delta(\bar{S})$ is such that (x, x) is a strict Nash equilibrium. Then $u(x, x) > u(y, x)$ for any $y \in \Delta(\bar{S}) - x$, implying that condition (i) for ESS holds. The above inequality also implies that there exists no $y \in \Delta(\bar{S}) - x$ such that $u(x, x) = u(y, x)$. Thus, condition (ii) (vacuously) holds, too. Therefore, x is an ESS. ■

We can also relate evolutionary stability to the weak dominance concept.

Proposition 6.5 *If a mixed strategy x in $\Delta(\bar{S})$ is an ESS, then x is weakly undominated.*

Proof Suppose x is an ESS that is weakly dominated. Then, there exists a strategy $z \in \bar{S}$ such that $u(z, y) \geq u(x, y)$ for all $y \in \bar{S}$. Since this is true for $y = x$ as well, z must be a best response to x. Also, x is a best response to x by condition (i) of ESS. Then, we must have $u(x, z) > u(z, z)$ by condition (ii) of ESS. On the other hand, the weak domination condition that $u(z, y) \geq u(x, y)$ for all $y \in \bar{S}$ implies that $u(z, z) \geq u(x, z)$ (as we can insert z for y above). Thus, a contradiction. Therefore, x must be weakly undominated if it is an ESS. ■

We should notice that the converse of the above proposition is not true. In the rock-scissors-paper game that we will analyse in Example 6.1, all three strategies of each player are weakly undominated while no strategy is an ESS.

6.2 Examples

Our first example illustrates that the set of ESS in a strategic form game may be empty.

Example 6.1 Consider the strategic form game in Table 6.1, which is known as the rock-scissors-paper game. It is easy to check that the above game has no pure-strategy Nash equilibrium. The unique Nash equilibrium is totally mixed, given by (σ, σ) where $\sigma = (1/3, 1/3, 1/3)$.

Let us check whether σ is an ESS. Since (σ, σ) is a Nash equilibrium, condition (i) in Proposition 6.1 is satisfied.

To check condition (ii), note that $u(\sigma, \sigma) = 1$. Clearly for any $y \in \{R, S, P\}$, we also have $u(y, \sigma) = 1$. Moreover, $u(\sigma, y) = 1 = u(y, y)$ for any $y \in \{R, S, P\}$. So, condition (ii) does not hold, implying that σ is not a ESS. Thus, the rock-scissors-paper game has no ESS. ■

Our next example illustrates a strategic form game with two (pure) ESS's.

Example 6.2 Reconsider the Stag Hunt Game in Table 1.5.

Player 1 \ Player 2	S	H
S	(s, s)	$(0, h)$
H	$(h, 0)$	(h, h)

Table 1.5 Reconsidered

where $0 < h < s$. This game has three Nash equilibria: Two pure strategy equilibria (S, S) and (H, H), and a totally-mixed strategy equilibrium $((\frac{h}{s})S + (1 -$

Table 6.1 The rock-scissors-paper game in Example 6.1

Player 1 \ Player 2	R	S	P
R	$(1, 1)$	$(2, 0)$	$(0, 2)$
S	$(0, 2)$	$(1, 1)$	$(2, 0)$
P	$(2, 0)$	$(0, 2)$	$(1, 1)$

$\frac{h}{s})H$, $(\frac{h}{s})S + (1 - \frac{h}{s})H)$. Note also that (S, S) and (H, H) are both strict Nash equilibria; therefore, by Proposition 6.4 the strategies S and H are both ESS. However, the totally-mixed strategy $(\frac{h}{s})S + (1 - \frac{h}{s})H$ cannot be an ESS, for otherwise Proposition 6.2 would require that it would be unique, contradicting the fact that both H and Y are ESS. To directly show that $\sigma = (\frac{h}{s})S + (1 - \frac{h}{s})H$ is not an ESS, note that it satisfies condition (i) of ESS, since (σ, σ) is a Nash equilibrium. Also note that $u(\sigma, \sigma) = u(H, \sigma) = h$. Then condition (ii) of ESS requires $u(\sigma, H) > u(H, H)$. But, $u(\sigma, H) = (1 - \frac{h}{s})h < h = u(H, H)$. Thus, condition (ii) is not satisfied, implying that σ is not an ESS. ■

Our last example illustrates a strategic form game with a unique ESS.

Example 6.3 Reconsider the Hawk-Dove Game in Table 1.7, under the assumption $0 < v < c$.

		Player 2	
		H	D
Player 1	H	$(\frac{v-c}{2}, \frac{v-c}{2})$	$(v, 0)$
	D	$(0, v)$	$(\frac{v}{2}, \frac{v}{2})$

Table 1.7 Reconsidered

It is easy to check that the mixed extension of the above game has a totally mixed Nash equilibrium (σ^*, σ^*) with $\sigma^* = (v/c, 1 - (v/c))$. Let us check whether σ^* is an ESS. Since (σ^*, σ^*) is a Nash equilibrium, σ^* is a best reply to itself. Thus, for σ^* condition (i) of ESS holds. Now, let us check condition (ii). Since σ^* is totally mixed, any $\sigma' \neq \sigma^*$ in the common strategy space of any player is also a best reply to σ^*. Thus, pick $\sigma' = (p, 1 - p)$ where $p \in [0, 1]$ and $p \neq v/c$. Then, the second-order best-reply condition for σ^* to be an ESS requires that $u(\sigma^*, \sigma') > u(\sigma', \sigma')$. Note that

$$u(\sigma^*, \sigma') = \frac{v}{c}\left(p\left(\frac{v-c}{2}\right) + (1-p)v\right) + \left(1 - \frac{v}{c}\right)\left((1-p)\frac{v}{2}\right)$$

and

$$u(\sigma', \sigma') = p\left(p\left(\frac{v-c}{2}\right) + (1-p)v\right) + (1-p)\left((1-p)\frac{v}{2}\right).$$

It then follows that

$$\begin{aligned} u(\sigma^*, \sigma') - u(\sigma', \sigma') &= \left(\frac{v}{c} - p\right)\left[p\left(\frac{v-c}{2}\right) + (1-p)v - (1-p)\frac{v}{2}\right] \\ &= \frac{(v - pc)^2}{2c} > 0, \end{aligned}$$

since $p \neq v/c$ by assumption. Thus, condition (ii) of ESS is also satisfied for σ^*. Hence, σ^* is an ESS. By Proposition 6.2, σ^* must be the unique ESS of the Hawk-Dove Game.

We should finally note that the Hawk-Dove Game becomes a Prisoner's Dilemma Game whenever $c \in (0, v)$. In that case, this game has a unique Nash equilibrium (H, H), which is a strict equilibrium. By Proposition 6.4, we then know that H is an ESS and by Proposition 6.3 and the uniqueness of the Nash equilibrium strategy profile we also infer that there can be no other strategy that is an ESS. ■

References

1. Maynard Smith J, Price GR (1973) The logic of animal conflict. Nature 246:15–18
2. Maynard Smith J (1974) The theory of games and the evolution of animal conflicts. J Theoret Biol 47(1):209–221

7 Correlated Equilibrium

Chapter 7 will show how the idea of Nash equilibrium can be extended to a strategic form game where the noncooperative strategies of players can be correlated by a mediator who picks a strategy profile (to be played in the game) randomly according to a commonly known probability distribution (over the set of all strategy profiles) and privately informs each player about the strategy recommended to play. The equilibrium of this game where each player finds it to her best interest to obey the recommendation of the mediator, under the belief that all other players will obey too, is called a correlated equilibrium. Chapter 7 will also illustrate that in some strategic form games some correlated equilibria lead to expected payoff profiles that players could not achieve by choosing their strategies independently.

In the previous chapters, we have assumed that the mixed strategies of players are independent. The implication of this assumption is that players cannot reach some probability distributions over the space of strategy profiles $S = \bigtimes_{i \in N} S_i$, possibly leading to the inefficiency that players cannot reach some expected payoff distributions that are Pareto undominated. In this chapter, we will see whether and how this inefficiency can be eliminated or reduced by a mediator who is able to coordinate the strategies of players. To illustrate this point, we will consider the following example.

Example 7.1 (*Traffic Control Game*) Suppose that two drivers are at an intersection point where one of them must stop, while the other continues to go. A driver who stops always derives a payoff of zero, whereas a driver who unilaterally drives has a payoff of 1. If both drivers decide to drive, they crash resulting in severe casualties, and each obtains a payoff of -9. We can represent the described game by Table 7.1.

I. Saglam, *Mastering Game Theory*, Classroom Companion: Economics,
https://doi.org/10.1007/978-3-031-82094-6_7

Table 7.1 The strategic form game in Example 7.1

		Player 2	
		GO	STOP
Player 1	GO	$-9, -9$	$1, 0$
	STOP	$0, 1$	$0, 0$

One can easily calculate that the mixed extension of the above strategic form game has three Nash equilibria:

$$(\text{GO,STOP}),\ (\text{STOP,GO}),\ \text{and}\ \ (0.1\ \text{GO} + 0.9\ \text{STOP},\ 0.1\ \text{GO} + 0.9\ \text{STOP}).$$

Player 1 prefers the equilibrium (GO,STOP) to obtain a payoff of 1, whereas player 2 prefers the equilibrium (STOP,GO) to obtain a payoff of 1. On the other hand, each player obtains a payoff of zero in the symmetric (and unpure) equilibrium (0.1 GO + 0.9 STOP, 0.1 GO + 0.9 STOP). One can argue that the symmetric unpure strategy Nash equilibrium compromises between the conflicting preferences of players on the two pure strategy Nash equilibria. However, there are two serious problems about this unpure equilibrium. First, the probability of crash is nonzero; it is $0.1 \times 0.1 = 0.01$. Moreover, the probability that both drivers must wait at the intersection is not zero. This probability is $0.9 \times 0.9 = 0.81$, indicating a substantial waste of time (and possibly a waste of gas, too).

Now, instead of mixed strategy profiles, let us consider a probability distribution function p, a mapping from the set of all possible strategy profiles $S = \{\text{GO,STOP}\} \times \{\text{GO,STOP}\}$ to the unit interval $[0, 1]$. For the traffic control game, the aforementioned two problems (the statistically expected crash and waste at the intersection) would be eliminated if the mixed strategies of the players would result in a probability distribution p^* where $p^*(\text{GO,STOP}) = 1/2$ and $p^*(\text{STOP,GO}) = 1/2$. Under such a probability distribution, neither car crash nor waste of time would ever occur, since $p^*(\text{GO,GO})$ and $p^*(\text{STOP,STOP})$ would be both zero. This probability distribution is nothing but a random version of the well-known traffic system with green (GO) and red (STOP) lights. Green light (to player 1) allows player 1 to go and requires player 2 to stop, whereas red light (to player 1) requires player 1 to stop and allows player 2 to go. One can check that p^* is the unique probability distribution that maximizes the sum of two players' payoffs, without favoring any one of them. That is,

$$u_1(p^*) + u_2(p^*) = 1/2 = \max_p u_1(p) + u_2(p) \ \ s.t.\ \ u_1(p) = u_2(p).$$

Unfortunately, it is impossible to obtain the described probability distribution p^* as a mixed strategy profile whenever the mixed strategies of players are independent.

Table 7.2 The probability distributions induced by the independent mixed strategies in Table 7.1

		Player 2	
		GO	STOP
Player 1	GO	$\sigma_{11}\sigma_{12}$	$\sigma_{11}(1-\sigma_{12})$
	STOP	$(1-\sigma_{11})\sigma_{12}$	$(1-\sigma_{11})(1-\sigma_{12})$

Table 7.2

To see this point, suppose towards a contradiction that a mixed strategy profile

$$\sigma = (\sigma_1, \sigma_2) = ((\sigma_{11}, 1-\sigma_{11}), (\sigma_{12}, 1-\sigma_{12}))$$

can generate the probability distribution p^*. For clarity, let us illustrate in Table 7.2 the probabilities generated by σ on the set of pure strategy profiles in the traffic control game.

As we desire $p^*(\text{GO,GO}) = 0$, the distribution p^* can be generated by σ only if $\sigma_{11} = 0$ or $\sigma_{12} = 0$, implying $\sigma_{11}\sigma_{12} = 0$. Then, we would have $\sigma_{11}(1-\sigma_{12}) = \sigma_{11}$. However, since we also desire $p^*(\text{GO,STOP}) = 1/2$ and $p^*(\text{GO,STOP}) = \sigma_{11}(1-\sigma_{12})$ must be true by supposition, it would follow that $\sigma_{11}(1-\sigma_{12}) = \sigma_{11} = 1/2$. This would further imply that $\sigma_{12} = 0$ since $p^*(\text{GO,GO}) = 0$. Then, we would have $(1-\sigma_{11})\sigma_{12} = 0$, implying that $p^*(\text{STOP,GO}) = 1/2$ cannot be equal to $(1-\sigma_{11})\sigma_{12}$, contradicting the supposition that p^* can be generated by some mixed strategy profile σ. ■

Example 7.1 showed that there are strategic form games where some payoff profiles cannot be obtained when players play their mixed strategies independently (regardless whether they play their best-response strategies or not). The solution to the traffic control game in Example 7.1 is the traffic lightning system which has been in use throughout the world for many decades. This solution, which was known as a folklore in game theory for years, simply hinges on the possibility to obtain any convex combination of two Nash equilibrium profiles, (GO,STOP) and (STOP,GO), using an external randomization device that produces p^* which puts probability 0.5 to each of these profiles.

The insufficiency of independent mixed strategies in strategic form games is a general phenomenon. This insufficiency is caused by the fact that the set of all independent mixed strategy profiles, $\Delta^N(S) = \times_{i \in N} \Delta(S_i)$, is not as large as the set of all probability distributions on S, which we denote by $\Delta(S)$. This insufficiency prevents the players in some strategic form games to coordinate on a mutually desirable strategy profile. As we have seen in the traffic control game, a remedy is to ensure players' coordination with the help of some externally given probability distribution on S, i.e., $p \in \Delta(S)$. This remedy, which was formally proposed by Robert Aumann in [1], will be presented in the next section.

7.1 Definitions

Given a strategic form game $G = \langle N, (S_i)_{i\in N}, (u_i)_{i\in N}\rangle$, a mediator turns a wheel of fortune where each strategy in $S = \times_{i\in N} S_i$ has the chance of being selected according to a probability distribution p on S; i.e., $p \in \Delta(S)$. This probability distribution is selected by a mediator in line with some social objective, which may be, for example, to maximize the total expected utility of players or the expected utility of the worst-off player. The mediator publicly announces the distribution p to the players before they make any decision. The mediator also announces that once he privately observes a strategy profile $s \in S$ selected by the wheel of fortune according to the distribution p, each player i will be privately recommended to play s_i. Since each player i is assumed to know the strategy space S in G, her utility function u_i, and the distribution p that was announced by the mediator, each player can calculate her expected utility from obeying any possible recommendation of the mediator before the mediator turns the wheel of fortune.

A probability distribution is called a correlated equilibrium if given this distribution each player finds it to her best interest to obey any recommendation of the mediator under the belief that the remaining players will also obey the recommendations of the mediator.

Definition 7.1 Given a finite strategic form game $G = \langle N, (S_i)_{i\in N}, (u_i)_{i\in N}\rangle$, a probability distribution $p \in \Delta(S)$ is a correlated equilibrium if for each $i \in N$ and, for each $r_i, t_i \in S_i$ with $\sum_{r_{-i}\in S_{-i}} p(r_{-i}|r_i) > 0$, we have

$$\sum_{r_{-i}\in S_{-i}} u_i(r_i, r_{-i})p(r_{-i}|r_i) \geq \sum_{r_{-i}\in S_{-i}} u_i(t_i, r_{-i})p(r_{-i}|r_i),$$

where $p(r_{-i}|r_i)$ denotes the conditional probability of r_{-i} given r_i.

According to this definition, each player $i \in N$ may think as follows: "Well, I do not know what strategy in S_i will be recommended to me; but if I am recommended r_i, then I can calculate my expected payoff both under r_i and under any alternative strategy t_i, and then decide whether to obey r_i or play another strategy. In calculating my expected payoff, I will use the probability distribution $p(.)$ and my private information r_i to calculate the conditional likelihoods with which the other players will play their strategies, i.e. the distribution $p(.|r_i)$." Under this reasoning, the expected utility of player i from obeying r_i is $\sum_{r_{-i}\in S_{-i}} u_i(r_i, r_{-i})p(r_{-i}|r_i)$, whereas her expected utility from playing an alternative strategy t_i is $\sum_{r_{-i}\in S_{-i}} u_i(t_i, r_{-i})p(r_{-i}|r_i)$. Given any recommendation r_i, player i should compare these two utility terms for each possible strategy t_i in deciding whether or not to obey r_i. Notice that when player i decides on whether or not to switch from r_i to t_i, she still uses the conditional probability $p(r_{-i}|r_i)$ in calculating the expected payoff from playing t_i. This is because she believes that all other players will obey the recommendations of the mediator even when she will not. Hence, in a correlated equilibrium, each player actually checks whether her obeying any possible recommendation is a best response to the remaining players' obedience to the strategies they will be privately recommended.

Also notice in the above definition that $p(r_{-i}|r_i)$ is defined if and only if $\sum_{r_{-i}\in S_{-i}} p(r_{-i}|r_i) > 0$; i.e., the marginal probability of r_i is positive. Whenever this condition is true, the inequality in Definition 7.1 can be written as

$$\sum_{r_{-i}\in S_{-i}} u_i(r_i, r_{-i})p(r_i, r_{-i}) \geq \sum_{r_{-i}\in S_{-i}} u_i(t_i, r_{-i})p(r_i, r_{-i}).$$

In fact, the last inequality is meaningfully defined even when $\sum_{r_{-i}\in S_{-i}} p(r_{-i}|r_i) = 0$, since $p(r_i, r_{-i})$ would be zero for any $r_{-i} \in S_{-i}$. Thus, we can rewrite Definition 7.1 as follows:

Definition 7.2 Given a finite strategic form game $G = \langle N, (S_i)_{i\in N}, (u_i)_{i\in N}\rangle$, a probability distribution $p \in \Delta(S)$ is a correlated equilibrium if, for each $i \in N$ and for each $r_i, t_i \in S_i$, we have

$$\sum_{r_{-i}\in S_{-i}} u_i(r_i, r_{-i})p(r_i, r_{-i}) \geq \sum_{r_{-i}\in S_{-i}} u_i(t_i, r_{-i})p(r_i, r_{-i}).$$

7.2 Theoretical Results

We will first relate a Nash equilibrium to a correlated equilibrium. However, we have to be cautious: a Nash equilibrium is defined on independent strategies, whereas correlated strategies are defined on correlated (interdependent) strategies. Thus, we have to first transform a Nash equilibrium strategy to a correlated strategy. Note that, for each independent strategy profile $\sigma \in \Delta^N(S)$, we can generate a probability distribution (correlated strategy vector) p^σ such that $p^\sigma(s) = \times_{i\in N} \sigma_i(s_i)$ for each $s \in S$.

Proposition 7.1 *Let $G = \langle N, (S_i)_{i\in N}, (u_i)_{i\in N}\rangle$ be a finite strategic form game and G^Δ be its mixed extension. For any Nash equilibrium σ^* of G^Δ, the induced probability distribution p^{σ^*} is a correlated equilibrium.*

Proof Let $\sigma^* \in \Delta^N(S)$ be a Nash equilibrium of a G^Δ. Then,

$$\tilde{u}_i(s_i, \sigma^*_{-i}) \geq \tilde{u}_i(s'_i, \sigma^*_{-i})$$

for all s_i with $\sigma_i^*(s_i) > 0$ and for all $s_i' \in S_i$. This inequality can be rewritten as

$$\sum_{s_{-i} \in S_{-i}} \sigma_{-i}^*(s_{-i}) u_i(s_i, s_{-i}) \geq \sum_{s_{-i} \in S_{-i}} \sigma_{-i}^*(s_{-i}) u_i(s_i', s_{-i})$$

for all s_i with $\sigma_i^*(s_i) > 0$ and for all $s_i' \in S_i$. It follows that

$$\sum_{s_{-i} \in S_{-i}} \sigma_i^*(s_i)\sigma_{-i}^*(s_{-i}) u_i(s_i, s_{-i}) \geq \sum_{s_{-i} \in S_{-i}} \sigma_i^*(s_i)\sigma_{-i}^*(s_{-i}) u_i(s_i', s_{-i})$$

for all $s_i, s_i' \in S_i$. Since p^{σ^*} is the probability distribution induced by σ^*, we have $p^{\sigma^*}(s_i, s_{-i}) = \sigma_i^*(s_i)\sigma_{-i}^*(s_{-i})$ for any $s \in S$. Thus, we can rewrite the last inequality above as

$$\sum_{s_{-i} \in S_{-i}} p^{\sigma^*}(s_i, s_{-i}) u_i(s_i, s_{-i}) \geq \sum_{s_{-i} \in S_{-i}} p^{\sigma^*}(s_i, s_{-i}) u_i(s_i', s_{-i})$$

for all $s_i, s_i' \in S_i$, implying that p^{σ^*} is a correlated equilibrium. ■

Since the mixed extension of a finite strategic form game has always a Nash equilibrium in mixed (pure or unpure) strategies, the above result implies that the set of correlated equilibria of a finite strategic form game is nonempty.

Proposition 7.2 For any finite strategic form game, the set of correlated equilibrium is nonempty.

Proof Let $G = \langle N, (S_i)_{i \in N}, (u_i)_{i \in N} \rangle$ be a finite strategic form game and G^Δ be its mixed extension. The set of Nash equilibria in G^Δ is nonempty by Nash's [2] existence theorem (Proposition 4.1). Moreover, for every Nash equilibrium in G^Δ, there exists a correlated equilibrium by Proposition 7.1. Therefore, the set of correlated equilibrium in G is nonempty. ■

The following result shows that if two probability distributions in a finite strategic form game are correlated equilibria, then any convex combination of them is also a correlated equilibrium.

Proposition 7.3 For any finite strategic form game, the set of correlated equilibria is convex.

Proof Let $G = \langle N, (S_i)_{i \in N}, (u_i)_{i \in N} \rangle$ be a finite strategic form game. By Proposition 7.2, the set of correlated equilibria, say $\mathcal{CE}$, is nonempty. If $\mathcal{CE}$ contains a unique equilibrium, then it is trivially convex. Otherwise, pick any two correlated equilibrium p and p' in $\mathcal{CE}$. By the definition of correlated equilibrium, we have

$$\sum_{s_{-i}\in S_{-i}} p(s_i, s-i)u_i(s_i, s_{-i}) \geq \sum_{s_{-i}\in S_{-i}} p(s_i, s-i)u_i(s_i', s_{-i}) \text{ for all } s_i, s_i' \in S_i$$

and

$$\sum_{s_{-i}\in S_{-i}} p'(s_i, s-i)u_i(s_i, s_{-i}) \geq \sum_{s_{-i}\in S_{-i}} p'(s_i, s-i)u_i(s_i', s_{-i}) \text{ for all } s_i, s_i' \in S_i.$$

Multiplying the first inequality by $\lambda \in [0, 1]$ and the second inequality by $1-\lambda$, and summing them up, we obtain

$$\sum_{s_{-i}\in S_{-i}} [\lambda p(s_i, s-i) + (1-\lambda)p'(s_i, s-i)]u_i(s_i, s_{-i}) \geq$$
$$\sum_{s_{-i}\in S_{-i}} [\lambda p(s_i, s-i) + (1-\lambda)p'(s_i, s-i)]u_i(s_i', s_{-i})$$

for all $s_i, s_i' \in S_i$. Thus, $\lambda p + (1-\lambda)p'$ is a correlated equilibrium, implying that the set $\mathcal{CE}$ is convex. ■

Geometrically, the set of correlated equilibrium distributions in a finite strategic form game $G = \langle N, (S_i)_{i\in N}, (u_i)_{i\in N}\rangle$ is a convex polytope. This polytope includes all probability distributions generated by the set of Nash equilibria. Interestingly, all Nash equilibria are on the boundary of this polytope, as demonstrated by Nau et al. [3]. We will illustrate this in the following example.

Example 7.2 Reconsider the Battle of Sexes Game in Table 1.4.

		Player 2	
		O	*F*
Player 1	*O*	(2, 1)	(0, 0)
	F	(0, 0)	(1, 2)

Table 1.4 Reconsidered

The game described above has three Nash equilibria: two pure strategy equilibria $\sigma_1^* = (O, O)$ and $\sigma_2^* = (F, F)$; and one unpure strategy equilibrium $\sigma_3^* = ((2/3)O + (1/3)F, (1/3)O + (2/3)F)$. These three equilibria yield the expected payoffs (2, 1), (1, 2), and (2/3, 2/3), respectively.

Now, we will find the set of correlated equilibria. Let $p = (p_{11}, p_{12}, p_{21}, p_{22})$ be a probability distribution on the set of strategy profiles $S = \{(O, O), (O, F), (F, O), (F, F)\}$ such that $p_{11} = p(O, O)$, $p_{12} = p(O, F)$, $p_{21} = p(F, O)$, $p_{22} = p(F, F)$. If p is a correlated equilibrium, then it must be true that

$$2p_{11} + 0p_{12} \geq 0p_{11} + 1p_{12}$$

$$0p_{21} + 1p_{22} \geq 2p_{21} + 0p_{22}$$

$$1p_{11} + 0p_{21} \geq 0p_{11} + 2p_{21}$$

$$0p_{12} + 2p_{22} \geq 1p_{12} + 0p_{22}$$

implying

$$2p_{11} \geq p_{12} \tag{7.1}$$

$$p_{22} \geq 2p_{21} \tag{7.2}$$

$$p_{11} \geq 2p_{21} \tag{7.3}$$

$$2p_{22} \geq p_{12}. \tag{7.4}$$

Thus, we have established that the probability distribution $p = (p_{11}, p_{12}, p_{21}, p_{22}) \in \Delta(S)$ is a correlated equilibrium if and only if it satisfies the inequalities (7.1)–(7.4). The set of correlated equilibrium distributions satisfying these inequalities is the pink-colored convex polytope in Fig. 7.1, as was earlier drawn by Nau et al. [3].

The pink polytope in Fig. 7.1 has 5 vertices and 6 facets. As demonstrated by Nau et al. [3], the (orange colored) Nash equilibrium distributions $\sigma_1^* = (O, O)$, $\sigma_2^* = (F, F)$; and $\sigma_3^* = ((2/3)O + (1/3)F, (1/3)O + (2/3)F)$ are on the boundary of this polytope. This convex polytope of correlated equilibrium distributions is tangent at the three Nash equilibria distributions to a blue-colored saddle, which portrays all possible profiles of independent strategies of the two players. Also note that the

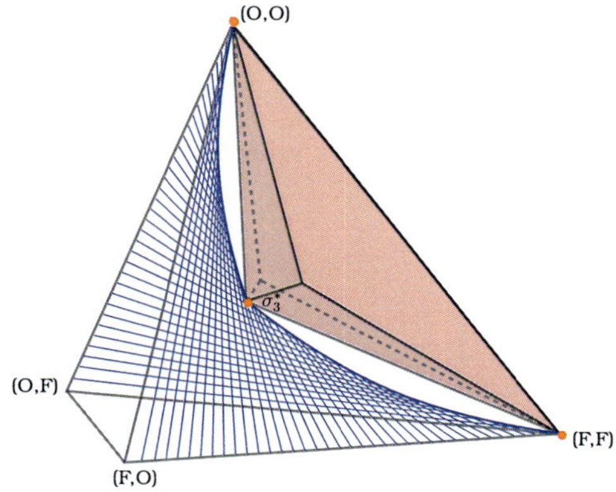

Fig. 7.1 Convex polytope of correlated equilibrium distributions

polytope of correlated equilibrium distributions is a part of a tetrahedron, a three dimensional convex shape with four vertices (O, O), (O, F), (F, O), and (F, F), representing the simplex of probability distributions on the outcomes of the Battle of Sexes. ■

7.3 Coordination on the Social Optimum

In the previous section, we have learned that the set of correlated equilibrium distributions in a finite strategic form game is convex. This set contains either a unique distribution as in the case of the Prisoner's Dilemma (where the unique Nash equilibrium is also the unique correlated equilibrium) or uncountably many distributions as in the case of the Battle of Sexes. In a game where there is more than one correlated equilibrium distributions, a question presents itself: Which correlated equilibrium should be in play? Under the assumption that the mediator can enforce any correlated equilibrium distributions, this question can be reformulated as which correlated equilibrium distribution the mediator should pick on behalf of the players. A mediator that acts as a socially benevolent planner may choose, for example, a correlated equilibrium that maximizes the sum of utilities of the players. In Example 7.2, we saw that the Battle of Sexes Game has uncountably many correlated equilibria. For this game, we will find below a correlated equilibrium that maximizes the players' total payoff.

Example 7.3 Reconsider the Battle of Sexes Game in Table 1.4 (which we studied in Example 7.2). The set of attainable expected payoff profiles under correlated strategies is the convex hull of the three payoff vectors, denoted by $\mathcal{A} = co\{(0, 0), (1, 2), (2, 1)\}$ and portrayed in Fig. 7.2.

We will later see that not every payoff in $\mathcal{A}$ is a correlated equilibrium payoff (contained by the set of payoffs generated by the pink polytope in Fig. 7.1). The payoffs in $\mathcal{A}$ are generated by the probability distributions in the tetrahedron

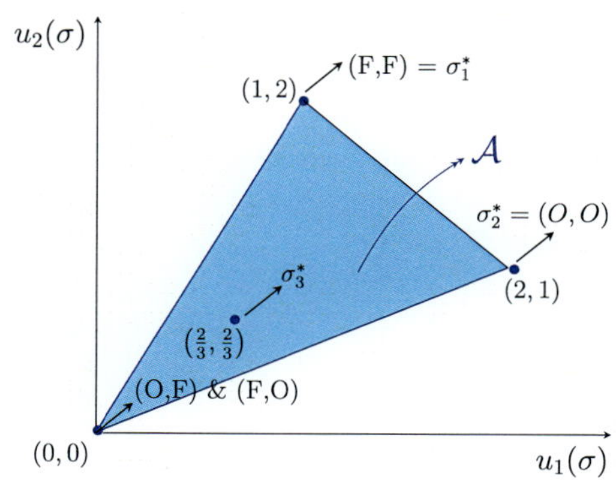

Fig. 7.2 Attainable payoff profiles under correlated strategies

in Fig. 7.1 (the simplex of probability distributions on the four outcomes of the Battle of Sexes).

Notice that the payoff profiles (2, 1) and (1, 2) corresponding, respectively, to pure strategy Nash equilibria $\sigma_1^* = (O, O)$ and $\sigma_2^* = (F, F)$ are Pareto undominated, whereas the payoff profile (2/3, 2/3) corresponding to the unpure Nash equilibrium $\sigma_3^* = ((2/3)O + (1/3)F, (1/3)O + (2/3)F)$ is not. Thus, σ_1^* and σ_2^* are on the boundary of $\mathcal{A}$, whereas σ_3^* is in the interior of it.

Now, recall from Example 7.2 that the probability distribution $p = (p_{11}, p_{12}, p_{21}, p_{22}) \in \Delta(S)$ is a correlated equilibrium if and only if it satisfies the inequalities (7.1)–(7.4). Let A^{corr} denote the set of expected payoffs corresponding to all possible correlated equilibria. Clearly, A^{corr} is a proper subset of $\mathcal{A}$. Below, we will show how we can portray A^{corr}.

For any $p \in \Delta(S)$, let us first calculate the players' total expected payoff as

$$V(p) = u_1(p) + u_2(p) = 3(p_{11} + p_{22}).$$

We will first calculate the correlated equilibrium that minimizes $V(p)$. Notice that

$$V(p) = 3(1 - p_{12} - p_{21}).$$

To minimize $V(p)$, we have to choose p_{12} and p_{21} are as high as possible. Thus, in the set of inequalities (7.1)–(7.4), we set $p_{12}/2 = 2p_{21} = p_{11} = p_{22}$, implying $p = (2/9, 4/9, 1/9, 2/9)$, which yields the expected payoff profile $(2/3, 2/3)$.

Next, we will calculate the correlated equilibrium that maximizes $V(p)$. Now, we have to choose p_{12} and p_{21} as low as possible and p_{11} and p_{22} as high as possible. Thus, in the set of inequalities (7.1)–(7.4), we set$p_{12} = p_{21} = 0$ and $p_{11} + p_{22} = 1$, implying $p = (p_{11}, 0, 0, 1 - p_{11})$ for any $p_{11} \in [0, 1]$, which yields the expected payoff profile $(1 + p_{11}, 2 - p_{11})$. For this correlated equilibrium, $V(p)$ attains its maximum level of 3.

Note that the expected payoff profile (2/3, 2/3) cannot Pareto dominate any other payoff profile corresponding to a correlated equilibrium, for otherwise the players' sum of expected payoffs under correlated strategies would not be minimized at the level of 4/3 ($= 3(2/9 + 2/9)$). This implies that the expected payoff profile $(2/3, 2/3)$ must be at the boundary of A^{corr}, i.e., the convex hull of correlated equilibrium payoff profiles. On the other hand, the expected payoff profile $(1 + p_{11}, 2 - p_{11})$ cannot be Pareto dominated by any other payoff profile corresponding to a correlated equilibrium, for otherwise the players' sum of expected payoffs under correlated strategies would not be maximized at the level of 3. This implies that for any $p_{11} \in [0, 1]$ the expected payoff profile $(1 + p_{11}, 2 - p_{11})$ must be at the Pareto boundary (frontier) of A^{corr}. In fact, the set of these payoff profiles is just the line segment connecting the pure strategy Nash equilibrium payoff profiles $(1, 2)$ and $(2, 1)$. Since A^{corr} is a convex set, it must be equal to the convex hull of the payoff profiles in $\{(1, 2), (2, 1), (2/3, 2/3)\}$, which we illustrate as the pink-colored region in Fig. 7.3.

Notice that the players' total payoff maximized (at the level of 3) on any payoff profile on the line segment connecting the profiles $(1, 2)$ and $(2, 1)$ in Fig. 7.3. On

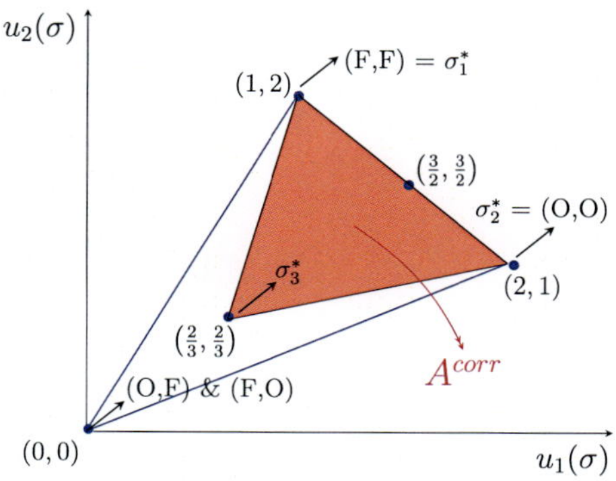

Fig. 7.3 Correlated equilibrium payoffs

this line segment, the expected payoff profile (3/2, 3/2) deserves a special attention. This profile is fair in the sense that it symmetrically compromises—in a symmetric game—between the two (Pareto efficient) Nash equilibrium payoff profiles (2, 1) and (1, 2) corresponding to the strategy profiles (O, O) and (F, F), respectively. If the mediator desires the players to enjoy the expected payoff profile (3/2, 3/2), she should enforce the correlated equilibrium distribution that puts probability 1/2 on each of the outcomes (O, O) and (F, F). ■

In the previous example, we have seen that the mediator's involvement to the Battle of Sexes Game by using a random device to correlate the strategies of players in a socially optimal way could not increase the social welfare over the level the players could independently achieve at a pure strategy Nash equilibrium. Nonetheless, the mediator's involvement with her correlation device allowed the players to enjoy a payoff profile which cannot be achievable at an unpure Nash equilibrium. In the next example, we will study a game where the mediator's involvement will have a positive effect on the social welfare, as well.

Table 7.3 The strategic form game in Example 7.4

		Player 2	
		L	R
Player 1	U	6, 6	2, 7
	D	7, 2	0, 0

Table 7.3

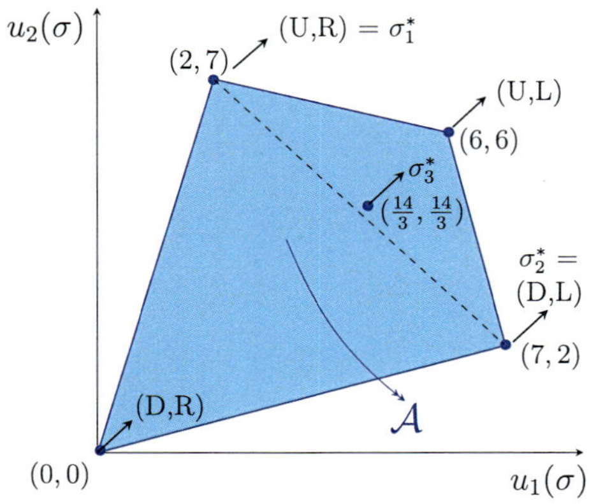

Fig. 7.4 Attainable payoff profiles under correlated strategies

Example 7.4 Consider the game in Table 7.3 earlier studied by Aumann [1].

The game has three Nash equilibria: two pure strategy equilibria $\sigma_1^* = (U, R)$ and $\sigma_2^* = (D, L)$; and one unpure strategy equilibrium $\sigma_3^* = ((2/3)U + (1/3)D, (2/3)L + (1/3)R)$. These three equilibria yield the expected payoffs (2, 7), (7, 2), and (14/3, 14/3), respectively. In Fig. 7.4, we plot the attainable expected payoff profiles under correlated strategies as the convex hull of the four payoff vectors corresponding to four pure strategy profiles; i.e., $\mathcal{A} = co\{(0, 0), (2, 7), (6, 6), (7, 2)\}$. As implied by Proposition 7.1, the Nash equilibrium (expected) payoff profiles are contained by the set $\mathcal{A}$. Notice that the payoff profiles corresponding to the pure strategy Nash equilibria σ_1^* and σ_2^* are Pareto undominated, whereas the payoff profile corresponding to the unpure Nash equilibrium σ_3^* is not.

Now, we will find the set of correlated equilibria and the corresponding set of expected payoff profiles. We assume that a mediator can implement with the help of a correlation device (a wheel of fortune) any correlated equilibrium outcome. Let $p = (p_{11}, p_{12}, p_{21}, p_{22})$ be a probability distribution on the set of strategy profiles $S = \{(U, L), (U, R), (D, L), (D, R)\}$ such that $p_{11} = p(U, L)$, $p_{12} = p(U, R)$, $p_{21} = p(D, L)$, $p_{22} = p(D, R)$. If p is a correlated equilibrium, then it must be true that

$$6p_{11} + 2p_{12} \geq 7p_{11} + 0p_{12}$$

$$7p_{21} + 0p_{22} \geq 6p_{21} + 2p_{22}$$

$$6p_{11} + 2p_{21} \geq 7p_{11} + 0p_{21}$$

$$7p_{12} + 0p_{22} \geq 6p_{12} + 2p_{22}$$

implying

$$2p_{12} \geq p_{11} \tag{7.5}$$

$$p_{21} \geq 2p_{22} \tag{7.6}$$

$$2p_{21} \geq p_{11} \tag{7.7}$$

$$p_{12} \geq 2p_{22}. \tag{7.8}$$

Thus, we have established that a probability distribution $p = (p_{11}, p_{12}, p_{21}, p_{22}) \in \Delta(S)$ is a correlated equilibrium if and only if it satisfies the inequalities (7.5)–(7.8). Let A^{corr} denote the set of expected payoffs corresponding to all possible correlated equilibria. Clearly, $A^{corr} \subseteq \mathcal{A}$. Below, we will deal with drawing the set A^{corr}.

For any $p \in \Delta(S)$, we can calculate the players' total expected payoff as

$$V(p) = u_1(p) + u_2(p) = 12p_{11} + 9p_{12} + 9p_{21}.$$

We will first calculate the correlated equilibrium that minimizes $V(p)$. Notice that

$$V(p) = 3p_{11} + 9(p_{11} + p_{12} + 9p_{21}) = 3p_{11} + 9(1 - p_{22}).$$

To minimize $V(p)$ we have to choose p_{22} as high as possible while p_{11} as low as possible. Thus, in the set of inequalities (7.5)–(7.8), we set $p_{11} = 0$, and $p_{12} = p_{21} = 2p_{22}$, implying $p = (0, 2/5, 2/5, 1/5)$, which yields the expected payoff profile $(18/5, 18/5)$.

Next, we will calculate the correlated equilibrium that maximizes $V(p)$. Now, we have to choose p_{22} as low as possible while p_{11} as high as possible. Thus, in the set of inequalities (7.5)–(7.8), we set $p_{22} = 0$ and $p_{11} = 2p_{12} = 2p_{21}$, implying $p = (1/2, 1/4, 1/4, 0)$, which yields the expected payoff profile $(21/4, 21/4)$.

Notice that the expected payoff profile $(18/5, 18/5)$ cannot Pareto dominate any other payoff profile corresponding to a correlated equilibrium, for otherwise the players' sum of expected payoffs under correlated strategies would not be minimized at the level of 36/5 $(= (18/5) + (18/5))$. Thus, the expected payoff profile $(18/5, 18/5)$ must be at the boundary of A^{corr}. On the other hand, the expected payoff profile $(21/4, 21/4)$ cannot be Pareto dominated by any other payoff profile corresponding to a correlated equilibrium, for otherwise the players' sum of expected payoffs under correlated strategies would not be maximized at the level of 21/2 $(= (21/4) + (21/4))$. This implies that the expected payoff profile $(21/4, 21/4)$ must be at the Pareto boundary of A^{corr}. We also know that the pure strategy Nash equilibrium payoff profiles (2, 7) and (7, 2) must be at the Pareto boundary of A^{corr}. This is because of the facts that (i) each Nash equilibrium induces a correlated equilibrium distribution, (ii) (2, 7) and (7, 2) are at the frontier of $\mathcal{A}$, and (iii) $A^{corr} \subseteq \mathcal{A}$.

Thus, we have established that all four payoff profiles in $\{(2, 7), (7, 2), (18/5, 18/5), (21/4, 21/4)\}$ are at the boundary of A^{corr}, with all of them, except for $(18/5, 18/5)$, also lying on the Pareto boundary. Since A^{corr} is a convex set, it must be equal to the convex hull of these four payoff profiles, which we illustrate as the pink-colored region in Fig. 7.5.

Notice that the maximum total expected payoff that can be achieved by players at a Nash equilibrium is 28/3. This payoff can be generated if the players play $\sigma_3^* = ((2/3, 1/3), (2/3, 1/3))$. The correlated equilibrium induced by σ_3^* is $p^{\sigma_3^*} = (4/9, 2/9, 2/9, 1/9)$. On the other hand, the maximum total expected payoff that can be achieved by players at a correlated equilibrium is $21/2$. This payoff is uniquely generated by the correlated equilibrium distribution $p = (1/2, 1/4, 1/4, 0)$. ■

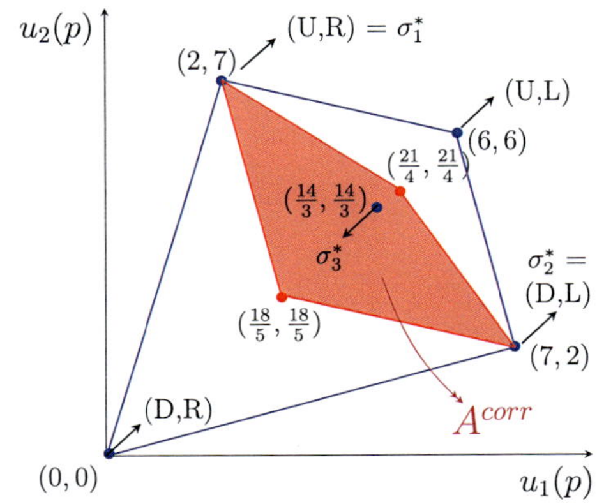

Fig. 7.5 Correlated equilibrium payoffs

References

1. Aumann RJ (1974) Subjectivity and correlation in randomized strategies. J Math Econ 1(1):67–96
2. Nash JF (1950) Equilibrium points in n-person games. PNAS 36(1):48–49
3. Nau R, Canovas SG, Hansen P (2003) On the geometry of Nash equilibria and correlated equilibria. Int J Game Theory 32:443–453

Part II

Static Noncooperative Games with Incomplete Information

In this part of the book, which consists of Chap. 8, we will study static noncooperative games with incomplete information, where some players have private information about certain aspects (e.g., the payoff structure, and the rules) of the game to be played. Clearly, the presence of private information held by any player creates uncertainty about certain aspects of the game from the viewpoint of all other players. To account for this uncertainty in strategic decision making, we will extend the Nash equilibrium to a solution concept called the Bayesian Nash equilibrium. We will present two versions of this extension, depending upon whether a player who has uncertainty about her opponents' payoffs completely knows her own payoffs or not when this player is about to choose a strategy in the game. We will also give the relationship between these two versions of Bayesian Nash equilibrium and apply each of them separately to a simple Bayesian game to highlight their difference.

Bayesian Nash Equilibrium

8

In Chap. 8, we will study strategic form games with incomplete information. These games involve at least one player who has uncertainty about the payoffs of other players. In real life, many strategic situations involve incomplete information. For example, the production cost, hence the operating profit, of a firm in an oligopolistic industry, or the valuation of a buyer about an object sold in an auction, is generally unknown to its competitors. We will call a strategic situation with incomplete information a (static noncooperative) Bayesian game. In Sect. 8.1, we will describe a Bayesian game and, in Sects. 8.2 and 8.3, we will, respectively, present the ex ante and interim Bayesian Nash equilibrium concepts with several propositions and examples.

8.1 Bayesian Game

A Bayesian game is a list $G^B = \langle N, (A_i)_{i\in N}, (T_i)_{i\in N}, (p_i)_{i\in N}, (u_i)_{i\in N}\rangle$. Here, $N = \{1, 2, \ldots, n\}$, with $n \geq 2$, denotes the set of players, A_i denotes the set of actions available for player i, T_i denotes the set of possible types of player i, p_i denotes a probability function representing the beliefs of player i about the types of other players, and u_i is the expected payoff (utility) function of player i.

We let $A = \bigtimes_{i\in N} A_i$ be the set of all action profiles in the game, $T = \bigtimes_{i\in N} T_i$ be the set of all possible type profiles in the game, and $T_{-i} = \bigtimes_{j\in N-i} T_j$ be the set of all possible types of the players excluding player i.

The function p_i of player i is a mapping from T_i to $\Delta(T_{-i})$, the set of probability distributions over T_{-i}. For any $t_i \in T_i$, the function p_i specifies a (subjective) probability distribution $p_i(.|t_i)$ over the set of type profiles T_{-i}, representing the beliefs of player i about the types of other players in case her own type is t_i.

I. Saglam, *Mastering Game Theory*, Classroom Companion: Economics,
https://doi.org/10.1007/978-3-031-82094-6_8

The elements of the Bayesian game G^B and the fact that each player knows her own type is common knowledge. The beliefs $(p_i)_{i \in N}$ in G^B are said to be *consistent* if there exists some common prior probability function p, from the set of possible profile types T to $\Delta(T)$, the set of probability distributions over T, such that for each $i \in N$ the belief function p_i of player i can be derived from the prior probability function p through the Bayes rule; i.e.,

$$p_i(t_{-i}|t_i) = \frac{p(t)}{\sum\limits_{t'_{-i} \in T_{-i}} p(t_i, t'_{-i})} \quad \text{for all } t \in T \text{ and } i \in N. \tag{8.1}$$

When the beliefs of players, $(p_i)_{i \in N}$, are not consistent, they cannot be common knowledge as shown by Aumann (1976). Therefore, in line with our earlier assumption that all elements of G^B, including the beliefs of the players, are common knowledge, we will assume that the beliefs of the players in N are consistent.

For each $i \in N$, the strategy of player i in the Bayesian game G^B is a function s_i that maps each possible type in T_i to an action in A_i. For each $i \in N$ and $t \in T$, we let $s_{-i}(t_{-i}) = (s_1(t_1), \ldots, s_{i-1}(t_{i-1}), s_{i+1}(t_{i+1}), \ldots, s_n(t_n))$. We let S_i to denote the set of all strategy functions of player i and $S = \times_{i \in N} S_i$ to denote the set of all strategy function profiles. We define the payoff function u_i of player i as a mapping from the domain $S \times T_i$ to the set of real numbers $\mathbb{R}$.

We can consider three possible states of the world: the ex ante state, the interim state, and the ex post state. In the ex ante state, no player knows about the type of any player, including herself. Each player believes that the probability of a type profile $t \in T$ is $p(t)$. (This probability is independent of the player since no player has private information in the ex ante state.) In the interim state, each player i knows her own type t_i and believes that the probability of any type profile $t_{-i} \in T_{-i}$ is $p_i(t_{-i}|t_i)$. (This probability is dependent on the identity of player, since each player has private information in the interim state.) In the ex post state, all types will be revealed to every player; hence each player knows the actual type profile $t \in T$ with certainty. In this chapter, we will only be interested in the ex ante and interim states of the world, since the ex post state coincides with the complete information situation that was studied in Chap. 4.

8.2 Ex Ante Bayesian Nash Equilibrium

In the ex ante state, no player knows any component of the actual type profile t in T. Each player has incomplete information about her own type and the type profile of other players. Therefore, given a strategy profile $(s_1(.), \ldots, s_n(.))$, the expected utility of player i in the ex ante state is

$$\sum_{t \in T} u_i((s_i(t_i), s_{-i}(t_{-i})), t_i)p(t).$$

Now, we are ready to define a Bayesian Nash equilibrium of the Bayesian game G^B in the ex ante state of the world.

Definition 8.1 A profile of pure strategy functions $\langle s_1^*(.), \ldots, s_n^*(.)\rangle$ is an ex ante Bayesian Nash equilibrium of a Bayesian game G^B if

$$\sum_{t \in T} u_i((s_i^*(t_i), s_{-i}^*(t_{-i})), t_i) p(t) \geq \sum_{t \in T} u_i((\tilde{s}_i(t_i), s_{-i}^*(t_{-i})), t_i) p(t) \tag{8.2}$$

for all $i \in N$ and for all $\tilde{s}_i(.) \in S_i$.

A profile of pure strategy functions $\langle s_1^*(.), \ldots, s_n^*(.)\rangle$ is an ex ante Bayesian Nash equilibrium if for each $i \in N$ the strategy function $s_i^*(.)$ of player i is a best response to the profile of strategy functions of other players, $s_{-i}^*(.)$, in the ex ante state of the world.

Example 8.1 Consider the following strategic form game (Table 8.1).

Player 1 privately observes the realization of a that takes the values 22 and 5 with equal probabilities, whereas player 2 privately observes the realization of b that takes the values 13 and 3 with equal probabilities.

The private type spaces of players 1 and 2 are $T_1 = \{t_{11}, t_{21}\}$ and $T_2 = \{t_{12}, t_{22}\}$, respectively. Let $T = T_1 \times T_2$. We define the strategy functions of players 1 and 2 as $s_1 : T_1 \to A_1 = \{U, D\}$ and $s_2 : T_2 \to A_2 = \{L, R\}$, respectively.

We are given the beliefs of player 1 as $p_1(t_{12}|t_1) = p_1(t_{22}|t_1) = 0.5$ for each $t_1 \in T_1$ and the beliefs of player 2 as $p_2(t_{11}|t_2) = p_2(t_{21}|t_2) = 0.5$ for each $t_2 \in T_2$. There is a common prior p such that

$$p_i(t_{-i}|t_i) = \frac{p(t_i, t_{-i})}{\sum_{t_{-i} \in T_{-i}} p(t_i, t_{-i})}.$$

This prior satisfies $p(t_1, t_2) = 0.25$ for all $(t_1, t_2) \in T_1 \times T_2$.

Let us now construct the Bayesian strategic form game where each player's strategy space will involve all possible strategy functions she may choose. Notice that each of the four type profiles in T is likely to occur with probability $p(t) = 0.25$. We can associate each $t \in T$ with a particular pair of (a, b). Thus, we have four possible payoff tables (Table 8.2):

Table 8.1 The strategic form game in Example 8.1

Player 1 \ Player 2	L	R
U	a, b	$7, 8$
D	$12, -2$	$12, 15$

Table 8.2 Four possible games associated with Table 8.1

Panel (i): Player 1 (rows), Player 2 (columns)

	L	R
U	22, 13	7, 8
D	12, −2	12, 15

$t = (t_{11}, t_{12}),\quad p(t) = 0.25$

$(a, b) = (22, 13)$

Panel (i)

Panel (ii): Player 1 (rows), Player 2 (columns)

	L	R
U	22, 3	7, 8
D	12, −2	12, 15

$t = (t_{11}, t_{22}),\quad p(t) = 0.25$

$(a, b) = (22, 3)$

Panel (ii)

Panel (iii): Player 1 (rows), Player 2 (columns)

	L	R
U	5, 13	7, 8
D	12, −2	12, 15

$t = (t_{21}, t_{12}),\quad p(t) = 0.25$

$(a, b) = (5, 13)$

Panel (iii)

Panel (iv): Player 1 (rows), Player 2 (columns)

	L	R
U	5, 3	7, 8
D	12, −2	12, 15

$t = (t_{21}, t_{22}),\quad p(t) = 0.25$

$(a, b) = (5, 3)$

Panel (iv)

The strategy functions of player 1 can be denoted by s_1^1, s_1^2, s_1^3, and s_1^4 such that

$$s_1^1(t_{11}) = U \text{ and } s_1^1(t_{21}) = U,$$
$$s_1^2(t_{11}) = U \text{ and } s_1^2(t_{21}) = D,$$
$$s_1^3(t_{11}) = D \text{ and } s_1^3(t_{21}) = U,$$
$$s_1^4(t_{11}) = D \text{ and } s_1^4(t_{21}) = D.$$

Similarly, the strategy functions of player 2 can be denoted by s_2^1, s_2^2, s_2^3, and s_2^4 such that

$$s_2^1(t_{12}) = L \text{ and } s_2^1(t_{22}) = L,$$
$$s_2^2(t_{12}) = L \text{ and } s_2^2(t_{22}) = R,$$
$$s_2^3(t_{12}) = R \text{ and } s_2^3(t_{22}) = L,$$
$$s_2^4(t_{12}) = R \text{ and } s_2^4(t_{22}) = R.$$

Table 8.3 The ex ante expected payoffs associated with Table 8.1

Player 2

Player 1	s_2^1	s_2^2	s_2^3	s_2^4
s_1^1	$\frac{54}{4}, \frac{32}{4}$	$\frac{41}{4}, \frac{42}{4}$	$\frac{41}{4}, \frac{22}{4}$	$\frac{28}{4}, \frac{32}{4}$
s_1^2	$\frac{68}{4}, \frac{12}{4}$	$\frac{53}{4}, \frac{34}{4}$	$\frac{53}{4}, \frac{24}{4}$	$\frac{38}{4}, \frac{46}{4}$
s_1^3	$\frac{34}{4}, \frac{12}{4}$	$\frac{36}{4}, \frac{34}{4}$	$\frac{36}{4}, \frac{24}{4}$	$\frac{38}{4}, \frac{46}{4}$
s_1^4	$\frac{48}{4}, \frac{-8}{4}$	$\frac{48}{4}, \frac{26}{4}$	$\frac{48}{4}, \frac{26}{4}$	$\frac{48}{4}, \frac{60}{4}$

We have obtained for each player $i = 1, 2$ the set of pure strategy functions $S_i = \{s_i^1, s_i^2, s_i^3, s_i^4\}$. Recall that the ex ante expected payoff of each player i at any profile $(s_1, s_2) \in S_1 \times S_2$ is

$$\sum_{t \in T} u_i\left((s_i(t_i), s_{-i}(t_{-i})), t_i\right) p(t).$$

Using the four equally likely payoff tables depicted earlier, we can calculate the ex ante expected payoff of each player at each possible profile of pure strategy functions to construct the following strategic form game.

To see how we have calculated the payoffs in Table 8.3, take for example the profile (s_1^2, s_2^3) and the corresponding payoff 24/4 received by player 2. Recall that according to the profile of strategy functions (s_1^2, s_2^3)

- player 1 plays the strategies $s_1^2(t_{11}) = U$ and $s_1^2(t_{21}) = D$, and
- player 2 plays the strategies $s_2^3(t_{12}) = R$ and $s_2^3(t_{22}) = L$.

So, we can calculate the expected payoff of player 2 at the profile (s_1^2, s_2^3) as

$$\begin{aligned}\tilde{u}_2(s_1^2, s_2^3) &= \sum_{(t_1,t_2)\in T} u_2\left(\left(s_1^2(t_1), s_2^3(t_2)\right), t_2\right) p(t_1, t_2)\\ &= u_2(U, R)(0.25) + u_2(U, L)(0.25) + u_2(D, R)(0.25) + u_2(D, L)(0.25)\\ &= (8)(0.25) + (3)(0.25) + (15)(0.25) + (-2)(0.25) = \frac{24}{4}.\end{aligned}$$

Given the (Bayesian) strategic form game constructed above, we will now find the Bayesian Nash equilibria in pure strategies, if any. In Table 8.4, we emphasize the best-response payoffs of players 1 and 2 by the red and blue colors, respectively.

We should observe that the unique Bayesian Nash equilibrium in pure strategies is the profile of strategy functions $\left(s_1^4(.), s_2^4(.)\right)$. ■

We say that a Bayesian game $G^B = \langle N, (A_i)_{i\in N}, (T_i)_{i\in N}, (p_i)_{i\in N}, (u_i)_{i\in N}\rangle$ is finite if it involves a finite number of players and each of these players has a finite number of types and actions, i.e., the set N is finite and, for each $i \in N$, the sets T_i and A_i are finite.

Table 8.4 The best-response payoffs in Table 8.3

Player 2

Player 1	s_2^1	s_2^2	s_2^3	s_2^4
s_1^1	$\frac{54}{4}, \frac{32}{4}$	$\frac{41}{4}, \frac{42}{4}$	$\frac{41}{4}, \frac{22}{4}$	$\frac{28}{4}, \frac{32}{4}$
s_1^2	$\frac{68}{4}, \frac{12}{4}$	$\frac{53}{4}, \frac{34}{4}$	$\frac{53}{4}, \frac{24}{4}$	$\frac{38}{4}, \frac{46}{4}$
s_1^3	$\frac{34}{4}, \frac{12}{4}$	$\frac{36}{4}, \frac{34}{4}$	$\frac{36}{4}, \frac{24}{4}$	$\frac{38}{4}, \frac{46}{4}$
s_1^4	$\frac{48}{4}, \frac{-8}{4}$	$\frac{48}{4}, \frac{26}{4}$	$\frac{48}{4}, \frac{26}{4}$	$\frac{48}{4}, \frac{60}{4}$

Proposition 8.1 ([1]) *Every finite Bayesian game has a mixed strategy Bayesian Nash equilibrium.*

Proof Consider a finite Bayesian game $G^B = \langle N, (A_i)_{i\in N}, (T_i)_{i\in N}, (p_i)_{i\in N}, (u_i)_{i\in N}\rangle$. Since T_i and A_i are both finite for each $i \in N$, there exists a finite set S_i containing all strategy functions s_i of player i mapping each type in T_i to an action in A_i. We let $\tilde{u}_i$ denote the (expected) payoff function of player i such that $\tilde{u}_i(s) = \sum_{t\in T} u_i((s_i(t_i), s_{-i}(t_{-i})), t_i)p(t)$ for each $s \in \times_{i\in N} S_i$. Thus, we have established that a finite Bayesian game G^B induces a finite strategic form game $\tilde{G} = \langle N, (S_i)_{i\in N}, (\tilde{u}_i)_{i\in N}\rangle$. We know by Nash's [2] existence theorem that the mixed extension of $\tilde{G}$ must have a mixed (pure or unpure) Nash equilibrium strategy profile. This strategy profile is also a Bayesian Nash equilibrium of the mixed extension of G^B. ■

Note that the Bayesian game in Example 8.1 contains a finite number of players, with each of them having a finite number of strategy functions. Therefore, appealing to Proposition 8.1 only (without making the calculations we did), one could have been sure that this game must have at least one mixed (pure or unpure) Bayesian Nash equilibrium profile of strategy functions.

8.3 Interim Bayesian Nash Equilibrium

Definition 8.2 A profile of pure strategy functions $\langle s_1^*(.), \ldots, s_n^*(.)\rangle$ is an interim Bayesian Nash equilibrium of a Bayesian game G^B if for all $i \in N$ and for all $t_i \in T_i$ occurring with positive probability

$$\sum_{t_{-i}\in T_{-i}} u_i((s_i^*(t_i), s_{-i}^*(t_{-i})), t_i)p_i(t_{-i}|t_i) \geq \sum_{t_{-i}\in T_{-i}} u_i((a_i, s_{-i}^*(t_{-i})), t_i)p_i(t_{-i}|t_i) \tag{8.3}$$

for all $a_i \in A_i$.

Table 8.5 The interim expected payoffs associated with four possible type profiles in Example 8.2

Panel (i): $t = (t_{11}, t_{12})$, $(a, b) = (22, 13)$ (rows: Player 1; columns: Player 2)

	s_2^1	s_2^2	s_2^3	s_2^4
s_1^1	$\frac{44}{2}, \frac{26}{2}$	$\frac{29}{2}, \frac{26}{2}$	$\frac{29}{2}, \frac{16}{2}$	$\frac{14}{2}, \frac{16}{2}$
s_1^2	$\frac{44}{2}, \frac{11}{2}$	$\frac{29}{2}, \frac{11}{2}$	$\frac{29}{2}, \frac{23}{2}$	$\frac{14}{2}, \frac{23}{2}$
s_1^3	$\frac{24}{2}, \frac{11}{2}$	$\frac{24}{2}, \frac{11}{2}$	$\frac{24}{2}, \frac{23}{2}$	$\frac{24}{2}, \frac{23}{2}$
s_1^4	$\frac{24}{2}, \frac{-4}{2}$	$\frac{24}{2}, \frac{-4}{2}$	$\frac{24}{2}, \frac{30}{2}$	$\frac{24}{2}, \frac{30}{2}$

Panel (ii): $t = (t_{11}, t_{22})$, $(a, b) = (22, 3)$ (rows: Player 1; columns: Player 2)

	s_2^1	s_2^2	s_2^3	s_2^4
s_1^1	$\frac{44}{2}, \frac{6}{2}$	$\frac{29}{2}, \frac{6}{2}$	$\frac{29}{2}, \frac{16}{2}$	$\frac{14}{2}, \frac{16}{2}$
s_1^2	$\frac{44}{2}, \frac{1}{2}$	$\frac{29}{2}, \frac{23}{2}$	$\frac{29}{2}, \frac{1}{2}$	$\frac{14}{2}, \frac{23}{2}$
s_1^3	$\frac{24}{2}, \frac{1}{2}$	$\frac{24}{2}, \frac{23}{2}$	$\frac{24}{2}, \frac{1}{2}$	$\frac{24}{2}, \frac{23}{2}$
s_1^4	$\frac{24}{2}, \frac{-4}{2}$	$\frac{24}{2}, \frac{30}{2}$	$\frac{24}{2}, \frac{-4}{2}$	$\frac{24}{2}, \frac{30}{2}$

Panel (iii): $t = (t_{21}, t_{12})$, $(a, b) = (5, 13)$ (rows: Player 1; columns: Player 2)

	s_2^1	s_2^2	s_2^3	s_2^4
s_1^1	$\frac{10}{2}, \frac{26}{2}$	$\frac{12}{2}, \frac{26}{2}$	$\frac{12}{2}, \frac{16}{2}$	$\frac{14}{2}, \frac{16}{2}$
s_1^2	$\frac{24}{2}, \frac{11}{2}$	$\frac{24}{2}, \frac{11}{2}$	$\frac{24}{2}, \frac{23}{2}$	$\frac{24}{2}, \frac{23}{2}$
s_1^3	$\frac{10}{2}, \frac{11}{2}$	$\frac{12}{2}, \frac{11}{2}$	$\frac{12}{2}, \frac{23}{2}$	$\frac{14}{2}, \frac{23}{2}$
s_1^4	$\frac{24}{2}, \frac{-4}{2}$	$\frac{24}{2}, \frac{-4}{2}$	$\frac{24}{2}, \frac{30}{2}$	$\frac{24}{2}, \frac{30}{2}$

Panel (iv): $t = (t_{21}, t_{22})$, $(a, b) = (5, 3)$ (rows: Player 1; columns: Player 2)

	s_2^1	s_2^2	s_2^3	s_2^4
s_1^1	$\frac{10}{2}, \frac{6}{2}$	$\frac{12}{2}, \frac{6}{2}$	$\frac{12}{2}, \frac{16}{2}$	$\frac{14}{2}, \frac{16}{2}$
s_1^2	$\frac{24}{2}, \frac{1}{2}$	$\frac{24}{2}, \frac{23}{2}$	$\frac{24}{2}, \frac{1}{2}$	$\frac{24}{2}, \frac{23}{2}$
s_1^3	$\frac{10}{2}, \frac{1}{2}$	$\frac{12}{2}, \frac{23}{2}$	$\frac{12}{2}, \frac{1}{2}$	$\frac{14}{2}, \frac{23}{2}$
s_1^4	$\frac{24}{2}, \frac{-4}{2}$	$\frac{24}{2}, \frac{30}{2}$	$\frac{24}{2}, \frac{-4}{2}$	$\frac{24}{2}, \frac{30}{2}$

The following result shows the equivalence between the ex ante and interim definitions of the Bayesian Nash equilibrium.

Proposition 8.2 *A profile of strategy functions $\langle s_1^*(.), \ldots, s_n^*(.)\rangle$ is an ex ante Bayesian Nash equilibrium of a Bayesian game G^B if and only if it is an interim Bayesian Nash equilibrium of G^B.*

Proof We will first consider the necessity (only if) part, which ensures that (8.2) holds only if (8.3) holds. Let a profile of strategy functions $\langle s_1^*(.), \ldots s_n^*(.)\rangle$ be a Bayesian Nash equilibrium of a Bayesian game G^B and suppose towards contradiction that there exists some player $i \in N$ and some $t_i' \in T_i$ occurring with positive probability such that the inequality in (8.3) does not hold. Then, there exists $a_i \in A_i$ such that

$$\sum_{t_{-i} \in T_{-i}} u_i((a_i, s_{-i}^*(t_{-i})), t_i') p_i(t_{-i}|t_i') > \sum_{t_{-i} \in T_{-i}} u_i((s_i^*(t_i'), s_{-i}^*(t_{-i})), t_i') p_i(t_{-i}|t_i').$$

Now, consider a strategy function $\tilde{s}$ such that $\tilde{s}(t_i') = a_i$ and $\tilde{s}(t_i) = s^*(t_i)$ if $t_i \in T_i \setminus \{t_i'\}$. For any $t_i \in T_i$ with $p(t_i) > 0$, define

$$D_i(t_i) = \sum_{t_{-i} \in T_{-i}} u_i((\tilde{s}(t_i), s_{-i}^*(t_{-i})), t_i) p_i(t_{-i}|t_i) - \sum_{t_{-i} \in T_{-i}} u_i((s_i^*(t_i), s_{-i}^*(t_{-i})), t_i) p_i(t_{-i}|t_i).$$

Since the beliefs of the players are consistent by assumption, there exists a prior belief p such that $p_i(t_{-i}|t_i) = p(t)/p(t_i)$ for any $t_i \in T_i$ with $p(t_i) > 0$. It then follows that for any $t_i \in T_i$ with $p(t_i) > 0$

$$p(t_i) D_i(t_i) \begin{cases} > 0 \text{ if } t_i = t_i' \\ = 0 \text{ if } t_i \in T_i \setminus \{t_i'\}. \end{cases}$$

Clearly, for any $t_i \in T_i$ with $p(t_i) = 0$, we have $p(t_i) D_i(t_i) = 0$. Therefore, we have

$$\sum_{t_i \in T_i} p(t_i) D_i(t_i) > 0,$$

which implies

$$\sum_{t \in T} u_i((\tilde{s}(t_i), s_{-i}^*(t_{-i})), t_i) p(t) > \sum_{t \in T} u_i((s_i^*(t_i), s_{-i}^*(t_{-i})), t_i) p(t),$$

contradicting that the strategy function s_i^* is a best response to the profile of strategy functions s_{-i}^*, or that the profile $\langle s_1^*(.), \ldots s_n^*(.)\rangle$ is an ex ante Bayesian Nash equilibrium. Thus, (8.2) cannot hold if (8.3) does not hold. This proves the only if part.

The sufficiency (if) part of the proposition is straightforward. If each player $i \in N$ can be sure that after she learns her type t_i in T_i it will be optimal to play the action $s_i^*(t_i)$ selected by the strategy function s_i^* (no matter what t_i is), provided that all other players will follow their strategy functions in s_{-i}^*, then there is simply no reason for any player to change her strategy function s_i^* before learning her type. ■

Example 8.2 Reconsider the Bayesian game in Example 8.1. Let us show that the unique ex ante Bayesian Nash equilibrium of this game, the profile of strategy functions (s_1^4, s_2^4), is also the unique interim Bayesian Nash equilibrium.

Recall that the beliefs of the two players are such that $p_1(t_{12}|t_1) = p_1(t_{22}|t_1) = 0.5$ for each $t_1 \in T_1$ and $p_2(t_{11}|t_2) = p_2(t_{21}|t_2) = 0.5$ for each $t_2 \in T_2$. For any $i \in \{1, 2\}$ and for any profile of strategy functions (s_i, s_{-i}), the interim expected utility of player i with type $t_i \in T_i$ is

$$\tilde{u}_i(s_i, s_{-i}, t_i) = \sum_{t_{-i} \in T_{-i}} u_i((s(t_i), s_{-i}(t_{-i})), t_i) p_i(t_{-i}|t_i).$$

Using Table 8.2, we calculate in Table 8.5 the interim expected utilities $\tilde{u}_1(s_1, s_2, t_1)$ and $\tilde{u}_2(s_2, s_1, t_2)$ of players 1 and 2 derived from each profile of strategy functions $(s_1, s_2) \in S_1 \times S_2$ at each type profile $(t_1, t_2) \in T_1 \times T_2$. The best-response payoffs of the two players reveal the set of Nash equilibria in each panel (or for each possible type profile). Let the strategic form game for the type profile $t \in T$ be denoted by $G(t)$ and the set of pure Nash equilibria of $G(t)$ be denoted by $NE[G(t)]$. Clearly, we have

$$\begin{aligned}
NE[G(t_{11}, t_{12})] &= \{(s_1^1, s_2^1), (s_1^1, s_2^2), (s_1^2, s_2^3), (s_1^3, s_2^4), (s_1^4, s_2^4)\} \\
NE[G(t_{11}, t_{22})] &= \{(s_1^1, s_2^3), (s_1^2, s_2^2), (s_1^3, s_2^4), (s_1^4, s_2^4)\} \\
NE[G(t_{21}, t_{12})] &= \{(s_1^2, s_2^3), (s_1^2, s_2^4), (s_1^4, s_2^3), (s_1^4, s_2^4)\} \\
NE[G(t_{21}, t_{22})] &= \{(s_1^2, s_2^2), (s_1^2, s_2^4), (s_1^4, s_2^2), (s_1^4, s_2^4)\}.
\end{aligned}$$

We should note that a profile of strategy functions (s_1, s_2) can be an interim Bayesian Nash equilibrium if and only if it is a common Nash equilibrium of the games $G(t)_{t \in T}$. Since $\bigcap_{t \in T} NE[G(t)] = \{(s_1^4, s_2^4)\}$, the unique interim Bayesian Nash equilibrium is (s_1^4, s_2^4), which is also the unique ex ante Bayesian Nash equilibrium. ■

Example 8.3 (*Stochastic Cournot Duopoly Game*) Consider a duopolistic industry with asymmetric information. The set of players is $N = \{1, 2\}$, where $i \in N$ denotes firm i. The action spaces of players 1 and 2 involve all possible nonnegative quantities, i.e., $Q_1 = [0, \infty)$ and $Q_2 = [0, \infty)$. Let $Q = Q_1 \times Q_2$. For any $q \in Q$, the price of the good produced by the two firms is $P(q_1 + q_2) = a - q_1 - q_2$ whenever $q_1 + q_2 \leq a$.

The cost of production of firm $i \in N$ is $C_i(q_i) = \theta_i q_i$ for any $q_i \in Q_i$. We assume that the (marginal cost) parameter θ_1 of firm 1 is common knowledge, whereas the parameter θ_2 is privately known by firm 2. However, it is common knowledge that the parameter θ_2 can take the values θ_2^L and θ_2^H with probabilities $p(\theta_2^L) \in (0, 1)$ and $p(\theta_2^H) = 1 - p(\theta_2^L)$, respectively. We assume that $0 < \theta_2^L < \theta_2^H < a$.

Given the above assumptions, the private type spaces of players 1 and 2 are $\Theta_1 = \{\theta_1\}$ and $\Theta_2 = \{\theta_2^L, \theta_2^H\}$, respectively. The beliefs of player 1 satisfy $p_1(\theta_2^L|\theta_1) = p(\theta_2^L)$ and $p_1(\theta_2^H|\theta_1) = p(\theta_2^H)$, whereas the beliefs of player 2 satisfy $p_2(\theta_1|\theta_2^L) = p_2(\theta_1|\theta_2^H) = 1$.

For any quantity (action) profile $q \in Q$, the payoff (operational profit) of firm $i \in N$ with cost parameter $\theta_i \in \Theta_i$ is

$$u_i(q, \theta_i) = \pi_i(q, \theta_i) = P(q_i + q_{-i})q_i - \theta_i q_i.$$

We define the strategy functions of firms 1 and 2 as $s_1 : \Theta_1 \to Q_1$ and $s_2 : \Theta_2 \to Q_2$, respectively. Suppose an interim Bayesian Nash equilibrium in pure strategy functions, say $s^* = (s_1^*, s_2^*)$, exists. Then, for all $i \in N$ and all $\theta_i \in \Theta_i$ it must be true that

$$s_i^*(\theta_i) = \text{argmax}_{q_i \in Q_i} \sum_{\theta_{-i} \in \Theta_{-i}} u_i((q_i, s_{-i}^*(\theta_{-i})), \theta_i) p_i(\theta_{-i}|\theta_i). \tag{8.4}$$

For player 1, the optimality condition in (8.4) can be written as:

$$s_1^*(\theta_1) = \text{argmax}_{q_1 \in Q_1} \pi_1^e((q_1, s_2^*), \theta_1) \tag{8.5}$$

where

$$\begin{aligned}
\pi_1^e((q_1, s_2^*), \theta_1) &= \pi_1((q_1, s_2^*(\theta_2^L)), \theta_1) p(\theta_2^L) + \pi_1((q_1, s_2^*(\theta_2^H)), \theta_1) p(\theta_2^H), \\
&= \left[\left(a - q_1 - s_2^*(\theta_2^L)\right) q_1 - \theta_1 q_1\right] p(\theta_2^L) + \\
&\quad \left[\left(a - q_1 - s_2^*(\theta_2^H)\right) q_1 - \theta_1 q_1\right] p(\theta_2^H) \\
&= \left(a - \theta_1 - q_1 - s_2^{*,e}\right) q_1
\end{aligned} \tag{8.6}$$

with

$$s_2^{*,e} = s_2^*(\theta_2^L) p(\theta_2^L) + s_2^*(\theta_2^H) p(\theta_2^H). \tag{8.7}$$

The first-order condition associated with (8.6) is

$$a - \theta_1 - 2q_1 - s_2^{*,e} = 0.$$

Since $s_1^*(\theta_1)$ must satisfy the above condition, we must have

$$a - \theta_1 - 2s_1^*(\theta_1) - s_2^{*,e} = 0,$$

implying

$$s_1^*(\theta_1) = \frac{a - \theta_1 - s_2^{*,e}}{2}. \tag{8.8}$$

For player 2, the optimality condition in (8.4) can be written as:

$$\begin{aligned}
s_2^*(\theta_2^L) &= \text{argmax}_{q_2 \in Q_2} \pi_2\left((q_2, s_1^*(\theta_1)), \theta_2^L\right) p_2\left(\theta_1 | \theta_2^L\right) \\
&= \left(a - s_1^*(\theta_1) - q_2\right) q_2 - \theta_2^L q_2
\end{aligned} \tag{8.9}$$

and

$$\begin{aligned}
s_2^*(\theta_2^H) &= \text{argmax}_{q_2 \in Q_2} \pi_2\left((q_2, s_1^*(\theta_1)), \theta_2^H\right) p_2\left(\theta_1 | \theta_2^H\right). \\
&= \left(a - s_1^*(\theta_1) - q_2\right) q_2 - \theta_2^H q_2
\end{aligned} \tag{8.10}$$

where we used $p_2\left(\theta_1 | \theta_2^L\right) = p_2\left(\theta_1 | \theta_2^H\right) = 1$. The first-order conditions associated with (8.9) and (8.10) are, respectively,

$$a - \theta_2^L - s_1^*(\theta_1) - 2q_2 = 0 \tag{8.11}$$

and

$$a - \theta_2^H - s_1^*(\theta_1) - 2q_2 = 0. \tag{8.12}$$

Since $s_2^*(\theta_2^L)$ and $s_2^*(\theta_2^H)$ must satisfy (8.11) and (8.12), respectively, we must have

$$a - \theta_2^L - s_1^*(\theta_1) - 2s_2^*(\theta_2^L) = 0$$

and

$$a - \theta_2^H - s_1^*(\theta_1) - 2s_2^*(\theta_2^H) = 0$$

implying

$$s_2^*(\theta_2^L) = \frac{a - \theta_2^L - s_1^*(\theta_1)}{2} \tag{8.13}$$

and

$$s_2^*(\theta_2^H) = \frac{a - \theta_2^H - s_1^*(\theta_1)}{2}. \tag{8.14}$$

We are ready to calculate the Bayesian Nash equilibrium of the Stochastic Cournot Duopoly using the equations (8.7), (8.8), (8.13), and (8.14). Multiplying (8.13) and (8.14) by $p(\theta_2^L)$ and $p(\theta_2^H)$, respectively, and adding them, we obtain

$$s_2^{*,e} = \frac{a - \theta_2^e - s_1^*(\theta_1)}{2}. \tag{8.15}$$

where

$$\theta_2^e = \theta_2^L p(\theta_2^L) + \theta_2^H p(\theta_2^H). \tag{8.16}$$

From (8.15), we obtain

$$s_1^*(\theta_1) = a - \theta_2^e - 2s_2^{*,e}. \tag{8.17}$$

Next, equating (8.8) and (8.17) yields

$$\frac{a - \theta_1 - s_2^{*,e}}{2} = a - \theta_2^e - 2s_2^{*,e}, \tag{8.18}$$

implying

$$s_2^{*,e} = \frac{a - 2\theta_2^e + \theta_1}{3}. \tag{8.19}$$

Inserting this into (8.17), we obtain

$$s_1^*(\theta_1) = \frac{a - 2\theta_1 + \theta_2^e}{3}. \tag{8.20}$$

Inserting (8.20) into (8.13) and (8.14), we obtain

$$s_2^*(\theta_2^L) = \frac{2a - 3\theta_2^L + 2\theta_1 - \theta_2^e}{6} \tag{8.21}$$

and

$$s_2^*(\theta_2^H) = \frac{2a - 3\theta_2^H + 2\theta_1 - \theta_2^e}{6}. \tag{8.22}$$

Thus, Eqs. (8.20)–(8.22) and (8.16) together characterize $(s_1^*(.), s_2^*(.))$, the unique Bayesian Nash equilibrium in pure strategy functions.

We should note that an increase in the expected marginal cost of firm 2, θ_2^e, increases the equilibrium output planned by firm 1 and reduces the equilibrium output planned by each cost type of firm 2. ■

References

1. Harsanyi JC (1967/1968) Games with incomplete information played by "Bayesian" players I, II, and III. Manage Sci 14:159–182, 320–334, 486–502
2. Nash JF (1950) Equilibrium points in n-person games. PNAS 36(1):48–49

Part III

Dynamic Noncooperative Games with Perfect or Imperfect Information

In this part of the book, we will study dynamic noncooperative games. The difference with the first two parts is that players will now be allowed to choose their actions sequentially (at least in some parts of the game) or repeatedly. Some real-life examples of dynamic noncooperative interactions involve market entry games, repeated games of oligopolistic competition, sequential auctions, and many parlor games involving chess, backgammon, bridge, and poker. In some of these games, like chess (and backgammon), the players know the history of actions chosen by their opponents (and the chance moves that led to those actions). We will call these strategic interactions "dynamic noncooperative games with perfect information" (with or without chance moves). In another set of dynamic games, some players have uncertainty about the state of play in some parts of the game, due to the asymmetric information caused by a randomization involved (like in poker or bridge) or due to simultaneous plays before or after sequential plays occur (like in market entry games). We will call these games "dynamic noncooperative games with imperfect information" (with or without chance moves). To isolate the role of perfectness or imperfectness of information in solving dynamic noncooperative games, we will restrict ourselves to situations with complete information (meaning that players have common knowledge about payoff structure, the sets of strategies, the external chance moves, and the rules of the game).

This part involves Chaps. 9–12. Chapter 9 introduces the model of extensive form games, represented by a tree structure. This model serves to study dynamic noncooperative games both with perfect information and with imperfect information. Chapter 9 also shows how players can randomize over their pure strategies in extensive form games and how these games can be represented as a strategic form game.

Chapter 10 introduces the subgame perfect Nash equilibrium as a refinement of Nash equilibrium in extensive form games. This equilibrium can be obtained by backward induction in extensive form games with perfect information and obtained by a generalized process of backward induction, called subgame perfection, in extensive form games with imperfect information.

Chapter 11 shows that the subgame perfect Nash equilibrium cannot always eliminate some "sequentially unreasonable" Nash equilibria in extensive form games with imperfect information. As a remedy, this chapter introduces two alternative refinements of the Nash equilibrium, known as the sequential equilibrium and the weak sequential equilibrium (or weak perfect Bayesian equilibrium).

Chapter 11 gives the relationship between these two refinements of the Nash equilibrium and their relations to the subgame perfect Nash equilibrium.

Finally, Chap. 12 studies extensive form games where a strategic form game is repeatedly played for a finite or infinite number of periods. This chapter shows whether and why the repetition of a strategic form game may have a subgame perfect Nash equilibria that cannot be obtained by repeatedly playing the Nash equilibria of that strategic form game.

Introduction to Extensive Form Games

9

In Chap. 9, we will introduce the model of extensive form games that will allow us to study games where players can play their actions sequentially or repeatedly. Section 9.1 will present the basic structures of the model, followed by some examples. Section 9.2 will discuss how an extensive form game can be represented as a strategic form game and Sect. 9.3 will present two alternative representations for the strategies of players in an extensive form game and a result that shows when these two representations, known as mixed and behavior strategies, are equivalent.

9.1 Extensive Form Games

We consider a group of $n \geq 2$ players denoted by the set $N = \{1, \ldots, n\}$. These players face an extensive form game that we represent by a tree structure. The tree of the game involves a set of branches, each of which connects two distinct points called nodes. We denote the set of all nodes in the game by $\mathcal{X}$. Each branch in the game is associated with a single action a lying in the set of possible actions $\mathcal{A}$. On each branch, we indicate the associated action.

The top node of the tree is called the root where the game starts. We denote the root by x_0. Usually, one of the players moves at the root. However, if an external player, called the Nature, exists in the game, then the game usually starts with the Nature playing at x_0. The Nature represents the incomplete information of the players about the game. Thus, it is assumed that the Nature plays each action available at x_0 with some probability in [0, 1]. The sum of the probabilities on the possible actions of the Nature is 1. We write these probabilities on the branches where the Nature plays its actions.

The nodes at the bottom of the tree are called terminal nodes. These are the nodes that are not followed by any branches. The set of terminal nodes is denoted by $\mathcal{X}^{\mathcal{T}}$.

I. Saglam, *Mastering Game Theory*, Classroom Companion: Economics,
https://doi.org/10.1007/978-3-031-82094-6_9

The remaining (nonterminal) nodes are called decision nodes. Given a decision node $x \in \mathcal{X} \setminus \mathcal{X}^{\mathcal{T}}$, we denote by $\mathcal{A}(x)$ (a subset of $\mathcal{A}$) the set of possible actions at x.

A profile of payoff functions $u = (u_1, \ldots, u_n)$ assigns a payoff to each player at each terminal node of the game, i.e. $u_i : \mathcal{X}^{\mathcal{T}} \to \mathbb{R}$. At each terminal node $x \in \mathcal{X}^{\mathcal{T}}$, we indicate the associated payoff profile $u(x) = (u_1(x), \ldots, u_n(x))$.

Each decision node $x \in \mathcal{X} \setminus \mathcal{X}^{\mathcal{T}}$ is assigned to an information set H. The set of all information sets form a partition $\mathcal{H}$ of the decision nodes $\mathcal{X} \setminus \mathcal{X}^{\mathcal{T}}$. Each player is assigned to at least one information set. The set of information sets of player i is denoted by $\mathcal{H}_i$. Clearly, $\mathcal{H}_1 \cup \mathcal{H}_2 \cup \ldots \mathcal{H}_n = \mathcal{H}$. We mark the set of nodes that are in the same information set by connecting these nodes by a dotted line segment. Any two nodes x_1 and x_2 are assumed to be in the same information set $H \in \mathcal{H}$ only if the player who plays at H cannot identify before the realization of payoffs whether the node x_1 or the node x_2 has been reached during the game. This implies that the number of branches and actions at x_1 and x_2 must be the same.

Example 9.1 Consider the extensive form game in Fig. 9.1, where at each terminal node the first payoff belongs to Amy and the second payoff belongs to Bob. The set of players is $N = \{Amy, Bob\}$. The set of all nodes is $\mathcal{X} = \{x_1, \ldots, x_{13}\}$, the set of terminal nodes is

$$\mathcal{X}^{\mathcal{T}} = \{x_6, x_7, x_8, x_9, x_{10}, x_{11}, x_{12}, x_{13}\},$$

and the set of decision nodes is

$$\mathcal{X} \setminus \mathcal{X}^{\mathcal{T}} = \{x_0, x_1, x_2, x_3, x_4, x_5\}.$$

Amy has three information sets $H^1_{Amy} = \{x_0\}$, $H^2_{Amy} = \{x_4\}$, and $H^3_{Amy} = \{x_5\}$. So, the set of information sets of Amy is

$$\mathcal{H}_{Amy} = \{H^1_{Amy}, H^2_{Amy}, H^3_{Amy}\}.$$

Bob has two information sets $H^1_{Bob} = \{x_1\}$ and $H^2_{Bob} = \{x_2, x_3\}$. So, the set of information sets of Bob is

$$\mathcal{H}_{Bob} = \{H^1_{Bob}, H^2_{Bob}\}.$$

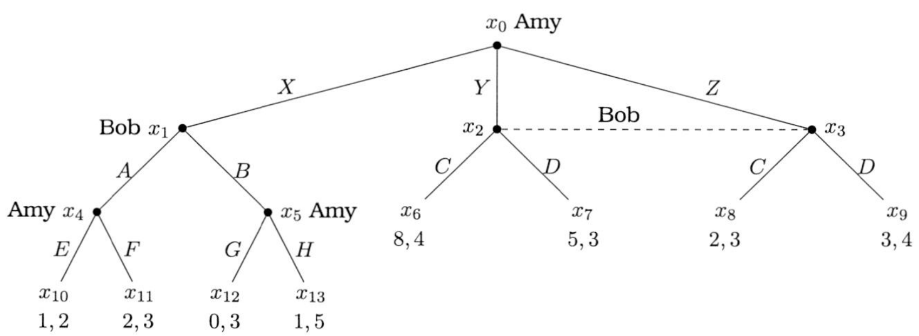

Fig. 9.1 An extensive form game

Thus, the set of all information sets in the game is

$$\mathcal{H} = \{\mathcal{H}_{Bob}, \mathcal{H}_{Amy}\} = \{\{x_0\}, \{x_1\}, \{x_2, x_3\}, \{x_4\}, \{x_5\}\}.$$

Note that $\mathcal{H}$ is a partition of $\mathcal{X} \setminus \mathcal{X}^{\mathcal{T}}$. The set of actions available in the game is

$$\mathcal{A} = \{X, Y, Z, A, B, C, D, E, F, G, H\}.$$

The set of actions available for Amy is

$$\mathcal{A}_{Amy} = \{X, Y, Z, E, F, G, H\},$$

whereas the set of actions available for Bob is

$$\mathcal{A}_{Bob} = \{A, B, C, D\}.$$

Recall that there are two nodes, x_2 and x_3, in Bob's information set H^2_{Bob}. At these two nodes the available actions for Bob are the same; i.e., $\mathcal{A}_{Bob}(x_2) = \mathcal{A}_{Bob}(x_3) = \{C, D\}$. The payoffs of Amy and Bob are shown below the terminal nodes. For example, $u(x_{13}) = (u_{Amy}(x_{13}), u_{Bob}(x_{13})) = (1, 5)$. ■

Note in the above example that whenever Bob knows that Amy did not play the action X at the node x_0, Bob does not know whether Amy played Y and Z. That is why we connect the nodes x_2 and x_3 by a dashed line to indicate that both of these nodes are in the Bob's information set (H^2_{Bob}) that is reached when Amy does not play X. In some games, each player always knows the previous moves of other players. We formally define these games as follows.

Definition 9.1 An extensive form game is a game with perfect information if each information set of each player contains a single decision node, i.e., for each $i \in N$ and $H_i \in \mathcal{H}_i$, there exists $x \in X \setminus \mathcal{X}^{\mathcal{T}}$ such that $H_i = \{x\}$.

If an extensive form game is not a game with perfect information, we will call it a game with imperfect information. Note that the game in Example 9.1 is an extensive form game with imperfect information. On the other hand, the game known as the chess is an extensive form game with perfect information. Below, we present a simple example for games with perfect information.

Example 9.2 Consider the extensive form game in Fig. 9.2. The set of players is $N = \{Amy, Bob\}$. At each terminal node the first payoff belongs to Amy and the second payoff belongs to Bob. The set of decision nodes is $\{x_0, x_1, x_2\}$. Amy has one information set $H^1_{Amy} = \{x_0\}$. The set of information sets of Amy is thus $\mathcal{H}_{Amy} = \{H^1_{Amy}\}$.

Bob has two information sets $H^1_{Bob} = \{x_1\}$ and $H^2_{Bob} = \{x_2\}$. The set of information sets of Bob is $\mathcal{H}_{Bob} = \{H^1_{Bob}, H^2_{Bob}\}$. The set of all information sets in the game is

$$\mathcal{H} = \{\mathcal{H}_{Bob}, \mathcal{H}_{Amy}\} = \{\{x_0\}, \{x_1\}, \{x_2\}\}.$$

Note that $\mathcal{H}$ is a partition of the set of decision nodes $\{x_0, x_1, x_2\}$. Also note that each information set of each player contains a single node. Therefore, this extensive form game is a game with perfect information. ■

We say that an extensive form game is finite if the number of players, the number of nodes and the number of actions at each information set in the game are finite. Note that the games in Examples 9.1 and 9.2 are finite.

9.2 Strategic Form Representation

We start by asking whether an extensive form game can be represented as a strategic form game. Consider the extensive form game in Example 9.2. In the strategic form representation of this game, Amy has three strategies corresponding to the three actions X, Y, and Z at her unique information set $\mathcal{H}^1_{Amy} = \{x_0\}$ in the extensive form game. Bob has four strategies, AA, AB, BA, and BB, as he can play in the extensive form game at two information sets $\mathcal{H}^1_{Bob} = \{x_1\}$ and $\mathcal{H}^2_{Bob} = \{x_2\}$ at both of which he can choose between two actions, A and B. (Thus, the strategy XY denotes for any $X, Y \in \{A, B\}$ that Bob plays the actions X and Y at the nodes x_1 and x_2, respectively.) Using the payoffs in Fig. 9.2, we can then construct the strategic form game in Table 9.1.

While we can find for each extensive form game a unique strategic form representation, the converse is not true. A strategic form game may represent more than one extensive form game. For example, the strategic form game in Table 9.1 represents all three extensive form games in Figs. 9.3, 9.4 and 9.5 below. Note that these three extensive form games look different. In particular, the first game involves perfect information, while the other two games involve imperfect information.

In the next chapter, we will consider several equilibrium concepts for the extensive form games that will exploit the sequential structure of the game as well as the information held by the players. As several extensive form games with different sequential and informational structures can possibly be represented by the same

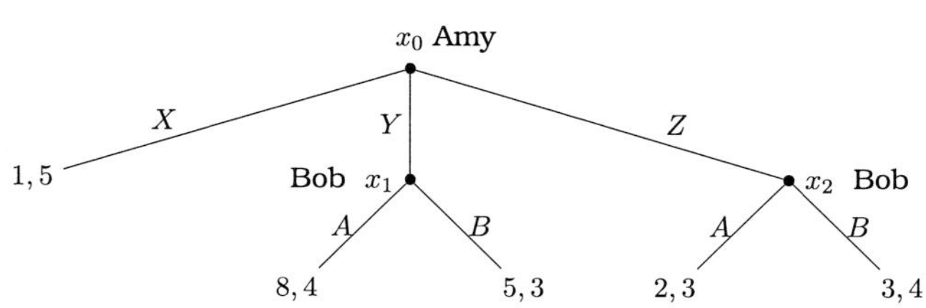

Fig. 9.2 An extensive form game with perfect information

Table 9.1 Strategic form game corresponding to the game in Fig. 9.2

		Bob			
		AA	AB	BA	BB
Amy	X	(1, 5)	(1, 5)	(1, 5)	(1, 5)
	Y	(8, 4)	(8, 4)	(5, 3)	(5, 3)
	Z	(2, 3)	(3, 4)	(2, 3)	(3, 4)

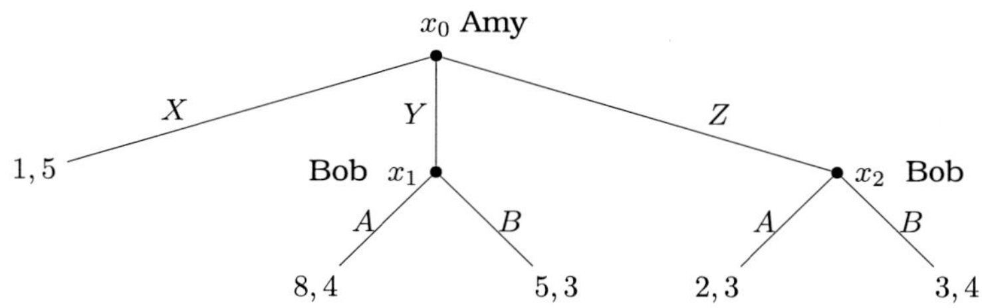

Fig. 9.3 An extensive form game that can be represented by Table 9.1

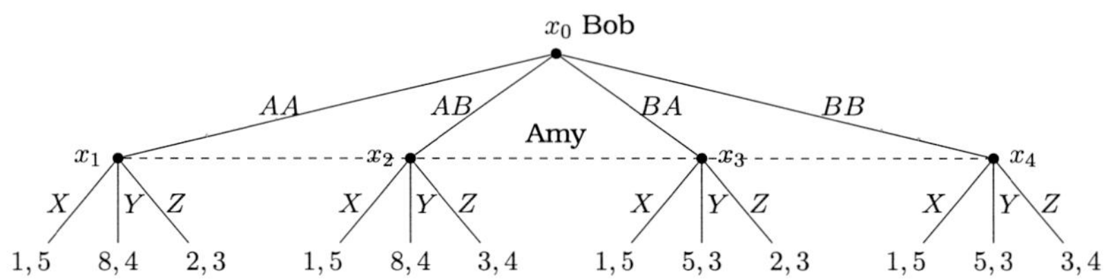

Fig. 9.4 A second extensive form game that can be represented by Table 9.1

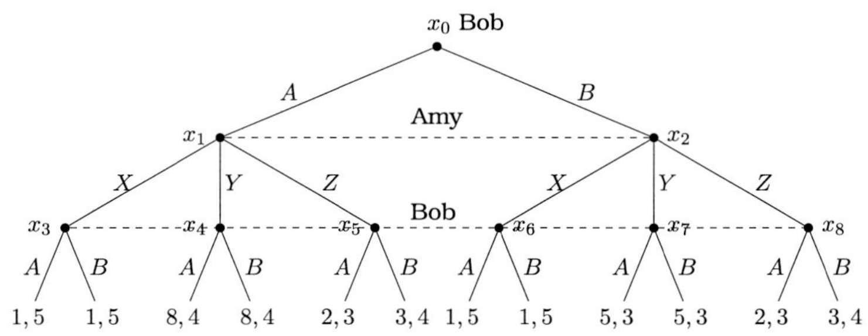

Fig. 9.5 A third extensive form game that can be represented by Table 9.1

(unique) strategic form game, it should be clear that many important details of extensive form games may be lost once they are transformed to a strategic form game. On the other hand, the strategic form game representation will be helpful for us in calculating the set of all Nash equilibria of an associated extensive form game.

If the extensive form game involves chance moves (due to the presence of the external player, called the Nature, located at the root x_0), then we should calculate the expected utilities of the players at each terminal node using the probabilities

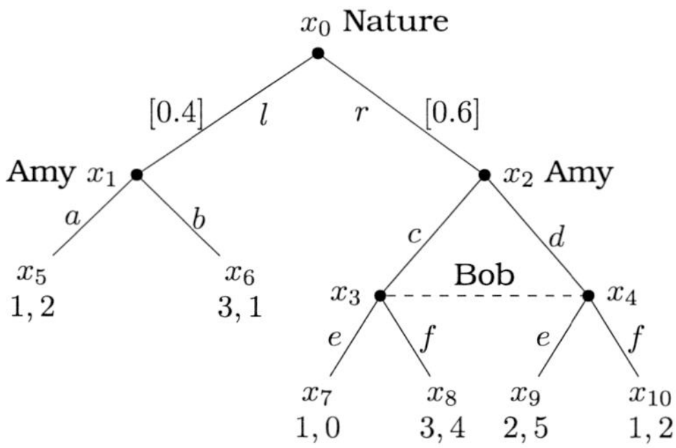

Fig. 9.6 An extensive form game with chance moves played by the Nature

assigned by the Nature to the possible actions at the root. Below, we present an example how to make such calculations.

Example 9.3 Consider the extensive form game in Fig. 9.6. In this game, the Nature plays the actions l and r with probabilities 0.4 and 0.6, respectively, (shown in brackets on the related branches of the game tree). In the strategic form representation of this game, Amy has four strategies ac, ad, bc, and bd, whereas Bob has two strategies e and f. These strategies lead to a 4×2 payoff matrix we present in Table 9.2.

The payoffs in the above table are calculated using the probabilities that the nodes x_1 and x_2 can be reached in the extensive form game. For example, consider the strategy profile (ad, e) in Fig. 9.6. The players should make the following calculations at this profile. With probability 0.4 the node x_1 will be reached and subsequently the action a will be played leading to the payoff profile at x_5, and with probability 0.6 the node x_2 will be reached and the action profile (d, e) will be played leading to the payoff profile at x_9. So, the expected payoffs of Amy and Bob at the strategy profile (ad, e) will, respectively, be

$$\begin{aligned} u^e_{Amy}(ad, e) &= \text{prob}(x_1) u_{Amy}(x_5) + \text{prob}(x_2) u_{Amy}(x_9) \\ &= (0.4)(1) + (0.6)2 = 1.6 \\ u^e_{Bob}(ad, e) &= \text{prob}(x_1) u_{Bob}(x_5) + \text{prob}(x_2) u_{Bob}(x_9) \\ &= (0.4)(2) + (0.6)5 = 3.8. \end{aligned}$$

Table 9.2 Strategic form game corresponding to the game in Fig. 9.6

	Bob e	Bob f
Amy ac	(1.0, 0.8)	(2.2, 3.2)
ad	(1.6, 3.8)	(1.0, 2.0)
bc	(1.8, 0.4)	(3.0, 2.8)
bd	(2.4, 3.4)	(1.8, 1.8)

We should note that the extensive form game portrayed above is a game with imperfect information. But, the informational imperfection is not caused by the chance moves of the Nature. It is rather caused by the fact that Bob's unique information set is non-singleton: it contains two nodes, x_3 and x_4. The strategic form representation we have derived calculates the ex ante payoffs, i.e., the payoffs expected by the players before the Nature chooses one of the two actions l and r with probabilities 0.4 and 0.6. However, after the Nature chooses one of these actions, both Amy and Bob will know in the extensive form game whether they are at the node x_1 or the node x_2. The only imperfection in the game is that if x_2 is ever reached, then Bob has to choose his action without knowing whether Amy has chosen the action c or the action d. ■

9.3 Behavior/Mixed Strategies

In the previous sections, we restricted the players, other than the Nature (the external player who has chance moves), to pure strategies involving a plan of actions. Now, we will extend our discussion to randomized strategies. We will consider two possibilities where all players will make randomization either over the possible pure strategies (plans of actions) they may be playing or over their actions available at each information set that may be reached in the extensive form game. The first type of randomization will yield strategies called mixed strategies with which we are familiar from earlier chapters, whereas the second type of randomization will yield strategies called the behavior strategies, which we formally define as follows.

Definition 9.2 A behavior strategy of player $i \in N$ in an extensive form game is a function b_i that maps each information set $H \in \mathcal{H}_i$ to a probability distribution over the set of available actions at H.

The following example will be helpful to understand the mixed and behavior strategies.

Example 9.4 Consider the extensive form game in Fig. 9.7, where the first payoff belongs to Amy and the second payoff belongs to Bob.

Amy's information sets are $H^1_{Amy} = \{x_0\}$ and $H^2_{Amy} = \{x_3, x_4\}$, whereas Bob's information sets are $H^1_{Bob} = \{x_1\}$ and $H^2_{Bob} = \{x_2\}$. A behavior strategy of Amy is a function b_{Amy} such that

$$b_{Amy}(H^1_{Amy}) = (b_A, b_B) \text{ where } b_A, b_B \in [0, 1] \text{ and } b_A + b_B = 1$$

and

$$b_{Amy}(H^2_{Amy}) = (b_X, b_Y) \text{ where } b_X, b_Y \in [0, 1] \text{ and } b_X + b_Y = 1.$$

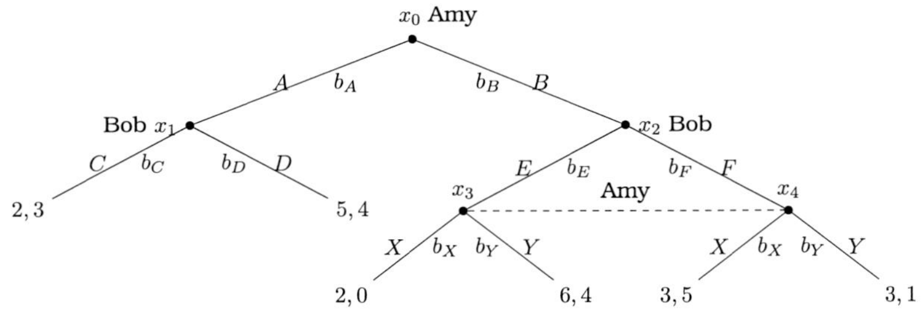

Fig. 9.7 Behavior strategies in an extensive form game

Likewise, a behavior strategy of Bob is b_{Bob} such that

$$b_{Bob}(H^1_{Bob}) = (b_C, b_D) \text{ where } b_C, b_D \in [0, 1] \text{ and } b_C + b_D = 1$$

and

$$b_{Bob}(H^2_{Bob}) = (b_E, b_F) \text{ where } b_E, b_F \in [0, 1] \text{ and } b_E + b_F = 1.$$

In the strategic form representation of the above extensive form game, the set of pure strategies of Amy is $\{AX, AY, BX, BY\}$ and the set of pure strategies of Bob is $\{CE, CF, DE, DF\}$. A mixed strategy of Amy is then a probability distribution σ_{Amy} such that

$$\sigma_{Amy} = (\sigma_{AX}, \sigma_{AY}, \sigma_{BX}, \sigma_{BY})$$

where

$$\sigma_{AX}, \sigma_{AY}, \sigma_{BX}, \sigma_{BY} \in [0, 1] \text{ and } \sigma_{AX} + \sigma_{AY} + \sigma_{BX} + \sigma_{BY} = 1.$$

Likewise, a mixed strategy of Bob is a probability distribution σ_{Bob} such that

$$\sigma_{Bob} = (\sigma_{CE}, \sigma_{CF}, \sigma_{DE}, \sigma_{DF})$$

where

$$\sigma_{CE}, \sigma_{CF}, \sigma_{DE}, \sigma_{DF} \in [0, 1] \text{ and } \sigma_{CE} + \sigma_{CF} + \sigma_{DE} + \sigma_{DF} = 1.$$ ■

Suppose the players in an extensive form game can freely choose to play either mixed or behavior strategies. One may ask whether there exists an equivalence between these two types of strategies with regard to the expected utilities they will generate. This question is important because if the answer is yes, then we can restrict ourselves to one of these two kinds of strategies without loss of generality. Before we give an answer to this question, let us define the following.

Definition 9.3 A mixed/behavior strategy profile is a profile of strategies $\tilde{\sigma} = (\tilde{\sigma}_i)_i$ where for each i the strategy $\tilde{\sigma}_i$ is either a mixed strategy or a behavior strategy of player i.

For two-person games, the above definition implies that if a strategy profile is a mixed/behavior strategy profile, then either (i) both players use mixed strategies, or (ii) both players use behavior strategies, or (iii) one player uses a mixed strategy, while the other player uses a behavior strategy. With any number of players, we should also note that both mixed and behavior strategies induce probabilities on each branch of a given extensive form game; therefore, we can calculate the probability that any decision or terminal node will be reached during the game. Formally, let $p(x|\tilde{\sigma})$ denote the probability that the node $x \in X$ will be reached under the mixed/behavior strategy profile $\tilde{\sigma}$ during the course of play. (Note that $p(x_0|\tilde{\sigma}) = 1$, trivially.) We are now ready to define the equivalence between mixed and behavior strategies.

Definition 9.4 Given an extensive form game, a behavior strategy b_i and a mixed strategy σ_i of player i are equivalent if for every node $x \in \mathcal{X}$ and for every mixed/behavior strategy profile $\tilde{\sigma}_{-i}$ of the players in $N - i$ it is true that $p(x|b_i, \tilde{\sigma}_{-i}) = p(x|\sigma_i, \tilde{\sigma}_{-i})$.

The above definition says that a player i regards a behavior strategy b_i and a mixed strategy σ_i to be equivalent in a given extensive form game if the probability of reaching any node $x \in X$ is the same under b_i and σ_i irrespective of the strategies of the remaining players and irrespective of whether any of them uses a behavior strategy or a mixed strategy.

Definition 9.5 Given an extensive form game, we say that a mixed strategy profile $\sigma = (\sigma_i)_{i \in N}$ and a behavior strategy profile $b = (b_i)_{i \in N}$ are equivalent to each other if for each $i \in N$ the strategies σ_i and b_i are equivalent for player i.

Given a mixed/behavior strategy profile $\tilde{\sigma}$ of the players in N, the expected utility of player j is $\tilde{u}_j(\tilde{\sigma})$. This expected utility changes to $\tilde{u}_j(\sigma_i, \tilde{\sigma}_{-i})$ if some player i uses the mixed strategy σ_i and it changes to $\tilde{u}_j(b_i, \tilde{\sigma}_{-i})$ if player i uses the behavior strategy b_i. (Note that players i and j need not be the same.) We should also note that for any $i, j \in N$

$$\tilde{u}_j(\sigma_i, \tilde{\sigma}_{-i}) := \sum_{x \in \mathcal{X}^T} u_j(x) p(x|b_i, \tilde{\sigma}_{-i})$$

and

$$\tilde{u}_j(b_i, \tilde{\sigma}_{-i}) := \sum_{x \in \mathcal{X}^T} u_j(x) p(x|\sigma_i, \tilde{\sigma}_{-i}).$$

If b_i and σ_i are equivalent for player i, then $p(x|b_i, \tilde{\sigma}_{-i}) = p(x|\sigma_i, \tilde{\sigma}_{-i})$ for each node $x \in X$, and in particular for each terminal node $x \in \mathcal{X}^T$. Therefore, if b_i and σ_i are equivalent for player i, then for any $j \in N$ the expected utility of player j when player i uses the mixed strategy σ_i and other players use their mixed/behavior strategies in $\tilde{\sigma}_{-i}$ must be the same as the expected utility of player j when player i

uses the behavior strategy b_i and other players use their mixed/behavior strategies in $\tilde{\sigma}_{-i}$; i.e.,

$$\tilde{u}_j(\sigma_i, \tilde{\sigma}_{-i}) = \tilde{u}_j(b_i, \tilde{\sigma}_{-i}) \text{ for all } j \in N.$$

Now consider any mixed strategy profile $\sigma = (\sigma_i)_{i \in N}$ and any behavior strategy profile $b = (b_i)_{i \in N}$ that are equivalent to each other. Pick any $j \in N$. Then, the above inequality implies that for any mixed/behavior strategy profile $\tilde{\sigma}_{-1}$ of the players in $N \setminus \{1\}$ we have

$$\tilde{u}_j(\sigma_1, \tilde{\sigma}_{-1}) = \tilde{u}_j(b_1, \tilde{\sigma}_{-1}).$$

Note that the last equality is true in particular for $\tilde{\sigma}_{-1} = \sigma_{-1}$, as well. Thus,

$$\tilde{u}_j(\sigma_1, \sigma_{-1}) = \tilde{u}_j(b_1, \sigma_{-1})$$

or equivalently

$$\tilde{u}_j(\sigma_1, \sigma_2, \ldots, \sigma_n) = \tilde{u}_j(b_1, \sigma_2, \ldots, \sigma_n). \tag{9.1}$$

Moreover, since σ_2 and b_2 are equivalent for player 2, we have

$$\tilde{u}_j(b_1, \sigma_2, \sigma_3, \ldots, \sigma_n) = \tilde{u}_j(b_1, b_2, \sigma_3, \ldots, \sigma_n).$$

Iterating in this way we can write for any $k = 1, \ldots, n-1$ the equation

$$\tilde{u}_j(b_1, \ldots, b_k, \sigma_{k+1}, \sigma_{k+2}, \ldots, \sigma_n) = \tilde{u}_j(b_1, \ldots, b_k, b_{k+1}, \sigma_{k+2}, \ldots, \sigma_n), \tag{9.2}$$

since the strategy profiles σ_k and b_k are equivalent for player k. Combining (9.1) with (9.2) yields

$$\tilde{u}_j(\sigma) = \tilde{u}_j(b).$$

We state this result below.

Proposition 9.1 *Given an extensive form game, pick any mixed strategy profile $\sigma = (\sigma_i)_{i \in N}$ and any behavior strategy profile $b = (b_i)_{i \in N}$ that are equivalent to each other. Then, $\tilde{u}_i(\sigma) = \tilde{u}_i(b)$ for all $i \in N$.*

The above result says that in an extensive form game each player's expected utility must be the same under a profile of behavior strategies and a profile of mixed strategies if these profiles are equivalent to each other. However, this result does not tell us whether we can always find for each behavior strategy profile in an extensive form game an equivalent mixed strategy profile or whether we can always find for each mixed strategy profile in an extensive form game an equivalent behavior strategy profile. The following example will show that the answers to both of these questions are negative; i.e., there exist games where the set of mixed strategy profiles is narrower than the set of behavior strategy profiles and games where the set of behavior strategy profiles is narrower than the set of mixed strategy profiles. We will see that these types of games fail to satisfy a property known *perfect recall* which requires that no player forgets in the game what she once knew.

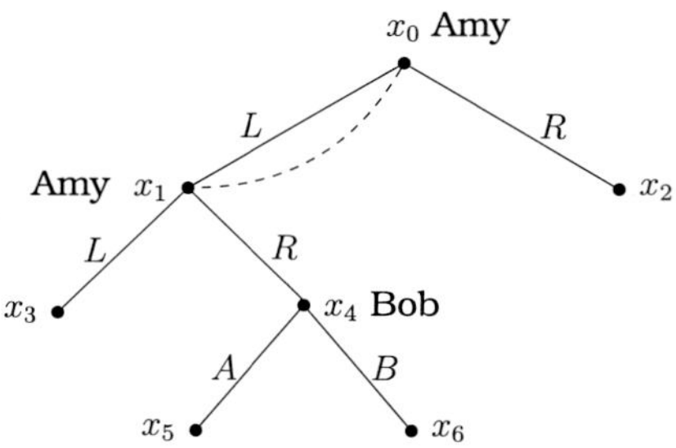

Fig. 9.8 A player (Amy) does not recall a previous node where she played

Example 9.5 Consider first the extensive form game in Fig. 9.8, where Amy has two strategies L and R and Bob has two strategies A and B. This game violates perfect recall since it involves a player, Amy, who forgets during the course of play the node where she previously was. Note that an information set of Amy, involving the nodes x_0 and x_1, reveals that if she plays L at the node x_0 and reaches to the node x_1, then she cannot recall that she earlier played at the node x_0.

Amy's unique information set is $H^1_{Amy} = \{x_0, x_1\}$, whereas Bob's unique information set is $H^1_{Bob} = \{x_4\}$. A behavior strategy of Amy is a function b_{Amy} such that

$$b_{Amy}(H^1_{Amy}) = (b_L, b_R) \text{ where } b_L, b_R \in [0, 1] \text{ and } b_L + b_R = 1.$$

Amy uses the behavior strategy probabilities (b_L, b_R) at both x_0 and x_1.

A behavior strategy of Bob is b_{Bob} such that

$$b_{Bob}(H^1_{Bob}) = (b_A, b_B) \text{ where } b_A, b_B \in [0, 1] \text{ and } b_A + b_B = 1.$$

Note that node x_5 is reached with probability $b_L(1 - b_L)b_A$ under the behavior strategy profile (b_{Amy}, b_{Bob}). This probability is positive if $b_L, b_A \in (0, 1)$.

In the strategic form representation of the above extensive form game, the set of pure strategies of Amy is $\{L, R\}$ and the set of pure strategies of Bob is $\{A, B\}$. For a given pure strategy profile (s_{Amy}, s_{Bob}) where $s_{Amy} \in \{L, R\}$ and $s_{Bob} \in \{A, B\}$, the players reach to x_3 if $s_{Amy} = L$ and they reach to x_2 if $s_{Amy} = R$. A mixed strategy of Amy is a probability distribution $\sigma_{Amy} = (\sigma_L, \sigma_R)$ where $\sigma_L, \sigma_R \in [0, 1]$ and $\sigma_L + \sigma_R = 1$. Likewise, a mixed strategy of Bob is a probability distribution $\sigma_{Bob} = (\sigma_A, \sigma_B)$ where $\sigma_A, \sigma_B \in [0, 1]$ and $\sigma_A + \sigma_B = 1$. A mixed strategy profile $(\sigma_{Amy}, \sigma_{Bob})$ leads to the node x_3 with probability σ_L and the node x_2 with probability $1 - \sigma_L$, where $\sigma_L \in [0, 1]$. So, the node x_5 cannot be reached under any mixed strategy profile. Thus, we have established that there is no mixed strategy profile that is equivalent to a behavior strategy profile that leads to x_5 with some positive probability; i.e., a profile $(b_{Amy}, b_{Bob}) = (b_L(L) + (1 - b_L)(R), b_A(A) + (1 - b_A)(B))$ where $b_L, b_A \in (0, 1)$.

Now, consider the extensive form game illustrated in Fig. 9.9. This game violates perfect recall since it involves a player, Amy, who forgets during the course of play a previous action she played. Note that an information set of Amy, involving the nodes x_4 and x_5, reveals that she does not recall her previous move at the node x_0, i.e., whether she chose there the action A or B.

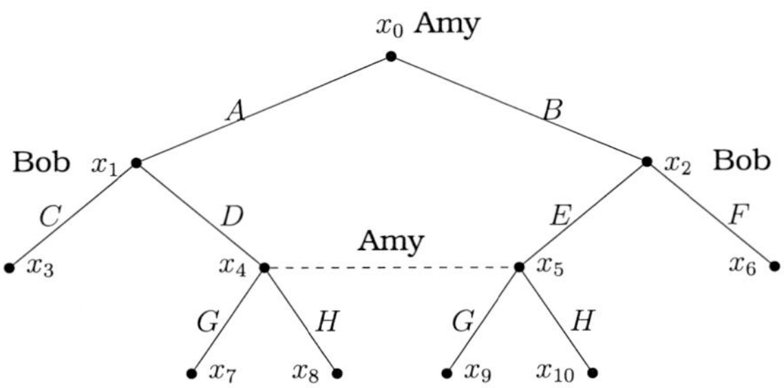

Fig. 9.9 A player (Amy) does not recall a previous action she played

In the strategic form representation of the above game, the pure strategies of Amy are AG, AH, BG, and BH, whereas the pure strategies of Bob are CE, CF, DE, and DF. So, consider the mixed strategy profile $\sigma = (\sigma_{Amy}, \sigma_{Bob})$ where

$$\sigma_{Amy} = \left(\frac{1}{2}(AG), 0(AH), 0(BG), \frac{1}{2}(BH)\right)$$

and

$$\sigma_{Bob} = \left(\frac{1}{4}(CE), \frac{1}{4}(CF), \frac{1}{4}(DE), \frac{1}{4}(DF)\right).$$

This mixed strategy profile implies that the nodes x_8 and x_9 are both reached with probability zero and the nodes x_7 and x_{10} are both reached with probability $1/4$. Note that any behavior strategy profile for the given extensive form game can be written as $b = (b_{Amy}, b_{Bob})$ where

$$b_{Amy}(\{x_0\}) = (b_A, 1 - b_A) \text{ and } b_{Amy}(\{x_4, x_5\}) = (b_G, 1 - b_G)$$

and

$$b_{Bob}(\{x_1\}) = (b_C, 1 - b_C) \text{ and } b_{Bob}(\{x_2\}) = (b_E, 1 - b_E)$$

with $b_A, b_G, b_C, b_E \in [0, 1]$. Now, suppose that there exists a behavior strategy profile b that is equivalent to the mixed strategy profile σ given above. Then, we must have $b_A > 0$ and $1 - b_C > 0$ since $p(x_7|\sigma) = 1/4 > 0$. This implies that $b_G = 1$ since $p(x_8|\sigma) = 0$. This would further imply that $p(x_{10}|b) = 0$. On the other hand, we know that $p(x_{10}|\sigma) = 1/4 > 0$. Thus, we conclude that no behavior strategy b can be equivalent to the selected mixed strategy profile σ.

Finally, consider the extensive form game illustrated in Fig. 9.10. This game violates perfect recall since it involves a player, Bob, who forgets during the course of play a previous action that his opponent, Amy, played.

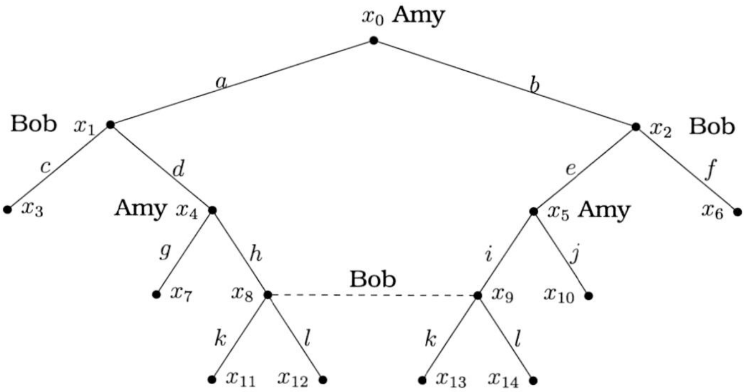

Fig. 9.10 A player (Bob) does not recall a previous action played by the opponent (Amy)

In the strategic form representation of the above game, the pure strategies of Amy are $agi, agj, ahi, ahj, bgi, bgj, bhi, bhj$, whereas the pure strategies of Bob are $cek, cel, cfk, cfl, dek, del, dfk, dfl$. So, consider the mixed strategy profile $\sigma = (\sigma_{Amy}, \sigma_{Bob})$ where

$$\sigma_{Amy} = 0(agi) + 0(agj) + 0(ahi) + \frac{1}{2}(ahj) + \frac{1}{2}(bgi) + 0(bgj) + 0(bhi) + 0(bhj)$$

and

$$\sigma_{Bob} = 0(cek) + \frac{1}{2}(cel) + 0(cfk) + 0(cfl) + 0(dek) + 0(del) + \frac{1}{2}(dfk) + 0(dfl).$$

This mixed strategy profile implies that the nodes x_{12} and x_{13} are both reached with probability zero and the nodes x_{11} and x_{14} are both reached with probability 1/4. Note that any behavior strategy profile for the given extensive form game can be written as $b = (b_{Amy}, b_{Bob})$ where

$$b_{Amy}(\{x_0\}) = (b_a, 1 - b_a),\ b_{Amy}(\{x_4\}) = (b_g, 1 - b_g),\ \text{and}\ b_{Amy}(\{x_5\}) = (b_i, 1 - b_i)$$

and

$$b_{Bob}(\{x_1\}) = (b_c, 1 - b_c),\ b_{Bob}(\{x_2\}) = (b_e, 1 - b_e),\ \text{and}\ b_{Bob}(\{x_8, x_9\}) = (b_k, 1 - b_k)$$

with $b_a, b_g, b_i, b_c, b_e, b_k \in [0, 1]$. Now, suppose that there exists a behavior strategy profile b that is equivalent to the mixed strategy profile σ given above. Then, we must have $b_a > 0$, $1 - b_c > 0$, $1 - b_g > 0$, and $b_k > 0$ since $p(x_{11}|\sigma) = 1/4 > 0$. This implies that $b_k = 1$ since $p(x_{12}|\sigma) = 0$. This would further imply that $p(x_{14}|b) = 0$. On the other hand, we know that $p(x_{14}|\sigma) = 1/4 > 0$. Thus, we conclude that no behavior strategy b can be equivalent to the selected mixed strategy profile σ. ■

In Example 9.5, we have seen three possibilities in which an extensive form game may violate perfect recall: a player may forget whether or not she earlier played or she may forget an action she earlier played or she may forget an action played by some other player (including the Nature) earlier. In case the first possibility holds in an extensive form game, there is a behavior strategy to which there exists no equivalent mixed strategy, whereas in case the last two possibilities hold, the extensive form game involves a mixed strategy to which there exists no equivalent behavior strategy.

For finite extensive games that satisfy perfect recall, we have the following result.

Proposition 9.2 [1,2] *In any finite extensive form game, if a player has perfect recall, then for every mixed strategy of this player there exists an equivalent behavior strategy.*

The above result implies that in any finite extensive form game where all players have perfect recall, players can ignore mixed strategies and play only behavior strategies without loss of generality. The benefit of this ignorance is a potentially huge computational simplicity. To give an example, suppose that an extensive form game involves a player with a total of eight information sets, $H_1, H_2, \ldots, H_{10}$, where for each $k = 1, 2, \ldots, 10$, the set H_k contains a single decision node x_k. This player has $2^{10} = 1024$ pure strategies in the strategic form representation. Let $\sigma = (\sigma_1, \sigma_2, \ldots, \sigma_{1024})$ be a mixed strategy of this player defined over the set of her pure strategies. Note that $\sum_{j=1}^{1024} \sigma_j = 1$. Thus, in order to cover all possible mixed strategies, the player has to consider 1023 independent parameters, say $\sigma_1, \ldots, \sigma_{1023}$, each of which can vary in [0, 1] as long as their sum is equal to 1. On the other hand, denoting by b the behavior strategy of the same player, we should see that for each $k = 1, 2, \ldots, 10$ $b(H_k) = (b_k^1, b_k^2)$ where $b_k^2 = 1 - b_k^1$. So, in order to cover all possible behavior strategies, our player has to consider only 10 independent parameters, say $b_1^1, b_2^1, \ldots, b_{10}^1$, each of which can vary in [0, 1]. If this player has perfect recall, using a behavior strategy defined by 10 independent parameters, she will be able to represent any mixed strategy defined by 1023 independent parameters.

Recall that Nash's [3] existence result (Proposition 4.1) ensures that the mixed extension of every finite strategic form game has a mixed strategy Nash equilibrium. Given this existing result, Kuhn's result on the equivalence of mixed and behavior strategies immediately implies that in finite extensive form games with perfect recall there exists a Nash equilibrium profile in behavior strategies.

References

1. Kuhn HW (1950) Extensive games. Proceed Nat Acad Sci 36(10):570–576
2. Kuhn HW (1953) Extensive games and the problem of information. In: Kuhn HW, Tucker AW (eds) Contributions to the theory of games, vol II. Princeton University Press. Princeton, NJ, pp 193–216
3. Nash JF (1950) Equilibrium points in n-person games. PNAS 36(1):48–49

Subgame Perfect Nash Equilibrium 10

This chapter introduces the subgame perfect Nash equilibrium as a refinement of Nash equilibrium in extensive form games. This equilibrium can be obtained by backward induction in extensive form games with perfect information and obtained by a generalized process of backward induction, called subgame perfection, in extensive form games with imperfect information.

From the previous chapter, we know that every finite extensive form game with perfect recall has a Nash equilibrium in behavior strategies (and in mixed strategies). In this chapter, we will show that we can eliminate some of the Nash equilibria in some extensive form games once we assume that the players are sequentially rational in the sense that they plan to behave "rationally" (playing their actions optimally) at every possible information set they may encounter during the game. Let us proceed with the following example to illustrate the existence of Nash equilibria with sequentially irrational plays.

Example 10.1 Consider Fig. 10.1, illustrating an extensive form game with perfect information. At each terminal node, the first payoff belongs to Amy and the second payoff belongs to Bob. This game has a strategic form representation given by Table 10.1.

The pure-strategy Nash equilibria of the above game are (X, BA), (X, BB), (Y, AA), and (Y, AB), implied by the colored best-response payoffs. Notice that not all of these equilibrium profiles are sequentially rational. In Fig. 10.1, rationality requires Bob to choose the action A at the node x_1 yielding the payoff profile $(8, 4)$ and to choose the action B at the node x_2 yielding the payoff profile $(3, 4)$. Therefore, Bob can be said to be playing rationally only if he plays the pure strategy AB as he would do under the Nash equilibrium profile (Y, AB).

I. Saglam, *Mastering Game Theory*, Classroom Companion: Economics,
https://doi.org/10.1007/978-3-031-82094-6_10

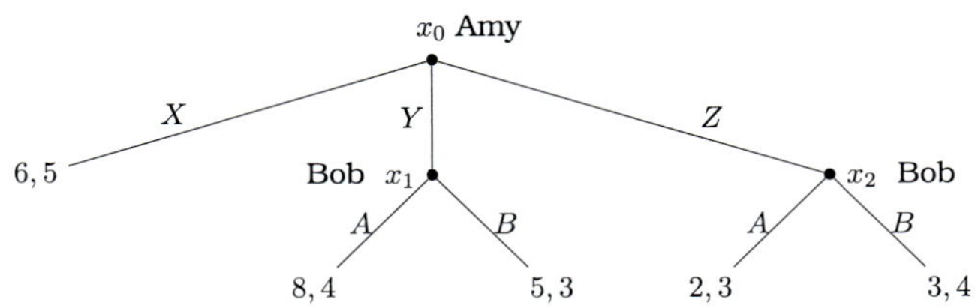

Fig. 10.1 An extensive form game with perfect information

Table 10.1 The strategic form representation of the extensive form game in Fig. 10.1

Bob

Amy	AA	AB	BA	BB
X	(6, 5)	(6, 5)	(6, 5)	(6, 5)
Y	(8, 4)	(8, 4)	(5, 3)	(5, 3)
Z	(2, 3)	(3, 4)	(2, 3)	(3, 4)

We will later check whether Amy is too playing rationally at the equilibrium profile (Y, AB). However, we can immediately say that the other three Nash equilibrium profiles violate sequential rationality since, for each of them, we can find at least one player (Bob) who is playing irrationally at a node (the node x_1 and/or the node x_2). ■

10.1 Backward Induction

The dynamic structure in an extensive form game may contain important details about what may have happened before each player chooses some action at an information set (below the root of the game tree). Once we transform an extensive form game to a strategic form game, we lose such details (if any) about the dynamic structure. For instance, for the extensive form game in Example 10.1 above, we used the strategic form representation to show that there exist four pure strategy Nash equilibria. But, the strategic form representation could not tell us anything about whether any of these equilibria would arise when the players in the game are sequentially rational. We are able to say this only by analyzing the players' moves in the extensive form game sequentially. We assumed that Bob should optimally play at each one of his information sets, $\{x_1\}$ and $\{x_2\}$, irrespective of whether any or both of this information sets would be reached after Amy plays at the root node x_0. If the informational structure of the game allows Amy to calculate the optimal actions of Bob at his information sets, then Amy can make these calculations before she makes her move

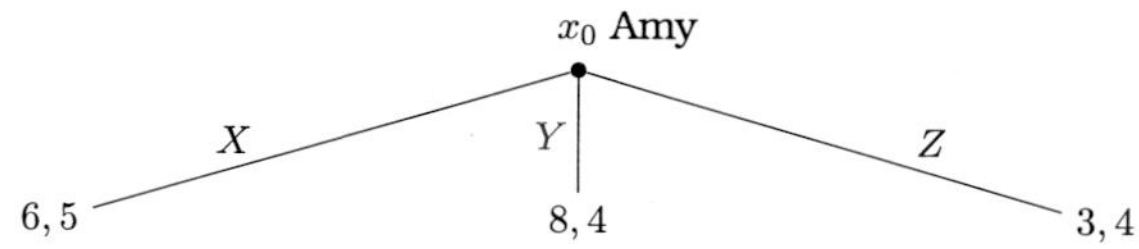

Fig. 10.2 The reduced game obtained from Fig. 10.1 using backward induction

to predict the outcomes that her possible actions may lead to. Using her predictions, Amy can then choose her optimal action at the root node. This solution process is known as the backward induction. Below, we apply this process to calculate the equilibrium of the extensive form game in Example 10.1.

Example 10.2 Reconsider the extensive form game in Example 10.1. Suppose that each player is rational and knows his/her own payoff, and moreover, Amy can make backward induction and has complete information about the rationality and payoffs of Bob. Then, Amy can predict that Bob will choose the action A at x_1 yielding the payoff profile $(8, 4)$ and that Bob will choose the action B at x_2 yielding the payoff profile $(3, 4)$. Using backward induction Amy can replace the subtrees starting at x_1 and x_2 with these payoff profiles.

In the reduced game portrayed by Fig. 10.2, the optimal strategy of Amy is to choose Y. This strategy would induce Bob to play at x_1 (on the equilibrium path of moves), optimally choosing A and leading to the equilibrium profile of payoffs $(8, 4)$. We can write the resulting equilibrium profile of strategies as

$$\sigma^* = ((\text{Amy plays } Y \text{ at } x_0), (\text{Bob plays } A \text{ at } x_1 \text{ and plays } B \text{ at } x_2))$$

or simply as $\sigma^* = (Y, AB)$. Note that σ^* is one of the four pure strategy Nash equilibria of the extensive form game in Example 10.1. Moreover, σ^* is the only pure strategy Nash equilibrium that satisfies sequential rationality. ■

Aumann [1] showed that common knowledge of rationality among players implies backward induction in games with perfect information. Conversely, if the players in such games do not play their equilibrium strategies obtained through backward induction, then the fact that these players are rational cannot be common knowledge. Below, we present another example to illustrate backward induction.

Example 10.3 Consider in Fig. 10.3 a sequential version of the Matching Pennies Game, where the first payoff at each terminal node belongs to Amy and the second payoff belongs to Bob. Suppose that both players are rational and also that their rationality and payoffs are both common knowledge.

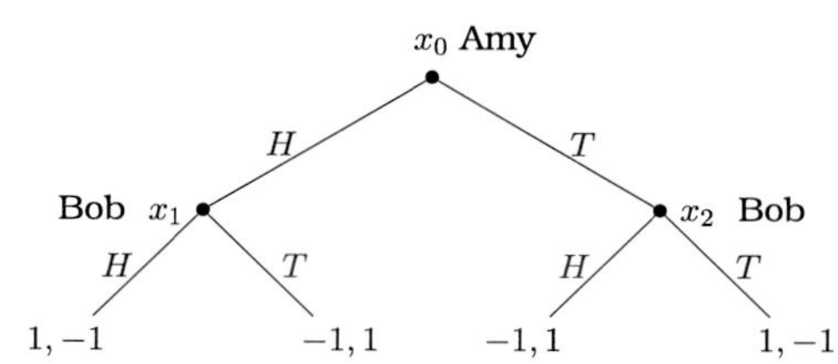

Fig. 10.3 A sequential matching pennies game

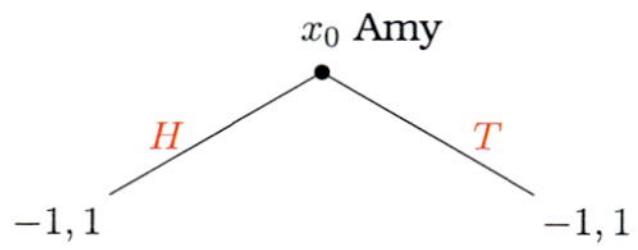

Fig. 10.4 The reduced game obtained from Fig. 10.3 using backward induction

Amy plays at the root node x_0 knowing that Bob will play at the node x_1 if Amy chooses H and Bob will play at x_2 if Amy chooses T. Rationality requires Bob to choose the action T at x_1 yielding the payoff profile $(-1, 1)$ and to choose the action H at x_2 yielding the payoff profile $(-1, 1)$. (We have highlighted these two actions at the nodes x_1 and x_2 by the blue color.) So, using backward induction Amy can replace the subtrees starting at x_1 and x_2 with these payoff profiles.

In the reduced game portrayed by Fig. 10.4, Amy is indifferent between choosing the action H or choosing the action T. Thus, there are two pure strategy profiles each of which can arise as an equilibrium using backward induction:

$$\sigma_1^* = ((\text{Amy plays } H \text{ at } x_0), (\text{Bob plays } T \text{ at } x_1 \text{ and plays } H \text{ at } x_2)),$$

$$\sigma_2^* = ((\text{Amy plays } T \text{ at } x_0), (\text{Bob plays } T \text{ at } x_1 \text{ and plays } H \text{ at } x_2)).$$

We can simply write these two equilibria as $\sigma_1^* = (H, TH)$ and $\sigma_1^* = (T, TH)$. Note that Bob, being the last player and making backward induction, should plan an optimal action at each information set where he could possibly be playing, irrespective of whether an information set would be reached after Amy makes her move. Therefore, Bob's equilibrium strategy is independent of Amy's equilibrium strategy. On the other hand, Amy, being the first player and making backward induction, should decide on her strategy only by predicting the moves of Bob at the nodes x_1 and x_2 and thereby reducing the game by backward induction. Note also that the equilibrium path of play is (H, T) under σ_1^* and (T, H) under σ_2^*, both leading to the payoff profile $(-1, 1)$. In the sequential version of the Matching Pennies Game where Amy is the leader (the first mover), Bob turns out to benefit from being the follower.

Now, let us calculate the set of all Nash equilibria of the Sequential Matching Pennies Game described above. Since any equilibrium obtained by backward induction is a Nash equilibrium, both σ_1^* and σ_2^* are a Nash equilibrium. To check whether the extensive form game we study has a Nash equilibrium other than these two equilibria, we will consider the associated strategic form representation in Table 10.2.
The pure-strategy Nash equilibria of the above game is (H, TH) and (T, TH), implied by the colored best-response payoffs of Amy and Bob. Clearly, these Nash equilibria, respectively, correspond to the equilibria σ_1^* and σ_2^*, obtained applying backward induction on the extensive form game that we study. We have seen that applying backward induction in the sequential matching pennies game did not eliminate any Nash equilibrium of the corresponding strategic form representation.

We should finally note that the set of equilibria obtained by applying backward induction on the Sequential Matching Pennies Game is sensitive to who starts the game. Suppose that the extensive form game in Fig. 10.3 is changed in such a way

Table 10.2 The strategic form representation of the extensive form game in Fig. 10.3

Amy \ Bob	HH	HT	TH	TT
H	$(1, -1)$	$(1, -1)$	$(-1, 1)$	$(-1, 1)$
T	$(-1, 1)$	$(1, -1)$	$(-1, 1)$	$(1, -1)$

that Bob is playing at the root node x_0, followed by Amy playing at the nodes x_1 and x_2 (with the first payoff at each terminal node still belonging to Amy). Then, the pure strategy equilibria obtained by backward induction would be

$$\sigma_1^* = ((\text{Bob plays } H \text{ at } x_0), (\text{Amy plays } H \text{ at } x_1 \text{ and plays } T \text{ at } x_2)),$$

$$\sigma_2^* = ((\text{Bob plays } T \text{ at } x_0), (\text{Amy plays } H \text{ at } x_1 \text{ and plays } T \text{ at } x_2)).$$

Both equilibria would lead to the payoff profile $(1, -1)$, at which Amy would enjoy the benefit of being the follower. ■

10.2 Subgame Perfect Nash Equilibrium

The extensive form games in Examples 10.1–10.3 are games with perfect information; these games do not involve any simultaneous play. Thus, when we applied backward induction on these games, the reduced game at each (singleton) information set was a one-player game (problem) of deciding between a number of actions with respect to their implied payoffs (under the belief that the rest of the game will be played as predicted). However, backward induction is of no direct use if the extensive form game is not a perfect information game. Consider, for example, the game

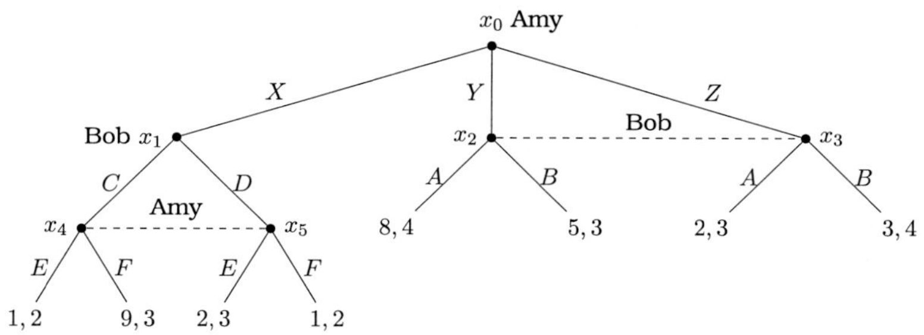

Fig. 10.5 An extensive form game with imperfect information

portrayed by Fig. 10.5, where at each terminal node the first payoff belongs to Amy and the second payoff belongs to Bob.

Now, we cannot apply the idea of backward induction in the way it was used to solve the games in the earlier examples. This is because the game in Fig. 10.5 is not a game with perfect information. The game involves at least one information set that contains more than one node. Note that the information sets of Amy are $\{x_0\}$ and $\{x_4, x_5\}$ and the information sets of Bob are $\{x_1\}$ and $\{x_2, x_3\}$. Both Amy and Bob have a non-singleton information set as well as a singleton information set. Consider, for example, Amy at the information set $\{x_4, x_5\}$. She does not know whether she is at x_4 or x_5. Moreover, neither E dominates F nor F dominates E with respect to the payoffs of Amy. Thus, the optimal action of Amy at the information set $\{x_4, x_5\}$ cannot be calculated without taking into consideration Bob's strategy at the information set $\{x_1\}$. By this example, it should be clear that to solve an extensive form game with imperfect information under the spirit of backward induction requires the players to deal with simultaneously-played smaller games that exist inside (or can be inductively derived from) the original game. This brings us to the wisdom of Reinhard Selten [2] who proposed the concept of subgame perfection that generalizes backward induction.

Selten [2] defines a subgame as a subtree, i.e., some part of the tree representing a given extensive form game Γ. Formally, given a decision node $x \in \mathcal{X} \setminus \mathcal{X}^T$, a subtree $\Gamma[x]$ of Γ is called a subgame if it satisfies the following conditions:

(i) x is the root of $\Gamma[x]$,

(ii) x is the unique node in the information set it belongs to,

(iii) If Γ contains a path from x to a terminal node $y \in \mathcal{X}^T$, then this path, involving all the nodes and branches (actions) on it, is also contained by $\Gamma[x]$,

(iv) For any decision node $y \in \mathcal{X} \setminus \mathcal{X}^T$ of the game Γ with $y \in \Gamma[x]$, any $i \in N$, and any information set $H \in \mathcal{H}$ in the game Γ, it is true that H is contained by $\Gamma[x]$ and assigned to player i in $\Gamma[x]$ if and only if H is assigned to player i in Γ.

According to the above definition, a subgame of an extensive form game is a subtree following a singleton information set, provided that the subtree does not break any information sets. Note that any extensive form game Γ has at least one subgame since the subtree $\Gamma[x_0] = \Gamma$ is a subgame according to the above conditions. The extensive form game Γ in Fig. 10.5 has two subgames $\Gamma[x_0] = \Gamma$ and $\Gamma[x_1]$, portrayed by Fig. 10.6.

Definition 10.1 Given an extensive form game Γ, a strategy profile σ is called a subgame perfect Nash equilibrium (SPNE) if it induces a Nash equilibrium in every subgame of Γ.

Since every extensive form game is a subgame of itself, the above definition implies that every SPNE is a Nash equilibrium. Therefore, to characterize the set of SPNE in an extensive form game, one can first calculate the set of Nash equilibria

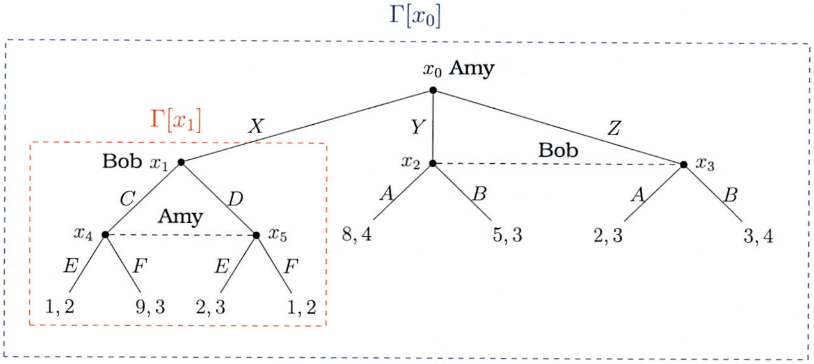

Fig. 10.6 Subgames in an extensive form game

and then check whether any of them is a SPNE applying Definition 10.1 on each Nash equilibrium separately. However, there is another way to find the set of SPNE. Before we present it, we say that an extensive form game Γ is minimal if the only subgame of Γ is itself.

A Procedure for Computing a SPNE:

[1] Given a finite extensive form game Γ, set $\Gamma^0 = \Gamma$.

[2] Run step-k described below iteratively for the values of $k = 1, 2, \ldots, K$ with some integer $K \geq 1$ to obtain a sequence of games $(\Gamma^1, \ldots, \Gamma^K)$ such that the game Γ^{K-1} is minimal.

Step-k:

- Identify in the game Γ^{k-1} all subgames that are minimal. Let M^{k-1} denote the set of these subgames.

- For each $M \in M^{k-1}$, calculate all Nash equilibrium strategy profiles. Let $NE^{k-1}(M)$ denote the set of these profiles.

- For each $M \in M^{k-1}$, pick a Nash equilibrium strategy profile $\sigma^{k-1}(M) \in NE^{k-1}(M)$.

- If Γ^{k-1} is not minimal, then replace each minimal subgame $M \in M^{k-1}$ with the payoff profile $u(\sigma^{k-1}(M))$ and call the resulting game as Γ^k. If Γ^{k-1} is minimal, then set $\Gamma^k = \emptyset$.

[3] For each player $i \in N$, let σ_i^* be the collection of the actions at the possible information sets he/she may play in step-k of the procedure when k is run from 1 to K. Then, $\sigma^* = (\sigma_1^*, \ldots, \sigma_n^*)$ is a profile of SPNE strategies.

We should note that using the iterative procedure described above we can find a SPNE of an extensive form game Γ after a finite number of steps, K, if Γ is a finite game. Also, since, at each step of the above procedure, there can be more than one Nash equilibrium strategy profile corresponding to each minimal subgame, by repeating the procedure for each possible Nash equilibrium strategy profile of each

Table 10.3 The strategic form game corresponding to the subgame $\Gamma[x_1]$ in Fig. 10.6

Amy \ Bob	C	D
E	(1, 2)	(2, 3)
F	(9, 3)	(1, 2)

possible subgame we can calculate all SPNE of a given extensive form game. The following example will illustrate how we can use the procedure described above.

Example 10.4 Reconsider the extensive form game Γ portrayed by Fig. 10.5. Let us apply the procedure described above on Γ.

[1] We set $\Gamma^0 = \Gamma$.

[2] We first run step-k for $k = 1$.

Step-1:

• We identify that Γ^0 has two subgames $\Gamma^0[x_0]$ and $\Gamma^0[x_1]$, as illustrated by Fig. 10.6. However, only $\Gamma^0[x_1]$ is minimal.

• The subgame $\Gamma^0[x_1]$ is represented by the strategic form game in Table 10.3. One can easily check that $\Gamma^0[x_1]$ has three Nash equilibria $\sigma_1 = (E, D)$, $\sigma_2 = (F, C)$, and $\sigma_3 = (\frac{1}{2}(E) + \frac{1}{2}(F), \frac{1}{9}(C) + \frac{8}{9}(D))$. The expected payoff profile, denoted by (u_1, u_2), at these three equilibria are, respectively, (2, 3), (9, 3), and (17/9, 5/2).

• If we replace the subgame $\Gamma^0[x_1]$ by the expected payoff profile (u_1, u_2) corresponding to any possible Nash equilibrium that we calculated above, the resulting extensive form game will be Γ^1, illustrated by Fig. 10.7.
Recall that we have to run step-k a total of $K \geq 1$ times to ensure that Γ^{K-1} is minimal. Since we know that Γ^0 is not minimal, K cannot be 1. Thus, we have to consider step-2 and check whether K is 2.

Step-2:

• We identify that Γ^1 has a unique subgame $\Gamma^1[x_0] = \Gamma^1$. Thus, $\Gamma^1[x_0]$ is minimal.

• Let us find the set of Nash equilibria in $\Gamma^1[x_0]$. The strategic form representation of this subgame is as follows:

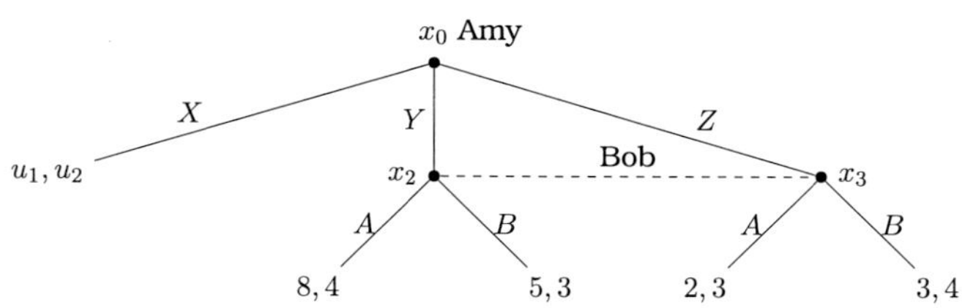

Fig. 10.7 The reduced game obtained from Fig. 10.6 using the computational procedure for SPNE

Table 10.4 The strategic form game corresponding to the extensive form game in Fig. 10.7

		Bob A	Bob B
Amy	X	(u_1, u_2)	(u_1, u_2)
	Y	(8, 4)	(5, 3)
	Z	(2, 3)	(3, 4)

Recall that the profile (u_1, u_2) depends on the Nash equilibrium we select for $\Gamma^0[x_1]$ in step-1. Depending on this profile, we will calculate the Nash equilibria of $\Gamma^1[x_0]$ and pick one of the equilibria as part of the SPNE profile. Since Γ^1 is minimal, we must then set $\Gamma^2 = \emptyset$ and complete the procedure at the end of step-2.

Let us now complete the missing part in step-2, the task of calculating the Nash equilibria of $\Gamma^1[x_0]$. We have to consider three cases:

Case 1: $(u_1, u_2) = (2, 3)$.

This profile arises if we pick $\sigma_1 = (E, D)$ in $\Gamma^0[x_1]$. In this case, the action Y becomes the dominant strategy for Amy in the game represented by Table 10.4, and Bob's best response to Y becomes A. So, the unique Nash equilibrium of the game $\Gamma^1[x_0]$ is (Y, A). Thus, we have found our first subgame perfect Nash equilibrium.

$$\begin{aligned} SPNE-1 \;=\; & (\text{Amy plays } Y \text{ at } x_0 \text{ and plays } E \text{ at both } x_4 \text{ and } x_5, \\ & \text{Bob plays } D \text{ at } x_1 \text{ and plays } A \text{ at both } x_2 \text{ and } x_3.) \end{aligned}$$

Case 2: $(u_1, u_2) = (9, 3)$.

This profile arises if we pick $\sigma_2 = (F, C)$ in $\Gamma^0[x_1]$. In this case, the action X becomes the dominant strategy for Amy in the game $\Gamma^1[x_0]$, and Bob's best response to X becomes to play A with any probability in [0, 1]. Thus, we have found our second subgame perfect Nash equilibrium.

$$\begin{aligned} SPNE-2 \;=\; & (\text{Amy plays } X \text{ at } x_0 \text{ and plays } F \text{ at both } x_4 \text{ and } x_5, \\ & \text{Bob plays } C \text{ at } x_1 \text{ and plays } \sigma_A(A) + (1-\sigma_A)(B) \\ & \text{for any } \sigma_A \in [0, 1] \text{ at both } x_2 \text{ and } x_3.) \end{aligned}$$

Case 3: $(u_1, u_2) = (17/9, 5/2)$.

This profile arises if we pick $\sigma_3 = (\frac{1}{2}(E) + \frac{1}{2}(F), \frac{1}{9}(C) + \frac{8}{9}(D))$ in $\Gamma^0[x_1]$. In this case, the action Y becomes the dominant strategy for Amy in the game $\Gamma^1[x_0]$, and Bob's best response to Y becomes to play A. Thus, we have found our third subgame perfect Nash equilibrium.

$$\begin{aligned} SPNE-3 \;=\; & (\text{Amy plays } Y \text{ at } x_0 \text{ and plays } \frac{1}{2}(E) + \frac{1}{2}(F) \text{ at both } x_4 \text{ and } x_5, \\ & \text{Bob plays } \frac{1}{9}(C) + \frac{8}{9}(D) \text{ at } x_1 \text{ and plays } A \text{ at both } x_2 \text{ and } x_3.) \end{aligned}$$

Table 10.5 The strategic form game corresponding to the extensive form game in Fig. 10.5

		Bob			
		CA	CB	DA	DB
Amy	XE	(1, 2)	(1, 2)	(2, 3)	(2, 3)
	XF	(9, 3)	(9, 3)	(1, 2)	(1, 2)
	YE	(8, 4)	(5, 3)	(8, 4)	(5, 3)
	YF	(8, 4)	(5, 3)	(8, 4)	(5, 3)
	ZE	(2, 3)	(3, 4)	(2, 3)	(3, 4)
	ZF	(2, 3)	(3, 4)	(2, 3)	(3, 4)

Thus, we have calculated all possible SPNEs of the extensive form game Γ in Fig. 10.5. Let us now calculate all pure-strategy Nash equilibria of Γ and see which of them are eliminated by subgame perfection. Note from Fig. 10.5 that the possible strategies of Amy are XE, XF, YE, YF, ZE, and ZF, whereas the possible strategies of Bob are CA, CB, DA, and DB. The strategic form representation of Γ is prsented in Table 10.5.

The best-response payoffs colored in Table 10.5 show that there are four pure-strategy Nash equilibria, (XF, CA), (XF, CB), (YE, DA), and (YF, DA). Of these four equilibria, (YE, DA) is SPNE-1, (XF, CA) is a member of SPNE-2 when $\sigma_A = 1$, and (XF, CB) is a member of SPNE-2 when $\sigma_A = 0$. We should observe that the Nash equilibrium (YF, DA) of the extensive form game Γ is eliminated by subgame perfection. ∎

Example 10.4 shows that a Nash equilibrium of an extensive form game does not have to be a SPNE (whereas each SPNE of any extensive form game must be a Nash equilibrium, as we argued earlier). The computational procedure we described earlier allows us to present the following existence result.

Proposition 10.1 *Any finite extensive form game has a SPNE.*

Proof Consider the procedure described earlier for the computation of a SPNE of any extensive form game. Recall that at each step of the procedure, the game is reduced in size (the number of nodes and branches) since each minimal subgame is replaced by a Nash equilibrium payoff profile. (Note that the finiteness of the game ensures that we can always identify the minimal subgames and each minimal subgame is also finite.) So, after a finite number of steps, the reduced game will be a minimal game (with the unique subgame being the reduced game itself) and the procedure will terminate. We should also note that Nash's [3] existence theorem ensures that there exists a Nash equilibrium for each (finite) minimal subgame considered at each step of the

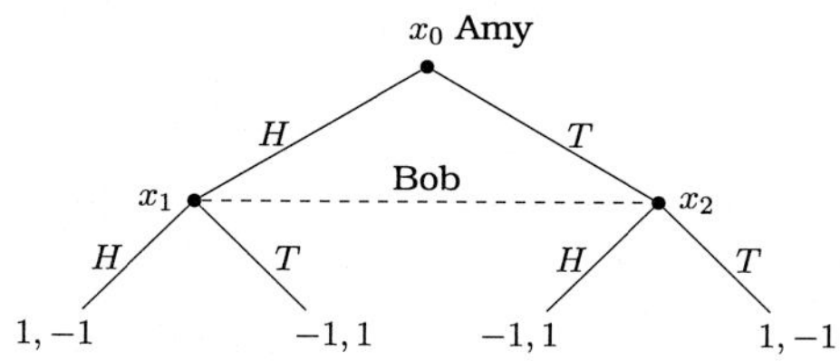

Fig. 10.8 Simultaneous matching pennies game in extensive form

computational procedure and a Nash equilibrium is trivially a SPNE for any subgame that is minimal. Thus, combining all Nash equilibrium profiles calculated for minimal subgames during the finite number of steps of the computational procedure will always yield a SPNE. ■

Every finite extensive form game has a SPNE, but not necessarily in pure strategies. The following example contains an extensive form game with a unique SPNE, which is found to be in unpure strategies.

Example 10.5 Consider the Matching Pennies Game in Fig. 10.8 represented in extensive form, where the first payoff at each terminal node belongs to Amy.

This game, say Γ, has a unique subgame, which is Γ itself. Therefore, the Nash equilibrium of Γ is a SPNE. We know that the strategic form representation of Γ has the unique Nash equilibrium $(\frac{1}{2}(H) + \frac{1}{2}(T), \frac{1}{2}(H) + \frac{1}{2}(T))$, which is unpure. Therefore, Γ has a unique SPNE, which is unpure. ■

The existence of a SPNE in pure strategies is ensured in games with perfect information, as we state in the next result.

Proposition 10.2 *Any finite extensive form game with perfect information has a SPNE in pure strategies. For any such game, the set of SPNE coincides with the set of equilibria obtained by backward induction. Moreover, there is a unique SPNE if no player has the same payoffs at any two terminal nodes.*

Proof The proof is straightforward. Let Γ^0 be a finite extensive form game with perfect information. By Proposition 10.1, the set of SPNE of Γ^0 is nonempty since Γ^0 is finite. Also, the assumption that Γ^0 is a game with perfect information ensures that every decision node of Γ^0 is the root of a subgame. So, consider the last decision nodes in Γ^0. (We can identify the last decision nodes since Γ^0 is finite.) Any SPNE of Γ^0 must specify optimal actions for the players who play at these nodes. Replacing the subgames starting at these nodes with the corresponding payoffs induced by the optimal actions, we obtain a reduced game, say Γ^1. If Γ^0 is minimal, then Γ^1 has no decision nodes, and we say $\Gamma^1 = \emptyset$. Otherwise, Γ^1 has at least one decision node. Then, we can repeat the above arguments made for Γ^0 for Γ^1 and obtain—using this induction going backward—a sequence of reduced games $\Gamma^2, \ldots, \Gamma^K$, where

$K \geq 1$, Γ^{K-1} is minimal, and $\Gamma^K = \emptyset$. In consequence, we conclude that any SPNE of Γ^0 must specify optimal actions for each player at each decision node she can play. This implies that (i) the set of SPNE of Γ^0, which we know to be nonempty, coincides with the set of equilibria obtained by backward induction, and (ii) the existence of a pure strategy equilibrium obtained by backward induction is always warranted. Clearly, if there is no player who has the same payoffs at any two terminal nodes, then backward induction cannot yield an equilibrium in unpure strategies, where a player mixes at least two actions with positive probabilities in one of the last decision nodes of one of the games in the sequence $(\Gamma^0, \Gamma^1, \ldots, \Gamma^K)$. ■

We should note that we appeal to the finiteness of an extensive form game twice in the proof of the above result. First, we use the finiteness assumption for the existence of SPNE of an extensive form game, as we earlier saw in the proof of Proposition 10.1. Second, we use the finiteness assumption to go the last decision nodes in the original game and in the sequence of reduced games while we calculate the equilibria through backward induction. However, we should also note that the finiteness of an extensive form game is only sufficient for Propositions 10.1 and 10.2, it is not necessary. Below, we present two examples to show that a SPNE or an equilibrium through backward induction can also exist in games that are not finite.

In our first example, the extensive form game contains a finite number of players (two players), but the number of decision nodes of the second player and the number of actions at each decision node of each player is uncountable.

Example 10.6 (*Stackelberg Duopoly Game*) The original version of this game was first considered by Stackelberg [4]. The story in our version is as follows: Two firms in a duopolistic industry compete in quantities. The production cost function of firm $i = 1, 2$ is $C_i(q_i) = cq_i$, where the variable q_i denotes the output of firm i and the constant $c \geq 0$ denotes the common marginal cost of the firms. The firms face an inverse demand function $P(Q) = a - Q$, where Q is the industry output, i.e., $Q = q_1 + q_2$, and $a > c$ is the maximal size of the demand for the good produced by the firms. The profit function of firm $i = 1, 2$, when it produces q_i and its rival produces q_j, is given by

$$\pi_i(q_i, q_j) = P(q_i + q_j)q_i - cq_i.$$

Firms 1 and 2 choose their outputs q_1 and q_2 sequentially (with firm 1 moving first and firm 2 moving second) to maximize their own profits.

For the described game, the set of players is $N = \{1, 2\}$ and the action spaces of players 1 and 2 are $Q_1 = [0, \infty)$ and $Q_2 = [0, \infty)$. For any $q \in Q_1 \times Q_2$, the players' payoffs are $u_1(q) = \pi_1(q_1, q_2)$ and $u_2(q) = \pi_2(q_2, q_1)$. Note that firm 1 plays only at the root while firm 2 can play at uncountably many decision nodes. We can associate each of these nodes with some action q_1 chosen by firm 1. So, pick any $q_1 \in Q_1$. At the decision node associated with q_1, firm 2 has uncountably many actions (as her action space is $Q_2 = [0, \infty)$.)

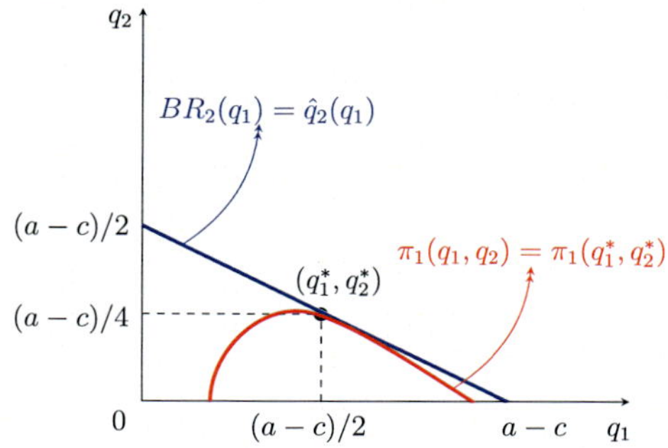

Fig. 10.9 Stackelberg duopoly game where firm 1 leads and firm 2 follows

The best-response correspondence of firm 2 at q_1 is

$$BR_2(q_1) = \{q_2 : \pi_2(q_2, q_1) \geq \pi_2(q_2', q_1) \text{ for all } q_2' \in Q_2\}.$$

In Example 4.10, we calculated this correspondence to be

$$BR_2(q_1) = \left\{q_2 = \frac{a - c - q_1}{2}\right\}.$$

Note that for each q_1, the best-response correspondence $BR_2(q_1)$ is single-valued, i.e., it is a function which we denote by $\hat{q}_2(q_1)$. We draw the best-response correspondence $BR_2(q_1)$ as the blue-colored curve in Fig. 10.9.

Now, using backward induction, firm 1 can first calculate $\hat{q}_2(q_1)$ for each $q_1 \in Q_1$ and then solve its maximization problem given by

$$\max_{q_1 \in Q_1} \pi_1(q_1, \hat{q}_2(q_1)) = \max_{q_1 \in Q_1} (a - q_1 - \hat{q}_2(q_1))q_1 - cq_1.$$

The first-order necessary condition for this problem is

$$a - 2q_1^* - \frac{a - c - q_1^*}{2} + \frac{q_1^*}{2} - c = 0,$$

implying $q_1^* = (a-c)/2$ and $q_2^* = \hat{q}_2(q_1^*) = (a-c)/4$. Notice that the second-order sufficiency condition is satisfied since $d^2\pi_1(q_1, \hat{q}_2(q_1))/d(q_1)^2 = -1 < 0$. So, the profit of firm 1, $\pi(q_1, \hat{q}_2(q_1))$ is maximized at $q_1^* = (a-c)/2$.

Let us call the curve $\pi_1(q_1, q_2) = \bar{\pi}$ the iso-profit curve of firm 1 associated with the profit level $\bar{\pi}$. We should observe that the red curve in Fig. 10.9 is the isoprofit curve of firm 1 associated with the highest attainable level of profit, $\pi_1(q_1^*, q_2^*)$, under the constraint that (q_1^*, q_2^*) must be on the best-response correspondence, $q_2 = BR_2(q_1)$, of firm 2. That is, why the red isoprofit curve is tangent to the blue curve, $BR_2(q_1)$, at the quantity level q_1^*.

To sum up, the SPNE of the described symmetric Stackelberg duopoly is a strategy profile σ^* where $\sigma_1^* = q_1^*$ and $\sigma_2^* = \hat{q}_2(q_1) = (a - c - q_1)/2$ for any $q_1 \in Q_1$. This strategy profile leads to the equilibrium path $q_1 = q_1^*$ and $q_2 = \hat{q}_2(q_1^*) = q_2^*$. ■

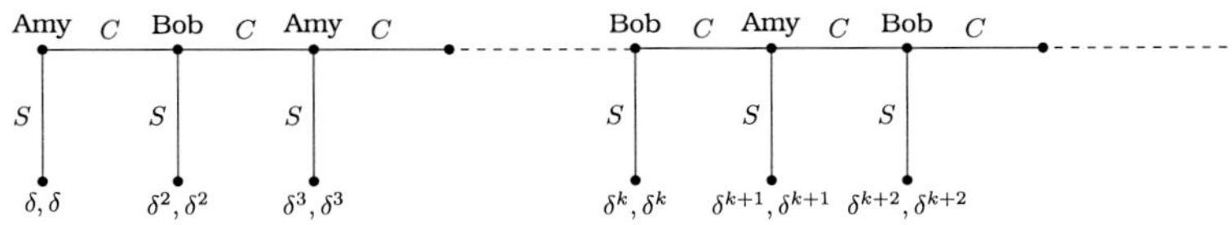

Fig. 10.10 An infinite extensive form game (with infinite number of nodes)

In our second example below, the extensive form game violates the finiteness assumption since the number of decision nodes is infinite.

Example 10.7 Consider the extensive form game in Fig. 10.10 where δ is a real number in (0, 1), and the first and second payoffs at each terminal node belong to Amy and Bob, respectively. The number of players is finite as there exist two players, Amy and Bob. Moreover, each player has only two actions, S (Stop) and C (Continue), at each decision node. However, there are infinitely many decision nodes in the game. Therefore, the game is not finite.

Let us call the described game Γ. We should note that Γ is a perfect information game. However, since the number of decision nodes is not finite, no decision node in Γ is a last decision node. So, for calculating the set of SPNE of Γ, we are not able to use backward induction, i.e., solving the simple optimization problems at the last decision nodes and coming backward till we reach to, and solve, the reduced problem at the root of Γ. Instead of backward induction, we will directly use the definition of SPNE, which requires that a strategy profile σ^* is a SPNE of Γ only if it induces a Nash equilibrium in each subgame of Γ.

So, pick any integer $j \geq 1$ and consider the node j in Γ where the payoff profile becomes (δ^j, δ^j) if the action S is chosen at that node. (Let us call the root of the game where Amy moves as the node 1. Then, Fig. 10.10 implies that the player who plays at the node j is Amy if j is odd and Bob if j is even.) Let us consider the decision problem of the player who moves at the node j. If she/he chooses the action S, she/he will get a payoff of δ^j, whereas if she/he chooses the action C, she/he will get a payoff not higher than δ^{j+1}. Therefore, the player at the node j must play S if she/he is rational, irrespective of what she/he or her/his opponent will play in any of the subsequent nodes. Since j was picked arbitrarily, this conclusion is true for all decision nodes of Γ. Thus, there exists a unique SPNE of Γ and in this SPNE each player's strategy is to play S at each decision node she/he is assigned in the game. The equilibrium path of this SPNE connects the root of Γ (with Amy playing the action S) to the terminal node with the payoff profile (δ, δ). ■

One may ask whether a SPNE must be Pareto dominant or Pareto undominated. An immediate answer is no, once we realize that the game known as the Prisoner's Dilemma can be represented in extensive form, as well. We illustrate our reasoning below.

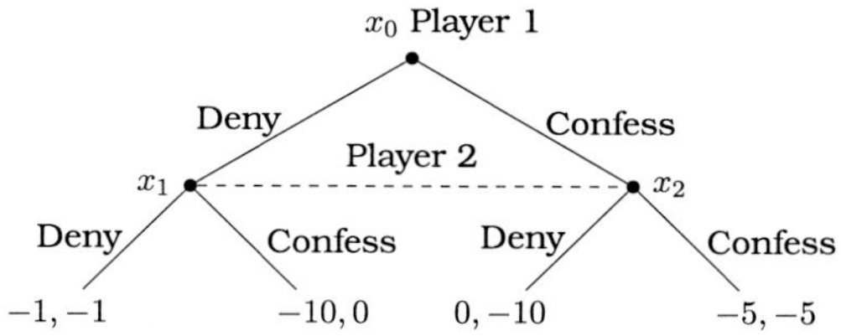

Fig. 10.11 Prisoner's dilemma game in extensive form

Example 10.8 Reconsider Prisoner's Dilemma in Table 1.3, which can be represented by the extensive form game in Fig. 10.11. The first and second payoffs at each terminal node belong to player 1 and player 2, respectively.

The above game, say Γ, has a unique subgame, which is Γ itself. Therefore, the Nash equilibrium of Γ is a SPNE. We know that the strategic form representation of Γ has the unique Nash equilibrium (Confess, Confess). Therefore, Γ has a unique SPNE. Moreover, this SPNE is Pareto dominated by the strategy profile (Deny, Deny). Thus, it is not a Pareto undominated profile of strategies (hence, it is not a Pareto dominant profile, either). ■

The extensive form game considered in Example 10.8 is a minimal game. Hence, it does not allow the application of backward induction. One may then ask whether the result we obtained in Example 10.11 could also be obtained if backward induction were applicable. The answer is yes. There exist extensive form games—with perfect information—where backward induction yields an equilibrium that is Pareto dominated.

Example 10.9 Consider the following extensive form game known as the Centipede Game. (The game has several versions; we use the version in MasCollel et al. [6] with slight modifications.) Two players, say Amy and Bob, each have 1 dollar at the beginning of the game. They alternate playing in a total of 198 decision nodes, $x_0, x_1, \ldots, x_{197}$, with Amy starting the game in the root node x_0. At each node there are two possible actions, S (Stop) and C (Continue). When a player chooses the action C, a referee (with external wealth) takes 1 dollar from this player and gives 2 dollars to her opponent. The game terminates either when the action S is chosen at a decision node or the action C is chosen by Bob in the last decision node x_{197}. The payoffs at the terminal nodes are as shown in Fig. 10.12. (The first and second payoffs at each terminal node belong to Amy and Bob, respectively.)

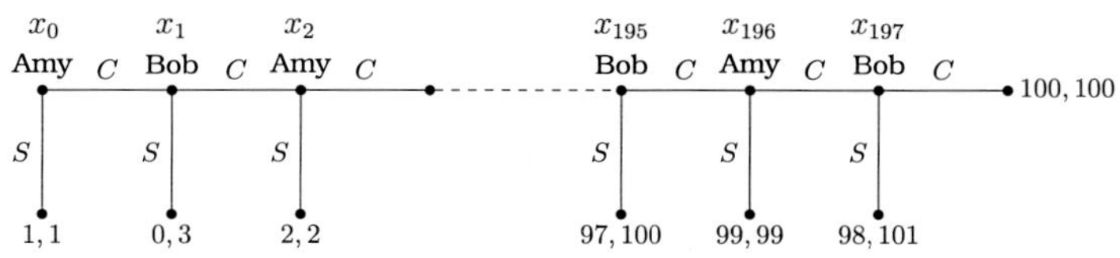

Fig. 10.12 Centipede game

Note that the above game has a finite number of decision nodes. Therefore, we can first go the last decision node, i.e., the node x_{197}. At this node, the optimal action for Bob is S, since it yields 101 dollars while the action C would yield 100 dollars. Thus, we can replace the node x_{197} by the payoff profile (98, 101). We can now go backward and consider the decision of Amy at the node x_{196}. She expects to get 98 dollars if she chooses C and 99 dollars if she chooses S. Thus, the optimal action of Amy at the node x_{196} is S. Going backward in this way, we can show that each player must choose S wherever she is called to play. Hence, the unique equilibrium obtained by backward induction (which is also a SPNE) is the strategy profile given by

$$\begin{aligned}\sigma^* &= (\text{Amy plays } S \text{ at the decision nodes } x_0, x_2, \ldots, x_{196},\\ &\quad \text{Bob plays } S \text{ at the decision nodes } x_1, x_3, \ldots, x_{197}.)\end{aligned}$$

Note that σ^* yields the payoff profile $(1, 1)$, and this profile is Pareto dominated by any payoff profile other than $(0, 3)$. Therefore, the equilibrium obtained by backward induction, σ^*, is Pareto dominated. Among many strategy profiles that Pareto dominate σ^*, consider the following profile:

$$\begin{aligned}\hat{\sigma} &= (\text{Amy plays } C \text{ at the decision nodes } x_1, x_3, \ldots, x_{197},\\ &\quad \text{Bob plays } C \text{ at the decision nodes } x_2, x_4, \ldots, x_{198}.)\end{aligned}$$

Note that $\hat{\sigma}$ would yield the payoff profile (100, 100), at which each player could enjoy to have 99 dollars in addition to the single dollar he/she would enjoy at σ^*. However, the players cannot expect $\hat{\sigma}$ to arise as an equilibrium under backward induction. The reason is that under backward induction Amy believes that Bob, when he is at the node x_{197}, would not choose C to get 100 dollars, because he could secure 101 dollars by choosing S instead. Amy can calculate that this action of Bob at the node x_{197} would yield 98 dollars for herself. Under this observation, the optimal action of Amy at the node x_{196} would be to play S to secure 99 dollars instead of playing C and getting 98 dollars at the node x_{197}. Since Bob can too make all these calculations, his optimal action at the node x_{195} would be to play S to secure 100 dollars instead of playing C and getting 99 dollars at the node x_{196}. These reasonings consequently require both players to choose the action S at each of their decision nodes as specified under σ^*. ■

Example 10.10 (*Rubinstein's* [5] *Infinite Horizon Bilateral Bargaining Game*)
Amy and Bob play the infinite extensive form game in Fig. 10.13, where they sequentially bargain to divide a cake of unit size 1 between themselves. The first and second payoffs at each terminal node belong to Amy and Bob, respectively, and $\delta \in (0, 1)$. Amy offers in periods $t = 0, 2, \ldots$ and Bob offers in periods $t = 1, 3, \ldots$. Both Amy and Bob discount future payoffs by the factor δ. In period 0, Amy makes an offer $(x_0, 1 - x_0)$. If Bob accepts this offer, the bargaining ends and players enjoy the payoffs $(x_0, 1 - x_0)$. On the other hand, if Bob rejects the offer of Amy, he makes the counteroffer $(y_1, 1 - y_1)$ in period 1. If Amy accepts the offer of Bob,

the discounted payoffs of the two players become $\delta y_1, \delta(1 - y_1)$. On the other hand, if Amy rejects the offer of Bob in period 1, she makes a counteroffer in period 2. Likewise, in any period k where Amy makes the offer $(x_k, 1 - x_k)$, the two players enjoy the discounted payoffs $(\delta^k x_k, \delta^k(1 - x_k))$ if Bob accepts this offer and in any period $k + 1$ where Bob makes the offer $(y_{k+1}, 1 - y_{k+1})$, the two players enjoy the discounted payoffs $(\delta^{k+1} y_{k+1}, \delta^{k+1}(1 - y_{k+1}))$ if Amy accepts this offer.

Notice that the number of players is finite as there exist two players, Amy and Bob, and each player has only two actions, O (Offer) and A (Accept) at each decision node. However, there are infinitely many decision nodes in the described extensive form game. Therefore, this game is not finite. We will show that this infinite extensive form game has a unique SPNE profile of strategies $s^* = (s_A^*, s_B^*)$ where

Amy's strategy s_A^* prescribes her to offer

$$(x_k^*, 1 - x_k^*) = \left(\frac{1}{1+\delta}, \frac{\delta}{1+\delta}\right)$$

in any period $k = 0, 2, \ldots$ and to accept any offer $(y_k, 1 - y_k)$ of Bob in any period $k = 1, 3, \ldots$ if $y_k \geq \delta x_{k+1}^* = \delta/(1 + \delta)$, and

Bob's strategy s_B^* prescribes him to offer

$$(y_k^*, 1 - y_k^*) = \left(\frac{\delta}{1+\delta}, \frac{1}{1+\delta}\right)$$

in any period $k = 1, 3, \ldots$ and to accept any offer $(x_k, 1 - x_k)$ of Amy in any period $k = 0, 2, \ldots$ if $1 - x_k \geq \delta(1 - y_{k+1}^*) = \delta/(1 + \delta)$.

The strategy profile s^* is a SPNE since it induces a Nash equilibrium (in each period) (or at each decision node of each player). In this SPNE, there is no delay: Amy and Bob reach an immediate agreement in period 0 (at the first decision node of Bob). In this agreement, Amy earns $1/(1 + \delta)$ and Bob earns $\delta/(1 + \delta)$. Since $1 > \delta$, Amy always has a higher equilibrium payoff. This is because of her first-mover advantage. However, it helps Bob if the players become more patient, for his equilibrium payoff is increasing in δ. As δ approaches to 1, the equilibrium payoff profile goes to $(1/2, 1/2)$, the symmetric distribution of the cake.

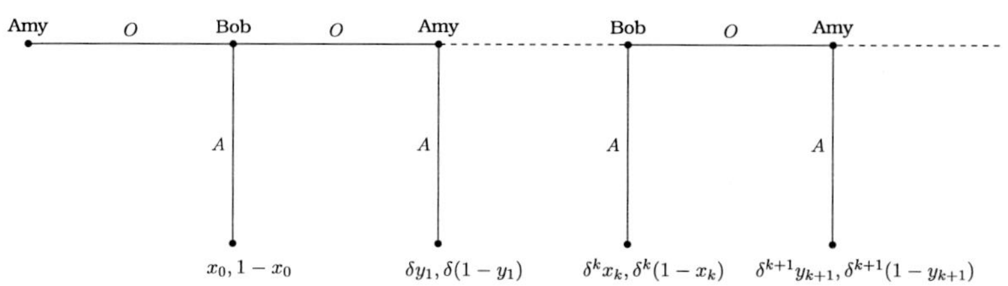

Fig. 10.13 Rubinstein's infinite horizon bilateral bargaining game

Below, we will show that the described strategy profile s^* is the unique SPNE of the extensive form game in Fig. 10.13. Let P^O and P^R, respectively, denote the players who make and receive an offer in any period. Also, let ν^{min} and ν^{max}, respectively, be the smallest and largest payoffs of P^O over all SPNE in all subgames. (These payoffs show the amounts before they are discounted to period zero values.)

Claim (i). $\nu^{min} \geq 1 - \delta\nu^{max}$.

This claim is true because the (one-period discounted) payoff of P^R will be at most $\delta\nu^{max}$ if s/he rejects the offer of P^O and makes a counteroffer in the next period.

Claim (ii). $\nu^{max} \leq 1 - \delta\nu^{min}$.

This claim is true because the payoff of P^R in the current period will be at least $\delta\nu^{min}$ since P^R would obtain at least ν^{min} in the next period where s/he could make a counteroffer. Therefore, the payoff of P^O cannot exceed $1 - \delta\nu^{min}$. Moreover, if P^R rejects the offer of P^O, s/he will not offer more than $\delta\nu^{max}$ in the next period. The present value of this amount in the current period is $\delta^2\nu^{max}$. Thus, ν^{max} cannot exceed $\delta^2\nu^{max}$, either. This last condition is automatically satisfied since $\delta < 1$.

Inserting claim (ii) into (i), we obtain

$$\nu^{min} \geq 1 - \delta\nu^{max} \geq 1 - \delta(1 - \delta\nu^{min}),$$

implying $\nu^{min} \geq 1/(1+\delta)$. Similarly, inserting claim (i) into (ii), we obtain

$$\nu^{max} \leq 1 - \delta\nu^{min} \leq 1 - \delta(1 - \delta\nu^{max}),$$

implying $\nu^{max} \leq 1/(1+\delta)$.

Since $\nu^{min} \leq \nu^{max}$ is true by definition, we have established that

$$\frac{1}{1+\delta} \leq \nu^{min} \leq \nu^{max} \leq \frac{1}{1+\delta},$$

implying that

$$\nu^{min} = \nu^{max} = \frac{1}{1+\delta}.$$

This proves that if a strategy profile is a SPNE, then P^O must make the offer $(\frac{1}{1+\delta}, \frac{\delta}{1+\delta})$ and P^R must accept an offer if and only if the payoff of P^R is not less than $\frac{\delta}{1+\delta}$. ■

References

1. Aumann RJ (1995) Backward induction and common knowledge of rationality. Games Econ Behav 8:6–19
2. Selten R (1975) Reexamination of the perfectness concept for equilibrium points in extensive games. Int J Game Theory 4:25–55
3. Nash JF (1950) Equilibrium points in n-person games. PNAS 36(1):48–49
4. von Stackelberg H (1934) Marktform und gleichgewicht. Springer, Berlin & Vienna
5. Rubinstein A (1982) Perfect equilibrium in a bargaining model. Econometrica 50:97–109
6. MasColell A, Whinston MD, Green JR (1995) Microeconomic theory. Oxford University Press

11 Sequential Equilibrium

In this chapter, we will first show that the subgame perfect Nash equilibrium cannot always eliminate some "sequentially unreasonable" Nash equilibria in extensive form games with imperfect information. As a remedy, we will then present two alternative refinements of the Nash equilibrium, known as the sequential equilibrium and the weak sequential equilibrium (or weak perfect Bayesian equilibrium). We will also give the relationship between these two refinements of the Nash equilibrium and their relations to the subgame perfect Nash equilibrium.

In the previous chapter, we saw that the idea of subgame perfection allows us to eliminate in extensive form games some Nash equilibria that are not reasonable. The set of Nash equilibria that remains is called subgame perfect because these equilibria induce a Nash equilibrium in every subgame, thereby ensuring plans of rational moves by each player even on subgames that are not on the equilibrium path of the play.

We also saw that in case an extensive form game is finite, we can calculate the set of subgame perfect Nash equilibria by using a generalized backward induction procedure. A basic step in this procedure—which reduces the game by replacing each minimal subgame in the game with one of the Nash equilibrium payoffs—is iteratively applied until the game cannot be reduced further. Given the dynamic nature of this procedure, one may tend to think that a subgame perfect equilibrium must be dynamically (sequentially) reasonable. But, this is not generally true as we illustrate in the next example.

Example 11.1 Consider the extensive form game Γ in Fig. 11.1. This game has no proper subgame. The only subgame is $\Gamma[x_0] = \Gamma$. Therefore, every Nash equilibrium of Γ is a SPNE. Below, we will first calculate the set of pure Nash equilibria of Γ and then discuss whether any of them is sequentially unreasonable.

I. Saglam, *Mastering Game Theory*, Classroom Companion: Economics,
https://doi.org/10.1007/978-3-031-82094-6_11

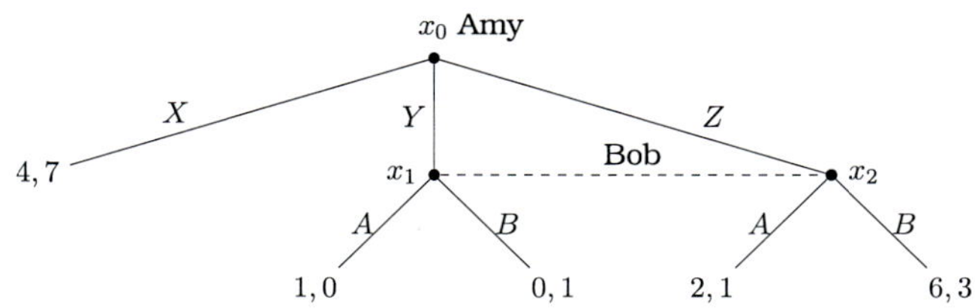

Fig. 11.1 An extensive form game with no proper subgame

Table 11.1 The strategic form game corresponding to the extensive form game in Fig. 11.1

Amy \ Bob	*A*	*B*
X	(4, 7)	(4, 7)
Y	(1, 0)	(0, 1)
Z	(2, 1)	(6, 3)

The strategic form representation of Γ is presented by Table 11.1. This game has two pure strategy Nash equilibria, (X, A) and (Z, B), and a set of unpure strategy Nash equilibria $(X, p(A) + (1 - p)(B))$ with $p \in [1/2, 1)$.

In the extensive form game, note that Bob can only play at the nodes x_1 and x_2, without knowing whether he is at x_1 or x_2. Thus, he has the same two actions, A and B, to choose at these two nodes. Also, note that the action A is strongly dominated for Bob by the action B. Despite this fact, the strategy profile (X, A) (where Bob plays A with probability 1) and the strategy profiles $(X, p(A) + (1 - p)(B))$ with $p \in [1/2, 1)$ (where Bob plays A with a nonzero probability) are both Nash equilibria. This is because of the fact that whenever Amy plays X, the game ends; and it becomes inconsequential for Bob whether he has earlier planned to play A or B. On the other hand, when Bob announces that he will play A with a probability not less than $1/2$ whenever he is called to play in the game, the best response of Amy is to play X at the node x_0. Since the game Γ has no subgame other than itself, subgame perfection is unable to prevent a sequentially irrational play of Bob at his information set $H_{Bob} = \{x_1, x_2\}$ and eliminate the resulting sequentially unreasonable equilibria. ■

What we have learned from the above example is that to eliminate a sequentially unreasonable/irrational SPNE from an extensive form game, one may need to put restrictions as to how players should move at each of their information sets. This idea was first proposed by Kreps and Wilson [1], leading to a very attractive equilibrium refinement of the Nash Equilibrium, called Sequential Equilibrium. For the clarity of presentation, we will first consider a weaker version of this equilibrium, called

Weak Perfect Bayesian Equilibrium (or Weak Sequential Equilibrium), which was introduced by Myerson [2].

11.1 Weak Perfect Bayesian Equilibrium

This equilibrium concept requires the specification of a strategy profile along with a system of beliefs.

Definition 11.1 Given an extensive form game Γ, we say that μ is a system of beliefs if for each information set $H \in \mathcal{H}$ and for each node $x \in H$, it is true that $\mu(x) \in [0, 1]$ and $\sum_{x \in H} \mu(x) = 1$.

Thus, a system of beliefs is a probability distribution on the decision nodes at each information set of the game. For any information set $H \in \mathcal{H}$, let $i[H]$ denote the player who can possibly play at H. Given a strategy profile σ, a belief system μ, and an information set $H \in \mathcal{H}$, we denote by $\tilde{u}_{i[H]}(\sigma|H, \mu)$ the expected utility of player $i[H]$ obtained from the part of the game starting at the information set H when this player and her opponents form their beliefs at their information sets according to μ and follow their strategies specified by σ. Now, given a belief system μ, we can define the rationality of a strategy profile σ at an information set H.

Definition 11.2 We say that a behavior strategy profile σ in an extensive form game Γ is rational at an information set $H \in \mathcal{H}$ relative to a belief system μ if the player $i[H]$ moving at this information set finds that playing the strategy $\sigma_{i[H]}$ is a best response to her opponents' strategy profile $\sigma_{-i[H]}$ conditional on H and μ; i.e.,

$$\tilde{u}_{i[H]}(\sigma_{i[H]}, \sigma_{-i[H]}|H, \mu) \geq \tilde{u}_{i[H]}(\tilde{\sigma}_{i[H]}, \sigma_{-i[H]}|H, \mu)$$

for every behavior strategy $\tilde{\sigma}_{i[H]}$ of player i.

When the above definition applies to all information sets in a game, we say that a behavior strategy profile σ is sequentially rational relative to the belief system μ.

Definition 11.3 A behavior strategy profile σ in an extensive form game Γ is sequentially rational relative to a belief system μ if σ is rational at every information set $H \in \mathcal{H}$ relative to μ.

Note in the above definition that the sequential rationality of a strategy profile σ is dependent on a given belief system μ. For the pair (σ, μ) to be reasonable, the belief system μ must be consistent with the strategy profile σ on every information set that is reached with positive probability under σ. Formally, we define this consistency requirement as follows.

Definition 11.4 Given an extensive form game Γ, a belief system μ is consistent with a behavior strategy profile σ if μ can be derived from σ using the Bayes rule whenever possible; i.e., for any $H \in \mathcal{H}$ with $\text{prob}(H|\sigma) > 0$ and any $x \in H$ it is true that

$$\mu(x) = \text{prob}(x|H, \sigma) \equiv \frac{\text{prob}(x|\sigma)}{\text{prob}(H|\sigma)}.$$

(In the above definition, $\text{prob}(x|H, \sigma)$ is the probability of reaching the node x under the behavior strategy σ conditional on that the information set H is reached.) Below, we present an example where a strategy profile σ in an extensive form game is sequentially rational relative to a belief system μ that is inconsistent with σ.

Example 11.2 Consider the extensive form game in Fig. 11.2, where the first and second payoffs at each terminal node belong to Amy and Bob, respectively.

Consider the strategy profile $\sigma =$ (Amy plays Y, Bob plays A) and the belief system μ such that $\mu(x_0) = 1$, $\mu(x_1) = 0.80$, and $\mu(x_2) = 0.20$. Let us check whether σ is sequentially rational relative to μ. Consider first the information set $H_{Amy} = \{x_0\}$ of Amy. We can calculate

$$\begin{aligned}\tilde{u}_{Amy}(X, \sigma_{Bob}|H_{Amy}, \mu) &= 1, \quad \text{and} \\ \tilde{u}_{Amy}(Y, \sigma_{Bob}|H_{Amy}, \mu) &= 2.\end{aligned}$$

The difference $\tilde{u}_{Amy}(Y, \sigma_{Bob}|H_{Amy}, \mu) - \tilde{u}_{Amy}(X, \sigma_{Bob}|H_{Amy}, \mu) = 1$ is positive. Thus, the strategy $\sigma_{Amy} = Y$ of Amy is rational at H_{Amy} relative to μ.

Now consider the information set $H_{Bob} = \{x_1, x_2\}$ of Bob. We can calculate

$$\begin{aligned}\tilde{u}_{Bob}(A, \sigma_{Amy}|H_{Bob}, \mu) &= (2)\mu(x_1) + (1)(1 - \mu(x_1)) = 1 + \mu(x_1) = 1.80, \quad \text{and} \\ \tilde{u}_{Bob}(B, \sigma_{Amy}|H_{Bob}, \mu) &= (1)\mu(x_1) + (2)(1 - \mu(x_1)) = 2 - \mu(x_1) = 1.20.\end{aligned}$$

The difference $\tilde{u}_{Bob}(A, \sigma_{Amy}|H_{Bob}, \mu) - \tilde{u}_{Bob}(B, \sigma_{Amy}|H_{Bob}, \mu) = 0.60$ is positive. Thus, the strategy $\sigma_{Bob} = A$ of Bob is rational at H_{Bob} relative to μ.

Thus, we have established that σ is sequentially rational relative to μ. Now, let us check whether the belief system μ is consistent with σ. When Amy plays $\sigma_{Amy} = Y$, the information set of Bob, H_{Bob}, is reached with probability 1, i.e.,

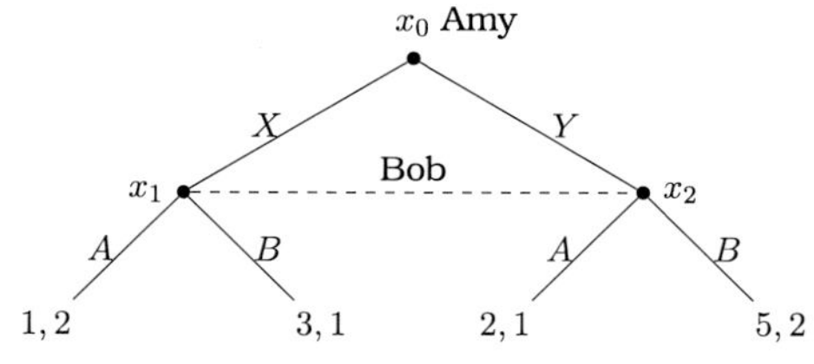

Fig. 11.2 An extensive form game to show inconsistency of some beliefs

$\text{prob}(H_{Bob}|\sigma) = 1$. So, the restriction of μ on the nodes of H_{Bob} must be derived from σ through the Bayes rule if μ is consistent with σ. We know that $\text{prob}(x_2|\sigma) = 1$. Thus,

$$\text{prob}(x_2|H_{Bob}, \sigma) = \frac{\text{prob}(x_2|\sigma)}{\text{prob}(H_{Bob}|\sigma)} = 1.$$

This value is not equal to $\mu(x_2)$, which is 0.20. Thus, the Bayes rule is violated at the node x_2. So, μ is inconsistent with σ. ■

Now, we present an example where a strategy profile σ in an extensive form game is sequentially rational relative to a belief system μ that is consistent with σ.

Example 11.3 Reconsider the extensive form game in Fig. 11.2. Now, consider the strategy profile $\sigma =$ (Amy plays Y, Bob plays B) and the belief system μ such that $\mu(x_0) = 1$, $\mu(x_1) = 0$, and $\mu(x_2) = 1$. Let us check whether σ is sequentially rational relative to μ. By arguments similar to those in Example 11.2, we can show that the strategy $\sigma_{Amy} = Y$ of Amy is rational at H_{Amy} relative to μ. Our calculations at the information set $H_{Bob} = \{x_1, x_2\}$ of Bob are as follows:

$$\begin{aligned} \tilde{u}_{Bob}(A, \sigma_{Amy}|H_{Bob}, \mu) &= (2)\mu(x_1) + (1)(1 - \mu(x_1)) = 1 + \mu(x_1) = 1, \quad \text{and} \\ \tilde{u}_{Bob}(B, \sigma_{Amy}|H_{Bob}, \mu) &= (1)\mu(x_1) + (2)(1 - \mu(x_1)) = 2 - \mu(x_1) = 2. \end{aligned}$$

The difference $\tilde{u}_{Bob}(B, \sigma_{Amy}|H_{Bob}, \mu) - \tilde{u}_{Bob}(A, \sigma_{Amy}|H_{Bob}, \mu) = 1$ is positive. Thus, the strategy $\sigma_{Bob} = B$ of Bob is rational at H_{Bob} relative to μ.

Thus, we have established that σ is sequentially rational relative to μ. Now let us check whether the belief system μ is consistent with σ. When Amy plays $\sigma_{Amy} = Y$, the information set of Bob, H_{Bob}, is reached with probability 1, i.e., $\text{prob}(H_{Bob}|\sigma) = 1$. So, the probability distribution on the nodes of H_{Bob} specified under μ must be derived from σ through the Bayes rule if μ is consistent with σ. We can calculate $\text{prob}(x_2|\sigma) = 1$, since $\sigma_{Amy} = Y$. Thus,

$$\text{prob}(x_2|H_{Bob}, \sigma) = \frac{\text{prob}(x_2|\sigma)}{\text{prob}(H_{Bob}|\sigma)} = 1.$$

This value is equal to $\mu(x_2)$. Since, H_{Bob} contains only two nodes, it is obvious that $\text{prob}(x_1|H_{Bob}, \sigma) = \mu(x_1)$ must be satisfied as well. Thus, we can derive the restriction of μ on the nodes of H_{Bob} from σ through the Bayes rule. So, μ is consistent with σ. ■

Definition 11.4 requires the consistency between a belief system μ and a strategy profile σ only on information sets that are reached with positive probability under σ. If a game Γ contains an information set $H \in \mathcal{H}$ that can never be reached under σ, i.e., $\text{prob}(H|\sigma) = 0$, then the probability distribution over the set of nodes in H can be arbitrarily assigned. Now, we are ready to present the Weak Perfect Bayesian Equilibrium concept.

Definition 11.5 A pair (σ, μ) involving a profile of strategies and a belief system is a Weak Perfect Bayesian Equilibrium (WPBE) of an extensive form game Γ if (i) σ is sequentially rational relative to μ and (ii) μ is consistent with σ.

In the following example, we will illustrate how we can use the above definition to calculate the set of WPBE of an extensive form game.

Example 11.4 Reconsider the extensive form game in Fig. 11.1.

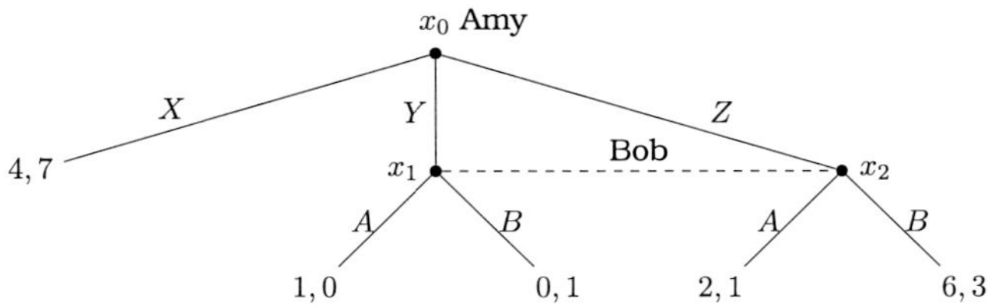

Figure 11.1 Reconsidered

Recall that the set of SPNE of this game involves the pure strategy profiles (X, A) and (Z, B), and the set of unpure strategy profiles $(X, p(A) + (1 - p)(B))$ with $p \in [1/2, 1)$. Let us check whether weak sequential rationality can eliminate any of these SPNE.

Let μ be a belief system. Amy has a unique information set $H_{Amy} = \{x_0\}$. Since H_{Amy} is a singleton set, we must have $\mu(x_0) = 1$. Note also that Bob has a unique information set $H_{Bob} = \{x_1, x_2\}$. Given the belief system μ, the probabilities at the nodes $x_1, x_2 \in H_{Bob}$ are $\mu(x_1) \in [0, 1]$ and $\mu(x_2) = 1 - \mu(x_1)$. Let σ_{Amy} be a strategy profile in $\Delta(A_{Amy})$, where $A_{Amy} = \{X, Y, Z\}$. We first calculate the expected payoffs of Bob from playing A and B conditional on H_{Bob} and μ:

$$\begin{aligned}\tilde{u}_{Bob}(A, \sigma_{Amy}|H_{Bob}, \mu) &= (0)\mu(x_1) + (1)(1 - \mu(x_1)) = 1 - \mu(x_1)\\ \tilde{u}_{Bob}(B, \sigma_{Amy}|H_{Bob}, \mu) &= 1\mu(x_1) + (3)(1 - \mu(x_1)) = 3 - 2\mu(x_1)\end{aligned}$$

for any $\sigma_{Amy} \in \Delta(A_{Amy})$. It follows that Bob prefers B to A for all values $\mu(x_1) \in [0, 1]$. Thus, a strategy profile σ can be a part of a WPBE only if $\sigma_{Bob} = B$. Given $\sigma_{Bob} = B$, we can calculate the expected payoffs of Amy from playing X, Y and Z:

$$\begin{aligned}\tilde{u}_{Amy}(X, \sigma_{Bob}|H_{Amy}, \mu) &= 4,\\ \tilde{u}_{Amy}(Y, \sigma_{Bob}|H_{Amy}, \mu) &= 0,\\ \tilde{u}_{Amy}(Z, \sigma_{Bob}|H_{Amy}, \mu) &= 6.\end{aligned}$$

Hence, Amy should play Z; i.e., $\sigma_{Amy} = Z$. In that case, the information set of Bob, H_{Bob}, will be reached with probability 1; i.e., $\text{prob}(H_{Bob}|\sigma) = 1$. Thus, μ must be derived from σ on H_{Bob} through the Bayes rule. We have $\text{prob}(x_1|\sigma) = 0$

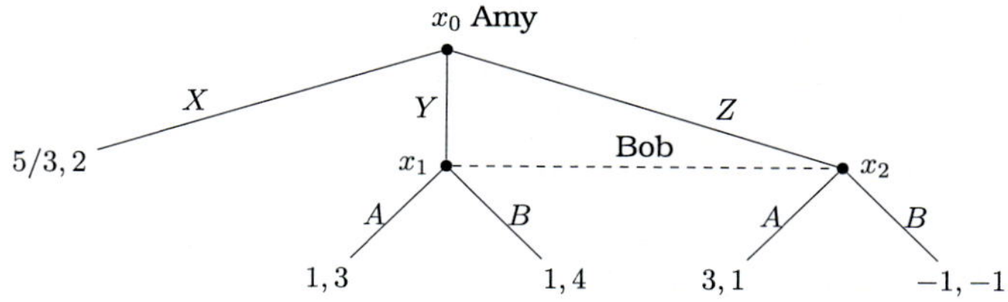

Fig. 11.3 An extensive form game where each SPNE is contained by a WPBE

and $\text{prob}(x_2|\sigma) = 1$ since $\sigma_{Amy} = Z$. Therefore, $\text{prob}(x_1|\sigma)/\text{prob}(H_{Bob}|\sigma) = 0$ and $\text{prob}(x_2|\sigma)/\text{prob}(H_{Bob}|\sigma) = 1$. So, the strategy profile σ can be consistent with a belief system μ only if $\mu(x_1) = 0$ and $\mu(x_2) = 1$. Thus, the extensive form game Γ has a unique WPBE (σ, μ), where $\sigma = (Z, B)$ and μ is such that $\mu(x_0) = 1$, $\mu(x_1) = 0$, and $\mu(x_2) = 1$.

Note that the unreasonable pure strategy SPNE profile (X, A) and the set of unreasonable unpure strategy SPNE profiles $(X, p(A) + (1 - p)(B))$ with $p \in [1/2, 1)$ are not a part of the unique WPBE, which only contains the reasonable SPNE profile (Z, B). ■

Below, we present another example, where calculating WPNE is more involved.

Example 11.5 Consider the extensive form game in Fig. 11.3, where the first and second payoffs at each terminal node belong to Amy and Bob, respectively.

This game has a single subgame, $\Gamma[x_0] = \Gamma$. Therefore, every Nash equilibrium of Γ is a SPNE. Let us calculate the set of pure Nash equilibria of Γ. The strategic form game associated with Γ is in Table 11.2 This game has two pure strategy Nash equilibria, (X, B), and (Z, A), and a set of unpure strategy Nash equilibria $(X, p(A) + (1 - p)(B))$ where $p \in (0, 2/3]$. Since the game Γ has no subgame other than itself, all Nash equilibria we calculated above is a SPNE of Γ. Let us check whether weak sequential rationality can eliminate any of SPNE.

Let μ be a belief system. Amy has a unique information set $H_{Amy} = \{x_0\}$. Since H_{Amy} is a singleton set, we must have $\mu(x_0) = 1$. Note also that Bob has a unique information set $H_{Bob} = \{x_1, x_2\}$. Given the belief system μ, the probabilities at

Table 11.2 The strategic form game corresponding to the extensive form game in Fig. 11.3.

Amy \ Bob	A	B
X	(5/3, 2)	(5/3, 2)
Y	(1, 3)	(1, 4)
Z	(3, 1)	(−1, −1)

the nodes $x_1, x_2 \in H_{Bob}$ are $\mu(x_1) \in [0, 1]$ and $\mu(x_2) = 1 - \mu(x_1)$. Let $\sigma_{Amy} \in \Delta(A_{Amy})$, where $A_{Amy} = \{X, Y, Z\}$. We first calculate the expected payoffs of Bob from playing A and B conditional on H_{Bob} and μ:

$$\begin{aligned}\tilde{u}_{Bob}(A, \sigma_{Amy}|H_{Bob}, \mu) &= 3\mu(x_1) + 1(1 - \mu(x_1)) = 1 + 2\mu(x_1)\\ \tilde{u}_{Bob}(B, \sigma_{Amy}|H_{Bob}, \mu) &= 4\mu(x_1) + (-1)(1 - \mu(x_1)) = -1 + 5\mu(x_1)\end{aligned}$$

for any $\sigma_{Amy} \in \Delta(A_{Amy})$. So, Bob

- prefers A to B if $1 + 2\mu(x_1) > -1 + 5\mu(x_1)$ or $\mu(x_1) < 2/3$,
- prefers B to A if $\mu(x_1) > 2/3$,
- becomes indifferent between A and B if $\mu(x_1) = 2/3$.

Now, we are ready to calculate the set of Weak Perfect Bayesian Equilibria (WPBE) by analyzing three possible cases.

Case 1: Suppose first $\mu(x_1) < 2/3$, which induces Bob to play A. In response, Amy should play Z. Then, the information set of Bob, H_{Bob}, is reached with probability 1 and the node x_1 is reached with probability 0 (since the node x_2 is reached with probability 1 when Amy plays Z). This is consistent with the supposition that $\mu_1 < 2/3$. Thus, we have calculated our first WPBE.

WPBE-1: $\sigma =$ (Amy plays Z, Bob plays$A$) and μ is such that $\mu(x_0) = 1$, $\mu(x_1) = 0$, and $\mu(x_2) = 1$.

Note that WPBE-1 contains the SPNE profile (Z, A).

Case 2: Suppose now $\mu(x_1) > 2/3$, which induces Bob to play B. In response, Amy should play X. In that case the information set of Bob, H_{Bob}, is never reached. Thus, any belief system μ is consistent with σ. The only restriction on μ is that $\mu(x_1) > 2/3$, which we supposed at the beginning of this case. Thus, we have calculated our second WPBE.

WPBE-2: $\sigma =$ (Amy plays X, Bob plays$B$) and μ is such that $\mu(x_0) = 1$, $\mu(x_1) = \alpha \in (2/3, 1]$, and $\mu(x_2) = 1 - \alpha$.

Note that WPBE-2 contains the SPNE profile (X, B).

Case 3: Finally, suppose $\mu(x_1) = 2/3$ inducing Bob to become indifferent between A and B. Then, Bob can randomize between A and B. Let $p \in [0, 1]$ denote the probability that Bob plays A. Now, we calculate the expected payoffs of Amy from playing X, Y and Z:

$$\begin{aligned}\tilde{u}_{Amy}(X, \sigma_{Bob}|H_{Amy}, \mu) &= 5/3,\\ \tilde{u}_{Amy}(Y, \sigma_{Bob}|H_{Amy}, \mu) &= 1(p) + 1(1-p) = 1,\\ \tilde{u}_{Amy}(Z, \sigma_{Bob}|H_{Amy}, \mu) &= 3(p) + (-1)(1-p) = -1 + 4p.\end{aligned}$$

Amy will be indifferent between X and Z if and only if $p = 2/3$. With $p = 2/3$, the expected payoffs of Amy from both X and Z will be $5/3$, which is higher than her expected payoff from Y. One can easily check that Amy will

- play X if $p < 2/3$,
- play Z if $p > 2/3$, and
- become indifferent between playing X and Z if $p = 2/3$.

So, we need to consider three subcases separately:

Case 3-i: Suppose $p < 2/3$. Then, $\sigma = (\text{Amy plays } X,\ \text{Bob plays } pA + (1-p)B)$. The information set of Bob, H_{Bob}, will never be reached, i.e., $\text{prob}(H_{Bob}|\sigma) = 0$. Thus, μ does not have to be derived from σ on H_{Bob} through the Bayes rule. Since Case 3-i arises under the supposition $\mu(x_1) = 2/3$, we must keep this restriction, though. Thus, we have calculated our third WPBE, where the equilibrium strategy profile varies with p.

WPBE-3: $\sigma(p) = (\text{Amy plays } X,\ \text{Bob plays } p(A) + (1-p)(B))$ where $p < 2/3$ and μ is such that $\mu(x_0) = 1$, $\mu(x_1) = 2/3$, and $\mu(x_2) = 1/3$.

Note that WPBE-3 contains the set of SPNE profiles $(X, p(A) + (1-p)(B))$ where $p \in [0, 2/3)$.

Case 3-ii: Suppose $p > 2/3$. Then, we have $\sigma(p) = (\text{Amy plays } Z,\ \text{Bob plays } p(A) + (1-p)(B))$. The information set of Bob, H_{Bob}, will be reached with probability 1; i.e., $\text{prob}(H_{Bob}|\sigma) = 1$. Thus, μ must be derived from σ on H_{Bob} through the Bayes rule. We calculate $\text{prob}(x_1|\sigma) = 0$ since $\text{prob}(x_2|\sigma) = 1$. Therefore, we have $\text{prob}(x_1|\sigma)/\text{prob}(H_{Bob}|\sigma) = 0$. But, we have $\mu(x_1) = 2/3$. Thus, $\mu(x_1)$ cannot be derived from σ through the Bayes rule, implying that (σ, μ) cannot be a WPBE if $p > 2/3$.

Case 3-iii: Suppose $p = 2/3$. Then,

$$\sigma(p) = (\text{ Amy plays } \sigma_X(X) + (1-\sigma_X)(Z),\ \text{Bob plays } \frac{2}{3}A + \frac{1}{3}B)$$

where $\sigma_X \in [0, 1]$. We now have two subcases to consider.

Case 3-iii-a: $\sigma_X = 1$. In this case $\text{prob}(H_{Bob}|\sigma) = 1 - \sigma_X = 0$, implying that the belief system μ, which satisfies $\mu(x_1) = 2/3$ and $\mu(x_2) = 1/3$, does not have to be derived from σ on H_{Bob} through the Bayes rule. Thus, the WPBE for this case is as follows:

WPBE-4: $\sigma = \big(\text{Amy plays } X, \text{Bob plays } \frac{2}{3}(A) + \frac{1}{3}(B)\big)$ and μ is such that $\mu(x_0) = 1$, $\mu(x_1) = 2/3$, and $\mu(x_2) = 1/3$.

Note that WPBE-4 contains the SPNE profile $(X, \frac{2}{3}(A) + \frac{1}{3}(B))$.

Case 3-iii-b: $\sigma_X \in [0, 1)$. In this case $\text{prob}(H_{Bob}|\sigma) = 1 - \sigma_X \in (0, 1]$, implying that μ must be derived from σ on H_{Bob} through the Bayes rule. We know that $\text{prob}(x_1|\sigma) = 0$. Thus, $\text{prob}(x_1|\sigma)/\text{prob}(H_{Bob}|\sigma) = 0$. On the other hand, we have $\mu(x_1) = 2/3$. Then, the equality $\text{prob}(x_1|\sigma)/\text{prob}(H_{Bob}|\sigma) = \mu(x_1)$ cannot hold, implying that there exists no WPBE in Case 3-iii-b.

We should note that each SPNE of Γ is contained by one of the four WPBE we calculated above. Weak sequential rationality did not eliminate any SPNE. ■

Earlier, in Example 11.4, we saw that not every SPNE of an extensive form game must be a part of a WPBE. A remaining question to answer is whether every WPBE of an extensive form game must contain a strategy profile that is a SPNE. Before answering this question, let us study the relationship between WPBE and Nash equilibrium. However, we need to make some preliminaries that will be used in the proof of Proposition 11.1.

We say that an information set $H_i \in \mathcal{H}_i$ of player i in an extensive form game is closest to the root node x_0 if every path from x_0 to a node in H_i does not pass through any other information set in $\mathcal{H}_i$. Let $\overline{\mathcal{H}}_i$ denote the set of player i's information sets that are closest to x_0. For example, for the game in Figure 10.5, $\overline{\mathcal{H}}_{Amy} = \{H^1_{Amy}\}$ where $H^1_{Amy} = \{x_0\}$ and $\overline{\mathcal{H}}_{Bob} = \{H^1_{Bob}, H^2_{Bob}\}$ where $H^1_{Bob} = \{x_1\}$ and $H^2_{Bob} = \{x_2, x_3\}$.

Let $\bar{p}(\sigma|\overline{\mathcal{H}}_i)$ denote the probability that a play of the game under the profile σ will not pass through any information set in $\overline{\mathcal{H}}_i$ and let $\bar{u}_i(\sigma|\overline{\mathcal{H}}_i]$ denote the expected payoff of player i when a play of the game under the profile σ does not pass through any information set in $\overline{\mathcal{H}}_i$.

We should note that for any $i \in N$, if player i does not play at the node x_0, then $\bar{p}(\sigma|\overline{\mathcal{H}}_i)$ and $\bar{u}(\sigma|\overline{\mathcal{H}}_i)$ need not be zero. However, $\bar{p}(\sigma|\overline{\mathcal{H}}_i)$ and $\bar{u}(\sigma|\overline{\mathcal{H}}_i)$ are independent of σ_i. Moreover, for any $H_i \in \overline{\mathcal{H}}_i$, the probability of reaching H_i under a behavior strategy profile σ, $\text{prob}(H_i|\sigma)$, is also independent of σ_i since by the definition of $\overline{\mathcal{H}}_i$ player i could not have ever played before H_i, or any other information set in $\overline{\mathcal{H}}_i$, was reached.

Now, we are ready to state and prove the following result.

Proposition 11.1 *Given an extensive form game, if a behavior strategy profile σ and a belief system μ constitute a WPBE, then σ is a Nash equilibrium.*

Proof See the appendix at the end of this chapter. ■

The converse of the above proposition is not true. There are extensive form games where some of the Nash equilibria in behavior strategies cannot be a part of a WPBE. Consider, for example, the extensive form game studied in Example 11.1. The behavior strategy $\sigma = (X, A)$ or explicitly

$$\sigma = (\text{Amy plays X at } H_{Amy} = \{x_0\}, \text{Bob plays } A \text{ at } H_{Bob} = \{x_1, x_2\})$$

is a Nash equilibrium as we earlier showed using Table 11.1. Also note that at the information set H_{Bob}, the action B of Bob strongly dominates the action A. Therefore, there is no belief system μ such that Bob's playing A can be rational at H_{Bob}. This implies that there is no belief system μ such that $\sigma = (X, A)$ is sequentially rational relative to μ. Therefore, the Nash equilibrium profile σ cannot be a part of any WPBE. Also recall that the extensive form game in Example 11.1 is minimal; i.e., the only subgame is the game itself. Therefore, every Nash equilibrium of that game, including $\sigma = (X, A)$ is a SPNE. Thus, Example 11.1 establishes that there exist extensive form games with SPNE that cannot be a part of any WPBE.

The general reason why a Nash equilibrium strategy profile σ does not have to be contained by a WPBE is that WPBE requires sequential rationality at all information sets in relative to a (consistent) belief system irrespective of whether an information set is reached under the profile σ. The sequential rationality implied by a Nash equilibrium is less restrictive. One can straightforwardly show that a strategy profile σ is a Nash equilibrium if and only if σ is sequentially rational at each information set H with $\text{prob}(H|\sigma) > 0$ relative to a belief system μ_σ that is derived, through the Bayes rule, from σ at each information set H with $\text{prob}(H|\sigma) > 0$.

So far, we have established that

(i) every WPBE contains a strategy profile that is a Nash equilibrium in behavior strategies,

(ii) not every Nash equilibrium in behavior strategies is part of a WPBE,

(iii) not every SPNE in behavior strategies is part of a WPBE.

A remaining question to answer is whether every WPBE contains a strategy profile that is a SPNE in behavior strategies. The answer is "no", as illustrated by the following example.

Example 11.6 Consider the extensive form game in Fig. 11.4.

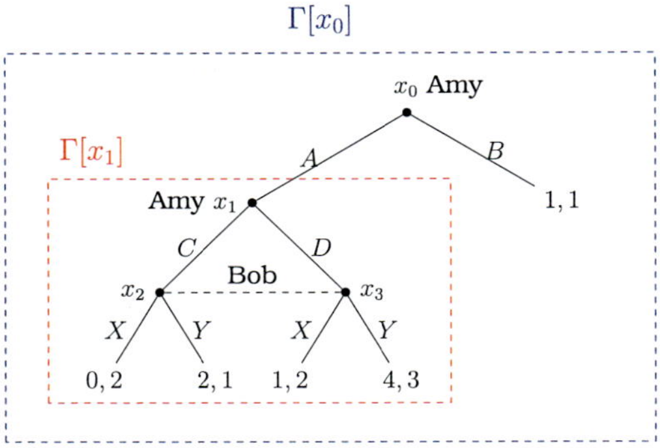

Fig. 11.4 An extensive form game where a WPBE is not a SPNE

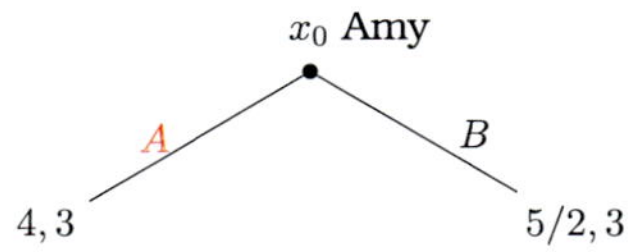

Fig. 11.5 The reduced game obtained from Fig. 11.4 replacing $\Gamma[x_1]$ with its unique Nash equilibrium payoff profile

This game has two subgames $\Gamma[x_0] = \Gamma$ and $\Gamma[x_1]$. Only $\Gamma[x_1]$ is minimal. To calculate the set of Nash equilibria in $\Gamma[x_1]$, we transform this subgame to its strategic form representation given by Table 11.3.

This strategic form game has a unique Nash equilibrium strategy profile, (D, Y), leading to the payoff profile (4, 3). If we replace the subgame $\Gamma[x_1]$ by the Nash equilibrium payoff profile (4, 3), then the reduced game becomes as in Fig. 11.5: Obviously, Amy should play A in the above game. Thus, the game Γ has a unique SPNE, $\sigma^* = ((A, D), Y)$, explicitly written as

$$\begin{aligned}\sigma^* = (&\text{Amy plays } A \text{ at the information set } H^1_{Amy} = \{x_0\} \text{ and}\\ &\text{plays } D \text{ at the information set } H^2_{Amy} = \{x_1\},\\ &\text{Bob plays } Y \text{ at the information set } H_{Bob} = \{x_2, x_3\})\,.\end{aligned}$$

(One can quickly check that the SPNE profile σ^* becomes a part of a WPBE under the belief system μ^* that satisfies $\mu^*(x_0) = 1$, $\mu^*(x_1) = 1$, $\mu^*(x_2) = 0$, and $\mu^*(x_3) = 1$.)

Table 11.3 The strategic form game corresponding to the subgame $\Gamma[x_1]$ in Figure 11.4

		Bob	
		X	Y
Amy	C	(0, 2)	(2, 1)
	D	(1, 2)	(4, 3)

Now, consider the strategy profile $\hat{\sigma} = ((B, D), X)$, explicitly written as

$$\begin{aligned}\hat{\sigma} = \big(&\text{Amy plays } B \text{ at the information set } H^1_{Amy} = \{x_0\} \text{ and}\\ &\text{plays } D \text{ at the information set } H^2_{Amy} = \{x_1\},\\ &\text{Bob plays } X \text{ at the information set } H^1_{Bob} = \{x_2, x_3\}\big)\end{aligned}$$

and the belief system $\hat{\mu}$ that satisfies

$$\hat{\mu}(x_0) = 1,\ \hat{\mu}(x_1) = 1,\ \hat{\mu}(x_2) = 1,\ \hat{\mu}(x_3) = 0.$$

We will prove that the pair $(\hat{\sigma}, \hat{\mu})$ is a WPBE.

Let us first show that $\hat{\sigma}$ is sequentially rational relative to $\hat{\mu}$. Consider Amy at her information set $H^1_{Amy} = \{x_0\}$. In the continuation game, she has four pure strategies to consider (A, C), (A, D), (B, C), and (B, D). For each of these strategies, we can calculate the expected payoffs of Amy when Bob plays $\hat{\sigma}_{Bob} = X$, conditional on H_{Amy} and $\hat{\mu}$:

$$\begin{aligned}\tilde{u}_{Amy}((A, C), \hat{\sigma}_{Bob}|H^1_{Amy}, \hat{\mu}) &= 0\\ \tilde{u}_{Amy}((A, D), \hat{\sigma}_{Bob}|H^1_{Amy}, \hat{\mu}) &= 1\\ \tilde{u}_{Amy}((B, C), \hat{\sigma}_{Bob}|H^1_{Amy}, \hat{\mu}) &= 1\\ \tilde{u}_{Amy}((B, D), \hat{\sigma}_{Bob}|H^1_{Amy}, \hat{\mu}) &= 1.\end{aligned}$$

Thus, the strategy $\hat{\sigma}_{Amy} = (B, D)$ is rational at H^1_{Amy} relative to $\hat{\mu}$.

Now, consider Amy at her information set $H^2_{Amy} = \{x_1\}$, where she has two strategies to consider, C and D. We can calculate

$$\begin{aligned}\tilde{u}_{Amy}(C, \hat{\sigma}_{Bob}|H^2_{Amy}, \hat{\mu}) &= 0\\ \tilde{u}_{Amy}(D, \hat{\sigma}_{Bob}|H^2_{Amy}, \hat{\mu}) &= 1.\end{aligned}$$

Thus, the strategy $\hat{\sigma}_{Amy} = (B, D)$ is rational at H^2_{Amy} relative to $\hat{\mu}$.

Finally, consider Bob at his unique information set $H^1_{Bob} = \{x_2, x_3\}$, where he has two strategies to consider, X and Y. We can calculate

$$\begin{aligned}\tilde{u}_{Bob}(X, \hat{\sigma}_{Amy}|H^1_{Bob}, \hat{\mu}) &= 2\\ \tilde{u}_{Bob}(Y, \hat{\sigma}_{Amy}|H^1_{Bob}, \hat{\mu}) &= 1.\end{aligned}$$

Therefore, the strategy $\hat{\sigma}_{Bob} = X$ is rational at H^1_{Bob} relative to $\hat{\mu}$.

Thus, we have established that the strategy profile $\hat{\sigma}$ is sequentially rational relative to $\hat{\mu}$. Now, we will show that $\hat{\mu}$ is consistent with $\hat{\sigma}$. Note that $\mu(x_0) = 1 = \text{prob}(x_0|H^1_{Amy}, \sigma)$ trivially holds. Also note that $\text{prob}(H^1_{Bob}|\sigma) = 0$ and $\text{prob}(H^2_{Amy}|\sigma) = 0$. Therefore, the belief system $\hat{\mu}$ does not have to satisfy the Bayes rule at the information sets H^1_{Bob} and H^2_{Amy}. Thus, $\hat{\mu}$ is consistent with $\hat{\sigma}$.

To conclude, we have proved that the pair $(\hat{\sigma}, \hat{\mu})$ is a WPBE, while $\hat{\sigma}$ is not a SPNE. ■

The extensive form games in Examples 11.4 and 11.5 are minimal. Hence, the set of SPNE in these examples coincides with the set of Nash equilibria. Since the strategy profile in a WPBE must be a Nash equilibrium, the set of WPBE in both examples contain strategy profiles that are SPNE. On the other hand, the extensive form game in Example 11.6 is not minimal, and it contains a Nash equilibrium $\hat{\sigma}$ that is not a SPNE. Moreover, there exists a belief system $\hat{\mu}$ such that $(\hat{\sigma}, \hat{\mu})$ is a WPBE. Thus, using Example 11.6 we have established that not every WPBE contains a strategy profile that is SPNE in behavior strategies. However, for extensive form games with perfect information, this negative result does not arise.

Given an extensive form game with perfect information, each information set is a singleton set. Then, let μ^{PI} denote the belief system such that for every information set $H = \{x\}$ in $\mathcal{H}$, $\mu^{PI}(x) = 1$. Using this belief system, we can state the following result.

Proposition 11.2 *In an extensive form game with perfect information, a behavior strategy profile σ is a SPNE if and only if the pair (σ, μ^{PI}) is a WPBE.*

Proof See the appendix at the end of this chapter. ■

11.2 Sequential Equilibrium

In the previous section, we saw that there are extensive form games that have a WPBE assessment (σ, μ) such that the strategy profile σ is not a SPNE. The reason underlying this result is simply that in a WPBE no consistency restrictions are placed on beliefs off the equilibrium path. That is, a belief system μ supporting a strategy profile σ as a WPBE can be arbitrarily chosen, i.e., it does not have to be obtained by Bayes rule, on information sets that are never reached under σ. In this section, we will study the effects of having a stronger consistency restriction on beliefs leaving no room for arbitrariness, as was originally proposed by Kreps and Wilson [1] in their definition of the Sequential Equilibrium Concept.

Definition 11.6 Given an extensive form game, a pair (σ, μ) involving a strategy profile and a belief system is a sequential equilibrium if it satisfies the following conditions:

(i) The strategy profile σ is sequentially rational relative to μ, and

(ii) There exists a sequence of completely mixed strategy profiles $(\sigma^k)_{k=1}^{\infty}$ with $\lim_{k\to\infty} \sigma^k = \sigma$ and a sequence of belief systems $(\mu^k)_{k=1}^{\infty}$ with $\lim_{k\to\infty} \mu^k = \mu$ such that μ^k is consistent with σ^k for all $k = 1, \ldots, \infty$.

Note in the above definition that for each k, the belief system μ^k must be derived from the strategy profile σ^k at each information set through the Bayes rule, since σ^k is completely unpure (mixed). Also note that each sequential equilibrium is a WPBE, since the limiting belief μ is consistent with σ; i.e., μ is equivalent to the belief that can be derived from σ on the play path of σ. But, the converse is not true. There are extensive form games where some WPBE is not a sequential equilibrium, as illustrated by the following example.

Example 11.7 Consider the extensive form game, Γ, in Example 11.6, which we reillustrate below.

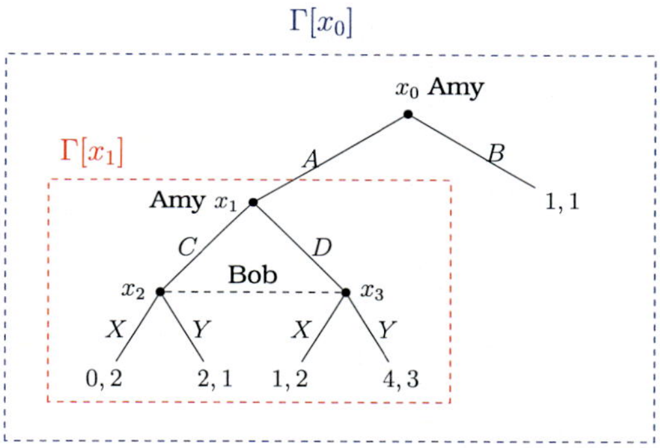

Figure 11.4 Reconsidered

We found that Γ has a unique SPNE, $\sigma^* = ((A, D), Y)$. We will show that Γ has a unique sequential equilibrium that contains σ^*.

Suppose a pair (σ, μ) is a sequential equilibrium of Γ. Then, we can pick a sequence of completely mixed strategy profiles $(\sigma^k)_{k=1}^{\infty}$ with $\lim_{k\to\infty} \sigma^k = \sigma$ and a sequence of belief systems $(\mu^k)_{k=1}^{\infty}$ with $\lim_{k\to\infty} \mu^k = \mu$ such that μ^k is consistent with σ^k for all $k \in \mathbb{N}$. Now, let us pick any k and consider σ^k and μ^k. Noting that $H^1_{Amy} = \{x_0\}$, $H^2_{Amy} = \{x_1\}$, and $H^1_{Bob} = \{x_2, x_3\}$, we can calculate the probability

$$\text{prob}(x_2 | H^1_{Bob}, \sigma^k) = \frac{\text{prob}(x_2|\sigma^k)}{\text{prob}(H^1_{Bob}|\sigma^k)}. \tag{11.1}$$

Using the Bayes rule we can write

$$\text{prob}(x_2|\sigma^k) = \text{prob}(x_2|H^2_{Amy}, \sigma^k)\text{prob}(H^2_{Amy}|\sigma^k), \quad \text{and} \tag{11.2}$$

$$\text{prob}(H^1_{Bob}|\sigma^k) = \text{prob}(H^1_{Bob}|H^2_{Amy}, \sigma^k)\text{prob}(H^2_{Amy}|\sigma^k). \tag{11.3}$$

Inserting (11.2) and (11.3) into (11.1) and making cancellations, we obtain

$$\text{prob}(x_2|H^1_{Bob}, \sigma^k) = \frac{\text{prob}(x_2|H^2_{Amy}, \sigma^k)}{\text{prob}(H^1_{Bob}|H^2_{Amy}, \sigma^k)}. \tag{11.4}$$

Note that $(H^1_{Bob}|H^2_{Amy}, \sigma^k) = 1$. Therefore,

$$\text{prob}(x_2|H^1_{Bob}, \sigma^k) = \text{prob}(x_2|H^2_{Amy}, \sigma^k). \tag{11.5}$$

Since μ^k is consistent with σ^k, we have $\mu^k(x_2) = \text{prob}(x_2|H^1_{Bob}, \sigma^k)$. From (11.5), it then follows that $\mu^k(x_2) = \text{prob}(x_2|H^2_{Amy}, \sigma^k)$.

Recall that $\lim_{k\to\infty} \sigma^k = \sigma$ and $\lim_{k\to\infty} \mu^k = \mu$. So, we must have $\mu(x_2) = \text{prob}(x_2|H^2_{Amy}, \sigma)$. We know that $\text{prob}(x_2|H^2_{Amy}, \sigma)$ is the probability that Amy plays the action C under her behavior strategy σ_{Amy} in the subgame $\Gamma[x_1]$. Therefore, given the belief system μ, the belief of Bob about the likelihood that the play of the game will pass through the node x_2 in the subgame $\Gamma[x_1]$ is exactly the likelihood that Amy will play C under her behavior strategy σ_{Amy} in the subgame $\Gamma[x_1]$.

Finally, since (σ, μ) is a sequential equilibrium, σ is sequentially rational relative to μ, implying that σ_{Amy} is rational at H^2_{Amy} relative to μ and σ_{Bob} is rational at H^1_{Bob} relative to μ. All of these facts ensure that σ must induce a Nash equilibrium in $\Gamma[x_1]$. We know from Example 11.6 that $\Gamma[x_1]$ has a unique Nash equilibrium (D, Y), yielding the expected payoff profile (4, 3). Since σ must be rational at H^1_{Amy} as well, Amy should play the action A at H^1_{Amy}, as it yields a higher expected payoff than obtained under the action B. Thus, we have established that (σ, μ) is a sequential equilibrium only if

$$\begin{aligned}\sigma = \big(&\text{Amy plays } A \text{ at the information set } H^1_{Amy} = \{x_0\} \text{ and}\\ &\text{plays } D \text{ at the information set } H^2_{Amy} = \{x_1\},\\ &\text{Bob plays } Y \text{ at the information set } H^1_{Bob} = \{x_2, x_3\}\big)\end{aligned}$$

and

$$\mu(x_0) = 1,\ \mu(x_1) = 1, \mu(x_2) = 0, \text{ and } \mu(x_3) = 1.$$

Thus, the profile (σ, μ) is a sequential equilibrium (and a WPBE), while σ is a SPNE (and a Nash equilibrium). ■

Recall from Example 11.6 that the game Γ contains a strategy profile $\hat{\sigma} = ((B, D), X)$, which arises in a WPBE. But, $\hat{\sigma}$ is not a SPNE and in Example 11.7 we establish that a behavior strategy profile, $\hat{\sigma}$, that is not subgame perfect is not contained by the singleton set of sequential equilibria. This result is not peculiar to the game in that example. As proposed by Kreps and Wilson [1], if an assessment (σ, μ) is a sequential equilibrium of an extensive form game, then σ is a subgame perfect Nash equilibrium (SPNE). Earlier, we mentioned that every sequential equilibrium is WPBE. Now, we know that an assessment (σ, μ) is a sequential equilibrium only if this assessment is WPBE and σ is SPNE.

Example 11.8 Consider the extensive form game in Example 11.5, which we reillustrate below.

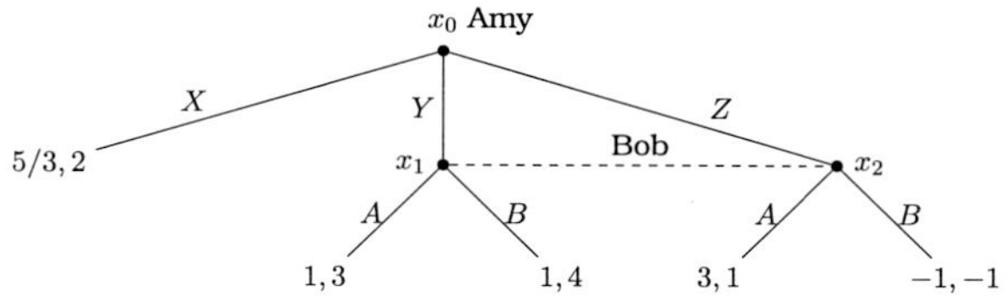

Figure 11.3 Reconsidered

We earlier found that the above game has four WPBEs. Let us check whether any of them is sequentially rational.

WPBE-1: $\sigma =$ (Amy plays Z, Bob plays A) and μ is such that $\mu(x_0) = 1$, $\mu(x_1) = 0$, and $\mu(x_2) = 1$.

Pick any $\delta \in (0, 1)$ and consider the sequence of belief systems $(\mu^k)_{k=1}^{\infty}$ where

$$\mu^k(x_0) = 1, \;\; \mu^k(x_1) = \delta^k, \;\; \text{and} \;\; \mu^k(x_2) = 1 - \delta^k$$

and the sequence of completely mixed strategy profiles $(\sigma^k)_{k=1}^{\infty}$ where

$$\sigma^k_{Amy} = \delta^k(X) + \delta^k(1 - \delta^k)(Y) + (1 - \delta^k)^2(Z)$$

and

$$\sigma^k_{Bob} = (1 - \delta^k)(A) + \delta^k(B).$$

Note that $\lim_{k\to\infty} \mu^k = \mu$ and $\lim_{k\to\infty} \sigma^k = (Z, A) = \sigma$. We know that σ is sequentially rational relative to μ since (σ, μ) is a WPBE. Also, μ^k is consistent with σ^k for all $k \in \mathbb{N}$ since we have

$$\text{prob}(x_0|\sigma^k, H_{Amy}) = 1 = \mu^k(x_0),$$

$$\text{prob}(x_1|\sigma^k, H_{Bob}) = \frac{\text{prob}(x_1|\sigma^k)}{\text{prob}(H_{Bob}|\sigma^k)} = \frac{\delta^k(1 - \delta^k)}{1 - \delta^k} = \delta^k = \mu^k(x_1)$$

and

$$\text{prob}(x_2|\sigma^k, H_{Bob}) = \frac{\text{prob}(x_2|\sigma^k)}{\text{prob}(H_{Bob}|\sigma^k)} = \frac{(1 - \delta^k)^2}{1 - \delta^k} = 1 - \delta^k = \mu^k(x_2).$$

Therefore, (σ, μ) is a sequential equilibrium.

WPBE-2: $\sigma = (\text{Amy plays } X, \text{ Bob plays} B)$ and μ is such that $\mu(x_0) = 1$, $\mu(x_1) = \alpha \in (2/3, 1]$, and $\mu(x_2) = 1 - \alpha$.

Pick any $\alpha \in (2/3, 1]$ and any $\delta \in (0, 1)$. Consider the sequence of belief systems $(\mu^k)_{k=1}^{\infty}$ where

$$\mu^k(x_0) = 1, \quad \mu^k(x_1) = \alpha \text{ and } \mu^k(x_2) = (1 - \alpha)$$

and the sequence of completely mixed strategy profiles $(\sigma^k)_{k=1}^{\infty}$ where

$$\sigma^k_{Amy} = (1 - \delta^k)(X) + \delta^k \alpha(Y) + \delta^k(1 - \alpha)(Z)$$

and

$$\sigma^k_{Bob} = \delta^k(A) + (1 - \delta^k)(B).$$

Note that $\lim_{k\to\infty} \mu^k = \mu$ and $\lim_{k\to\infty} \sigma^k = (X, B) = \sigma$. We know that σ is sequentially rational relative to μ since (σ, μ) is a WPBE. Also, μ^k is consistent with σ^k for all $k \in \mathbb{N}$ since we have

$$\text{prob}(x_0|\sigma^k, H_{Amy}) = 1 = \mu^k(x_0),$$

$$\text{prob}(x_1|\sigma^k, H_{Bob}) = \frac{\text{prob}(x_1|\sigma^k)}{\text{prob}(H_{Bob}|\sigma^k)} = \frac{\delta^k \alpha}{\delta^k} = \alpha = \mu^k(x_1)$$

and

$$\text{prob}(x_2|\sigma^k, H_{Bob}) = \frac{\text{prob}(x_2|\sigma^k)}{\text{prob}(H_{Bob}|\sigma^k)} = \frac{\delta^k(1 - \alpha)}{\delta^k} = 1 - \alpha = \mu^k(x_2).$$

Therefore, (σ, μ) is a sequential equilibrium.

WPBE-3: $\sigma(p) = (\text{Amy plays } X, \text{ Bob plays } p(A) + (1 - p)(B))$ where $p < 2/3$ and μ is such that $\mu(x_0) = 1$, $\mu(x_1) = 2/3$, and $\mu(x_2) = 1/3$.

Pick any $p \in [0, 2/3)$ and any $\delta \in (0, 1 - p)$. Consider the sequence of belief systems $(\mu^k)_{k=1}^{\infty}$ where

$$\mu^k(x_0) = 1, \quad \mu^k(x_1) = 2/3 \text{ and } \mu^k(x_2) = 1/3$$

and a sequence of completely mixed strategy profiles $(\sigma^k)_{k=1}^{\infty}$ where

$$\sigma^k_{Amy} = (1 - \delta^k)(X) + \frac{2\delta^k}{3}(Y) + \frac{\delta^k}{3}(Z)$$

and

$$\sigma^k_{Bob} = (p + \delta^k)(A) + (1 - p - \delta^k)(B).$$

Note that $\lim_{k\to\infty} \mu^k = \mu$ and $\lim_{k\to\infty} \sigma^k = (X, p(A) + (1-p)(B)) = \sigma$. We know that σ is sequentially rational relative to μ since (σ, μ) is a WPBE. Also, μ^k is consistent with σ^k for all $k \in \mathbb{N}$ since we have

$$\text{prob}(x_0|\sigma^k, H_{Amy}) = 1 = \mu^k(x_0),$$

$$\text{prob}(x_1|\sigma^k, H_{Bob}) = \frac{\text{prob}(x_1|\sigma^k)}{\text{prob}(H_{Bob}|\sigma^k)} = \frac{2\delta^k/3}{\delta^k} = \frac{2}{3} = \mu^k(x_1)$$

and

$$\text{prob}(x_2|\sigma^k, H_{Bob}) = \frac{\text{prob}(x_2|\sigma^k)}{\text{prob}(H_{Bob}|\sigma^k)} = \frac{\delta^k/3}{\delta^k} = \frac{1}{3} = \mu^k(x_2).$$

Therefore, (σ, μ) is a sequential equilibrium.

WPBE-4: $\sigma = \left(\text{Amy plays } X, \text{Bob plays } \frac{2}{3}(A) + \frac{1}{3}(B)\right)$ and μ is such that $\mu(x_0) = 1$, $\mu(x_1) = 2/3$, and $\mu(x_2) = 1/3$.

Pick any $\delta \in (0, 1)$. Consider the sequence of belief systems $(\mu^k)_{k=1}^{\infty}$ where

$$\mu^k(x_0) = 1, \quad \mu^k(x_1) = 2/3, \quad \text{and} \quad \mu^k(x_2) = 1/3$$

and a sequence of completely mixed strategy profiles $(\sigma^k)_{k=1}^{\infty}$ where

$$\sigma^k_{Amy} = (1-\delta^k)(X) + \frac{2\delta^k}{3}(Y) + \frac{\delta^k}{3}(Z)$$

and

$$\sigma^k_{Bob} = \frac{2}{3}(A) + \frac{1}{3}(B).$$

Note that $\lim_{k\to\infty} \mu^k = \mu$ and $\lim_{k\to\infty} \sigma^k = (X, \frac{2}{3}(A) + \frac{1}{3}(B)) = \sigma$. We know that σ is sequentially rational relative to μ since (σ, μ) is a WPBE. Also, μ^k is consistent with σ^k for all $k \in \mathbb{N}$ since we have

$$\text{prob}(x_0|\sigma^k, H_{Amy}) = 1 = \mu^k(x_0),$$

$$\text{prob}(x_1|\sigma^k, H_{Bob}) = \frac{\text{prob}(x_1|\sigma^k)}{\text{prob}(H_{Bob}|\sigma^k)} = \frac{2\delta^k/3}{\delta^k} = \frac{2}{3} = \mu^k(x_1)$$

and

$$\text{prob}(x_2|\sigma^k, H_{Bob}) = \frac{\text{prob}(x_2|\sigma^k)}{\text{prob}(H_{Bob}|\sigma^k)} = \frac{\delta^k/3}{\delta^k} = \frac{1}{3} = \mu^k(x_2).$$

Therefore, (σ, μ) is a sequential equilibrium. ■

For the extensive form games in Examples 11.7 and 11.8, we showed that sequential equilibria exist. However, this existence result is not peculiar to these games. A proposition of Kreps and Wilson [1] shows that every extensive form game has at least one sequential equilibrium.

11.3 Chapter Appendix

Proof of Proposition 11.1 Let Γ be an extensive form game. Suppose a behavior strategy profile σ and a belief system μ constitute a WPBE of Γ. The expected payoff of player i at σ is given by

$$\tilde{u}_i(\sigma) = \bar{p}(\sigma|\overline{\mathcal{H}}_i)\bar{u}(\sigma|\overline{\mathcal{H}}_i) + \sum_{H_i \in \overline{\mathcal{H}}_i} \text{prob}(H_i|\sigma)\tilde{u}_i(\sigma_i, \sigma_{-i}|H_i, \mu). \tag{11.6}$$

If player i deviates to another behavior strategy σ_i', her expected payoff becomes

$$\begin{aligned}\tilde{u}_i(\sigma_i', \sigma_{-i}) &= \bar{p}(\sigma_i', \sigma_{-i}|\overline{\mathcal{H}}_i)\bar{u}(\sigma_i', \sigma_{-i}|\overline{\mathcal{H}}_i) \\ &\quad + \sum_{H_i \in \overline{\mathcal{H}}_i} \text{prob}(H_i|\sigma_i', \sigma_{-i})\tilde{u}_i(\sigma_i', \sigma_{-i}|H_i, \mu).\end{aligned}$$

Since $\bar{p}(\sigma_i', \sigma_{-i}|\overline{\mathcal{H}}_i) = \bar{p}(\sigma|\overline{\mathcal{H}}_i)$, $\bar{u}(\sigma_i', \sigma_{-i}|\overline{\mathcal{H}}_i) = \bar{u}(\sigma|\overline{\mathcal{H}}_i)$, and $\text{prob}(H_i|\sigma_i', \sigma_{-i}) = \text{prob}(H_i|\sigma)$ for each $H_i \in \overline{\mathcal{H}}_i$, the above equation becomes

$$\tilde{u}_i(\sigma_i', \sigma_{-i}) = \bar{p}(\sigma|\overline{\mathcal{H}}_i)\bar{u}(\sigma|\overline{\mathcal{H}}_i) + \sum_{H_i \in \overline{\mathcal{H}}_i} \text{prob}(H_i|\sigma)\tilde{u}_i(\sigma_i', \sigma_{-i}|H_i, \mu). \tag{11.7}$$

Because (σ, μ) is a WPBE, σ is sequentially rational relative to μ. Thus, σ is rational at every $H_i \in \mathcal{H}_i$ such that $\text{prob}(H_i|\sigma) > 0$, i.e.,

$$\tilde{u}_i(\sigma_i, \sigma_{-i}|H_i, \mu) \geq \tilde{u}_i(\sigma_i', \sigma_{-i}|H_i, \mu). \tag{11.8}$$

The fact that $\overline{\mathcal{H}}_i \subseteq \mathcal{H}_i$, the inequality (11.8), and the Eqs. (11.6)–(11.7) together imply $\tilde{u}_i(\sigma) \geq \tilde{u}_i(\sigma_i', \sigma_{-i})$, proving that σ is a Nash equilibrium. ■

Proof of Proposition 11.2 Let Γ be an extensive form game with perfect information. First suppose (σ, μ^{PI}) is a WPBE. Consider any subgame Γ' of Γ. Obviously, the restriction of (σ, μ^{PI}) on Γ' is a WPBE. Then, by Proposition 11.1, the restriction of σ on Γ' is a Nash equilibrium of Γ'. Since this is true for any subgame Γ' in Γ, σ is a SPNE of Γ.

Now, suppose that σ is a SPNE of Γ. Pick any decision node $x \in \mathcal{X} \setminus \mathcal{X}^T$ and consider the subgame $\Gamma[x]$. Define the information set $H[x] = \{x\}$. The player assigned to $H[x]$ is $i[H[x]]$. Since σ is a SPNE of Γ, σ induces a Nash equilibrium on $\Gamma[x]$. Then,

$$\tilde{u}_i(\sigma_{i[H[x]]}, \sigma_{-i[H[x]]}|H[x], \mu^{PI}) \geq \tilde{u}_i(\sigma'_{i[H[x]]}, \sigma_{-i[H[x]]}|H[x], \mu^{PI}) \tag{11.9}$$

for any $\sigma'_{i[H[x]]}$, implying that σ is rational at $H[x]$ relative to μ^{PI}. Since this is true for all $x \in \mathcal{X} \setminus \mathcal{X}^T$, σ is rational at each information set $H \in \mathcal{H}$ relative to μ^{PI}.

Obviously, μ^{PI} is consistent with σ. If σ passes trough an information set $H = \{x\} \in \mathcal{H}$, then $\text{prob}(x|\sigma) = \text{prob}(H|\sigma)$. Thus, $\text{prob}(x|H, \sigma) = \text{prob}(x|\sigma)/\text{prob}(H|\sigma) = 1$. This is ensured by μ^{PI}, since $\mu^{PI}(x) = 1$ for any $x \in \mathcal{X} \setminus \mathcal{X}^T$. On the other hand, If σ does not pass through an information set $H = \{x\} \in \mathcal{H}$, then the Bayes rule does not need not to apply. Therefore, μ^{PI} is consistent with σ. This proves that (σ, μ^{PI}) is a WPBE. ∎

References

1. Kreps DM, Wilson RB (1982) Sequential equilibria. Econometrica 50:863–894
2. Myerson RB (1991) Game theory: analysis of conflict. Harvard University Press, Cambridge, Massachusetts

Repeated Games

12

In this chapter, we will study extensive form games where a strategic form game (the stage game) is repeatedly played for a finite or infinite number of periods. A well-known real-life example is the repeated strategic interaction of firms competing in an oligopoly. After formally defining finitely, and infinitely, repeated games, we will show that whether and why the repetition of a strategic form game may have subgame-perfect Nash equilibria that cannot be obtained by repeatedly playing the Nash equilibria of that strategic form game. We will also state and prove a result that characterizes the set of expected payoffs in a strategic form game that can be obtained as the limit average payoff of some subgame-perfect Nash equilibrium of the infinite repetition of that game.

12.1 Finitely Repeated Games

Consider a strategic form game, $G = \langle N, (A_i)_{i \in N}, (u_i)_{i \in N} \rangle$, where $N = \{1, 2, \ldots, n\}$ with $n \geq 2$ is the set of players, A_i is the set of actions of player $i \in N$, and u_i is the utility function of player $i \in N$ mapping each action $a_i \in A_i$ to a real number. Let $A = \times_{i \in N} A_i$. Suppose that the strategic form game G is played in each period $t = 1, 2, \ldots, T$, where T is a finite integer greater than 1. We assume that each player discounts the future payoffs by a common discount factor $\delta \in (0, 1]$. (Since T is finite, we do not need to assume $\delta < 1$ to make the discounted sum of the T-period payoffs bounded.) We will call the game G as the stage game and let $G^\delta[T]$ denote the associated T-period repeated game with discount factor δ.

We assume that at the end of each period t, all players can perfectly observe (monitor) the action profile $a = (a_1, a_2, \ldots a_n)$ chosen. We denote by h^t a history of actions up to (but excluding) period t. That is, for each $t = 1, \ldots, T$,

$$h^t = (a^1, \ldots, a^{t-1}),$$

I. Saglam, *Mastering Game Theory*, Classroom Companion: Economics,
https://doi.org/10.1007/978-3-031-82094-6_12

where $a^\tau \in A$ for each $\tau = 1, \ldots, t-1$. Thus, the set of possible histories up to period t is

$$H^t = \underbrace{A \times A \times \ldots A}_{t-1 \text{ times}} = A^{t-1},$$

where $H^1 = \emptyset$. We denote by $H[T]$ the set of possible histories that can occur in the game $G^\delta[T]$; i.e.,

$$H[T] = \cup_{t=1}^{T} H^t.$$

We represent the strategy of player $i \in N$ by a function $s_i : H[T] \to A_i$. We write this strategy as $s_i = (s_i^1, s_i^2, \ldots, s_i^T)$, where $s_i^t : H^t \to A_i$ for each $t = 1, \ldots, T$.

Given a strategy profile $s = (s_1, s_2, \ldots, s_n)$, the induced $T-$period outcome is a profile $y(s) = (y^1(s), y^2(s), \ldots, y^T(s))$, where, for each $i \in N$, we have $y_i^1(s) = s_i(\emptyset)$ and $y_i^t(s) = s_i(y_i^{t-1})$ for each $t = 2, \ldots, T$. Given a strategy profile s, we denote by $V_i^{\delta,T}(s)$ the discounted sum of the T-period payoffs of player i, which is given by

$$V_i^{\delta,T}(s) = \sum_{t=1}^{T} \delta^{t-1} u_i(y_i^t(s)).$$

Henceforth, we will assume that N is finite and the stage game G is a mixed extension of a finite game, i.e., for each $i \in N$, there exists some finite set of actions F_i such that $A_i = \Delta(F_i)$. Thus, the action space of each player contains mixed (pure or unpure) actions. Let $NE(G) = \{a_1^*, \ldots, a_k^*\}$ denote the set of Nash equilibrium profiles of G. Since G is a mixed extension of a finite game, we know by Nash's [1] existence result (Proposition 4.1) that $NE(G) \neq \emptyset$.

Finally, for any action profile $a \in A$, let $sec(a, t-1) = (a, a, \ldots, a)$ denote a constant sequence where the profile a appears $t-1$ times.

To show that a particular strategy profile is a subgame-perfect Nash equilibrium of a (finitely or infinitely) repeated game, we will check whether the following property is satisfied.

One-Period Deviation Principle [2]. A strategy profile in a repeated (or extensive form) game satisfies the one-period deviation principle if no player can increase his payoff in any subgame through a one-period deviation from the given strategy profile.

This principle offers an enormous computational benefit for repeated games due to a result (inspired by the work of Blackwell, [2]) that a strategy profile can be a subgame-perfect Nash equilibrium if and only if it satisfies the one-period deviation principle. (Moreover, in a finitely repeated game this result holds even when the players do not discount the future payoffs, i.e., $\delta = 1$.) The proof of this result hinges upon the observations that (i) the discounted sum of the lifetime payoffs of players in a repeated game are additively separable over the stages (periods) of the game and (ii) any deviation from a strategy profile can be decomposed into a sequence of one-period deviations. Therefore, if there exists no one-period deviation that can increase the payoff of a deviant player, then there cannot exist any profitable deviation from the given strategy profile.

Now, we are ready to present the first result of this chapter.

Proposition 12.1 *Let $(a^1, a^2, \dots, a^T)$ be a T-period sequence of actions where $a^t \in NE(G)$ for each $t = 1, 2, \dots, T$ and let $s = (s_1, s_2, \dots, s_n)$ be a strategy profile such that $s_i^t(h^t) = a_i^t$ for each $i \in N$, $t = 1, 2, \dots, T$, and $h^t \in H^t$. Then, the profile s is a subgame-perfect Nash equilibrium of the T-period repeated game $G^\delta[T]$ for any $\delta \in (0, 1]$.*

Proof To prove this proposition, it is sufficient to check whether the strategy profile s satisfies the one-period deviation principle. Note that in any period $t = 1, 2, \dots, T$, the action profile a^t played is a Nash equilibrium profile of the stage game G. Therefore, no player $i \in N$ can increase its period t payoff by using a one-period deviation in period t. Moreover, the one-period deviation of any player i from the strategy s_i in any period t before the last period T cannot affect the future actions of her opponents when they follow their equilibrium strategies in s_i, since the equilibrium strategy of each player in any period is independent of the history of the game. Since s satisfies one-period deviation principle, it is a subgame-perfect Nash equilibrium of $G^\delta[T]$ for any $\delta \in (0, 1]$. ■

The above proposition states that a strategy profile s where the players play in any period one of the Nash equilibrium action profiles of the stage game G is a Nash equilibrium of the T-period repeated game. Since each period starts a subgame in the extensive form representation of a repeated game, the described strategy profile is also subgame perfect.

For example, if the stage game G has two Nash equilibrium profiles a and a', then the strategy profiles that lead to the sequences of actions (a, a), (a, a'), (a', a), and (a', a') are all (subgame-perfect) Nash equilibria of the 2-period repeated game $G^\delta[2]$ for any $\delta \in (0, 1]$.

Proposition 12.1 does not state or imply that the Nash equilibria of the T-period repeated game must only involve the strategy profiles under which the players play in each period one of the Nash equilibrium action profiles. We will later show that there are finitely repeated games that have (additional) Nash equilibria where players play in each period some action profiles that are not in the set of the Nash equilibria of the stage game. We may call these repeated games with additional Nash equilibria (uncovered by Proposition 12.1) nontrivial (or rich). The next result shows that when the stage game has a unique Nash equilibrium profile, the induced finitely repeated game must be trivial.

Proposition 12.2 *Let G be a stage game that has a unique Nash equilibrium profile of actions a^* and let $s = (s_1, s_2, \dots, s_n)$ be a strategy profile such that $s_i^t(h^t) = a_i^*$ for each $i \in N$, $t = 1, 2, \dots, T$, and $h^t \in H^t$. Then, the profile s is the unique (subgame-perfect) Nash equilibrium of the T-period repeated game $G^\delta[T]$ for any $\delta \in (0, 1]$.*

Proof Proposition 12.1 implies that the strategy profile s is a (subgame-perfect) Nash equilibrium of the T-period repeated game $G^\delta[T]$ for any $\delta \in (0, 1]$. Now, suppose that there is a strategy profile $\hat{s} \neq s$ that is also a Nash equilibrium of the T-period repeated game $G^\delta[T]$ for any $\delta \in (0, 1]$. Consider the last period $t = T$. Since the stage game G has a unique Nash equilibrium profile, a^*, the profile $\hat{s}$ can be an equilibrium only if $\hat{s}_i^T(h^T) = a_i^*$ for each $i \in N$ and $h^T \in H^T$. If this condition was violated for any player i, she could increase its period T payoff by deviating from $\hat{s}_i^T$. Since T is finite, using backward induction we can make similar arguments for periods $T - 1, T - 2$, and so on, up to period 1, contradicting that $\hat{s}$ is different from s. ■

Below, we give an example of a nontrivial repeated game, which involves a Nash equilibrium uncovered by Proposition 12.1.

Example 12.1 Consider a stage game, G, which is equal to the mixed extension of the strategic form game in Table 12.1.

Let us first find the Nash equilibria of G. Note that the action A is strongly dominated by C for both players. Therefore, a mixed action profile that puts a positive probability on A for any player cannot be a Nash equilibrium. Once we eliminate the action A for both players from G, the reduced game has three Nash equilibria: the action profiles (B, B), (C, C), and $(\frac{1}{5}B + \frac{4}{5}C, \frac{1}{5}B + \frac{4}{5}C)$, leading to the expected payoffs $(4, 4)$, $(1, 1)$, and $(4/5, 4/5)$, respectively.

Now, we will check whether for any value of $\delta \in (0, 1]$ the two-period repeated game $G^\delta[2]$ has a Nash equilibrium where players play in period 1 an action profile that is not a Nash equilibrium of the stage game G.

Consider a strategy profile s for $G^\delta[2]$ such that $s_i^1(\emptyset) = A$ for each $i = 1, 2$ and

$$s_i^2(h^2) = \begin{cases} B \text{ if } h^2 = ((A, A)) \\ C \text{ if } h^2 \neq ((A, A)). \end{cases}$$

The described strategy profile s induces the outcome profile $y(s) = (y^1(s), y^2(s))$ where, for each $i \in N$, we have $y_i^1(s) = A$ and $y_i^2(s) = B$. So, if all players stick to their strategies in s, then the discounted sum of the 2-period payoffs of player $i \in N$ becomes

$$V_i^{\delta,2}(s) = \sum_{t=1}^{2} \delta^{t-1} u_i(y^t(s))$$

$$= u_i((A, A)) + \delta u_i((B, B)) = 5 + \delta(4).$$

Table 12.1 The stage game in Example 12.1

Player 1 \ Player 2	A	B	C
A	(5, 5)	(−1, 0)	(0, 6)
B	(0, −1)	(4, 4)	(0, 0)
C	(6, 0)	(0, 0)	(1, 1)

Note that no player has an incentive to deviate from s in period 2, since s prescribes the players to play a Nash equilibrium action profile of G in period 2. So, if any player finds deviation from s profitable, this deviation must be in period 1. Also note that since each player obtains a payoff of 5 under s in period 1, any player can increase her period 1 payoff only by switching to C, which would yield a payoff of 6. However, this deviation would be punished, as prescribed by s, in the second period since her opponent would switch to play C instead of B. Because of this punishment, the deviant player would obtain a payoff of 1 in the second period (instead of a payoff of 4 that she would obtain in case of no deviation). Therefore, if a player $i \in N$ deviates from $s_i = (A, s_i^2)$ to $s_i' = (C, s_i^2)$, then the discounted sum of the 2-period payoffs of player i becomes

$$V_i^{\delta,2}(s_i', s_{-i}) = \sum_{t=1}^{2} \delta^{t-1} u_i(y^t(s))$$

$$= u_i((C, A)) + \delta u_i((C, C)) = 6 + \delta(1).$$

Note that $V_i^{\delta,2}(s_i, s_{-i}) \geq V_i^{\delta,2}(s_i', s_{-i})$ if and only if

$$5 + 4\delta \geq 6 + \delta$$

or $\delta \geq 1/3$. This implies that the profile s will be a Nash equilibrium of $G^\delta[2]$ if and only if $\delta \in [1/3, 1]$. ■

12.2 Infinitely Repeated Games

In this section, we assume that $T = \infty$, i.e., the stage game is infinitely repeated starting from period one. For any strategy profile s in the infinitely repeated game, we define the discounted sum of the infinite-period (lifetime) payoffs of player $i \in N$ as

$$V_i^{\delta,[1,\infty]}(s) = \sum_{t=1}^{\infty} \delta^{t-1} u_i(y^t(s))$$

whenever $\delta \in (0, 1)$ is the common discount factor of players in N. (Note that to ensure that $V_i^{\delta,[1,\infty]}(s)$ is bounded, we have now assumed that $\delta < 1$.) Likewise, we will define the discounted sum of the payoffs from any period τ to ∞ as

$$V_i^{\delta,[\tau,\infty]}(s) = \sum_{t=\tau}^{\infty} \delta^{t-1} u_i(y^t(s)).$$

We retain all other assumptions, definitions, and structures in the previous section where we dealt with finitely repeated games.

An immediate question is whether our results obtained for the case of finitely repeated games extend to the case of infinitely repeated games. For Proposition 12.1, we can easily see that the answer is "yes". A strategy profile s where the players play in any period one of the Nash equilibrium action profiles of the stage game G is a Nash equilibrium of the infinitely repeated game $G^{\delta}[\infty]$ for all values of $\delta \in (0, 1)$. The reason is that under such a strategy profile s no player can increase her payoff in any period by deviating in that period from the action profile specified by s for that period since that action profile is a Nash equilibrium of the stage game G. Also, no player can affect, by deviating in any period from s, the future actions of her opponents and her own future payoffs since the equilibrium strategy of each player in any period is independent of the history of the game.

On the other hand, Proposition 12.2 is not valid for infinitely repeated games. That is, we can find a stage game G with a unique Nash equilibrium profile $a^* \in A$ and a discount factor $\delta \in (0, 1)$ such that the associated infinitely repeated game $G^{\delta}[\infty]$ has a Nash equilibrium strategy profile under which players do not play the action profile a^* in some (even in any) period. The following example illustrates this result.

Example 12.2 Consider a stage game, G, which is equal to the mixed extension of the strategic form game in Table 12.2.

The stage game G is a Prisoner's Dilemma Game. It has a unique Nash equilibrium, which is the action profile (B, B), and this equilibrium is the unique action profile that is not Pareto efficient. Given a common discount factor $\delta \in (0, 1)$ for the two players, we can now consider the repeated game $G^{\delta}[\infty]$ associated with the stage game G.

Consider a strategy profile s for $G^{\delta}[\infty]$ such that $s_i^1(\emptyset) = A$ and for each $t = 2, 3, \ldots, \infty$

$$s_i^t(h^t) = \begin{cases} A \text{ if } h^t = sec((A, A), t-1) \\ B \text{ if } h^t \neq sec((A, A), t-1). \end{cases}$$

The described strategy profile s induces the outcome profile $y(s) = (y^1(s), y^2(s), \ldots, y^T(s))$ where, for each $t = 1, 2, \ldots, T$, we have $y^t(s) = (A, A)$. Note that if a strategy profile is a subgame-perfect Nash equilibrium, it must satisfy the one-period deviation principle. So, let us check whether s satisfies this principle. Pick any $i \in N$ and period $\tau \geq 1$. If player i deviates in period τ from the strategy profile s, then she should deviate to play the action B in period τ, which would yield a payoff of 4. However, this deviation would be punished, as prescribed by s, from the period $\tau + 1$ onwards since her opponent would switch to play the action B prescribed by the unique Nash equilibrium of G. Because of this punishment, the deviant player

Table 12.2 The stage game in Example 12.2

		Player 2: A	Player 2: B
Player 1	A	3, 3	0, 4
	B	4, 0	1, 1

would need to play the action B starting from the period $\tau + 1$ onwards and obtain a payoff of 1 (instead of a payoff of 3 that she would obtain in case of no deviation). Therefore, if a player $i \in N$ deviates from $s_i = (s_i^1, \ldots, s_i^{\tau-1}, s_i^{\tau}, s_i^{\tau+1}, \ldots, s_i^{\infty})$ to $s_i' = (s_i^1, \ldots, s_i^{\tau-1}, B, s_i^{\tau+1}, \ldots, s_i^{\infty})$, then the discounted payoff sum of player i starting from period τ onwards becomes

$$V_i^{\delta,[\tau,\infty]}(s_i', s_{-i}) = \sum_{t=\tau}^{\infty} \delta^{t-1} u_i(y^t((s_i', s_{-i}))) = \delta^{\tau-1} u_i((B, A)) + \delta^{\tau} \frac{u_i((B, B))}{1-\delta}$$

$$= \delta^{\tau-1}\left(4 + \delta\frac{1}{1-\delta}\right) = \delta^{\tau-1}\left(\frac{4-3\delta}{1-\delta}\right).$$

On the other hand, if all players stick to their strategies in s, then the discounted payoff sum of player i starting from period τ onwards becomes

$$V_i^{\delta,[\tau,\infty]}(s) = \sum_{t=1}^{\infty} \delta^{t-1} u_i(y^t(s)) = \delta^{\tau-1}\left(\frac{u_i((A, A))}{1-\delta}\right)$$

$$= \delta^{\tau-1}\left(\frac{3}{1-\delta}\right).$$

Note that one-period deviation is not profitable if and only if

$$V_i^{\delta,[\tau,\infty]}(s_i, s_{-i}) \geq V_i^{\delta,[\tau,\infty]}(s_i', s_{-i})$$

or equivalently

$$\delta^{\tau-1}\left(\frac{3}{1-\delta}\right) \geq \delta^{\tau-1}\left(\frac{4-3\delta}{1-\delta}\right)$$

implying $\delta \geq 1/3$. Since $\delta \in (0, 1)$, this condition implies that the profile s will be a Nash equilibrium of $G^{\delta}[\infty]$ if and only if $\delta \in [1/3, 1)$.

Note that s is also subgame perfect since the restriction of s on the subgames that are off the equilibrium path (where the punishment strategies of players get into the play) is also a Nash equilibrium (with the action profile (B, B) being played in each period of these subgames). ■

In the above example, the stage game is a Prisoner's Dilemma Game. We know that this game has a unique Nash equilibrium. Recall from the previous section, if a stage game G with a unique Nash equilibrium action profile a is repeated for a finite number of periods, then the repeated game has a unique subgame-perfect Nash equilibrium strategy profile and in that equilibrium strategy profile the players must play the unique Nash equilibrium action profile a in each period. On the other hand, the above example shows that when the same stage game is repeated for an infinite number of periods, the resulting repeated game may have additional subgame-perfect

Nash equilibria. The reason why these additional equilibria may arise in an infinitely, but not finitely, repeated Prisoner's Dilemma Game is simply because that there is no last period when the number of periods, T, is infinite.

When T is finite, the players must play in the last period of the game the unique Nash equilibrium of the Prisoner's Dilemma Game, since otherwise always one of the players would have an incentive to deviate from the action profile prescribed by their strategy profile. This means that the players cannot condition their strategies in period T on the history of their actions before period T. Thus, the strategies of players in period T cannot be used as a punishment device to prevent the players from any possible deviation from their original strategies in period $T-1$. Consequently, the players must play in period $T-1$, and by the same argument in every period of the game, the unique Nash equilibrium of the Prisoner's Dilemma game.

On the other hand, when T is infinite, there is always a future period (more precisely infinitely many future periods) during the game. Therefore, when T is infinite, there is no period where players must play a Nash equilibrium of the stage game unconditionally. For each period of the game (after the first period), the players find it possible to condition their strategies on the past history of the action profiles. This conditioning allows the players in each period after the first period to punish a player who unilaterally deviates from an equilibrium strategy profile in the future periods of the game. Whether this punishment can be deterring for a potential deviant depends, of course, on the payoff structure of the stage game and the discount factor of players.

In the above example, there are other (uncountably many) subgame-perfect Nash equilibria that we have not calculated. The existence of these additional equilibria will become apparent after we present below a folk (anonymous) theorem that deals with the characterization of the set of all subgame perfect Nash equilibria of any infinitely repeated game. But, first we have to introduce some preliminaries.

Given a stage game G, let $CP(G)$ denote the convex polytope associated with G; i.e. the convex hull of the set of pure action payoffs in G. To formally define $CP(G)$, recall that we have described a stage game G by the tuple $G = \langle N, (A_i)_{i\in N}, (u_i)_{i\in N}\rangle$, where each A_i satisfies $A_i = \Delta(F_i)$ for some finite set of actions F_i. Let $F = \times_{i\in N} F_i$. For convenience, let us denote F by the finite set $\{f^1, f^2, \ldots, f^k\}$ for some integer k. The set of pure action payoff profiles in G is the set

$$Z = \{z \in \mathbb{R}^n : z = u(f) \text{ for some } f \in F\}.$$

The convex polytope of G is the convex hull of Z; therefore

$$CP(G) = \{v \in \mathbb{R}^n : \exists(\alpha^1, \ldots, \alpha^k) \in \mathbb{R}^k_+ \text{ with } \sum_{j=1}^{k} \alpha^j = 1$$

$$\text{such that } v = \sum_{j=1}^{k} \alpha^j u(f^j)\}.$$

Fig. 12.1 Convex Polytope of G in Example 12.2

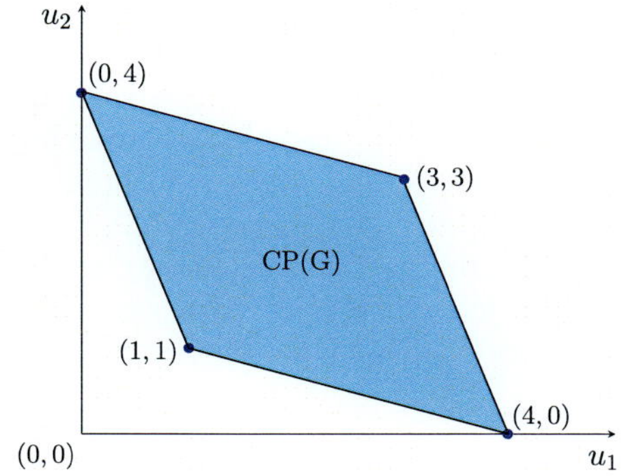

For two-player games, the convex polytope of G is also called the polygon of G. In Fig. 12.1, we plot the convex polytope $CP(G)$ of the Prisoner's Dilemma Game G in Example 12.2 using the pure action payoffs in Table 12.2.

Given a stage game $G = \langle N, (A_i)_{i\in N}, (u_i)_{i\in N}\rangle$, consider a sequence of action profiles $(a^1, a^2, \ldots, a^\infty)$ where $a^t \in A = \times_{i\in N} A_i$ for each $t \in \{1, 2, \ldots, \infty\}$. For this sequence, we define the limit average payoff of player i as $\lim_{T\to\infty} \frac{1}{T}\sum_{t=1}^{T} u_i(a^t)$. Note that for each element a^t of the given sequence, the utility profile $(u_1(a^t), \ldots, u_n(a^t))$ is in $CP(G)$. Therefore, the limit average payoff is also in $CP(G)$.

Also note that since the set $CP(G)$ is the convex hull of the pure action payoffs in G, for every payoff profile u in $CP(G)$ we can find a sequence of action profiles $(a^1, a^2, \ldots, a^\infty)$ with $a^t \in A$ for each $t \in \{1, 2, \ldots, \infty\}$ such that $u = \lim_{T\to\infty} \frac{1}{T}\sum_{t=1}^{T} u_i(a^t)$.

For a sequence of action profiles $(a^1, a^2, \ldots, a^\infty)$ with $a^t \in A$ for each $t \in \{1, 2, \ldots, \infty\}$, we will define the average discounted sum of the payoffs from any period τ to period ∞ as

$$\tilde{V}_i^{\delta,[\tau,\infty]}(s) = (1-\delta)\sum_{t=\tau}^{\infty} \delta^{t-1} u_i(a^t).$$

Proposition 12.3 (Folk Theorem) *Let $G = \langle N, (A_i)_{i\in N}, (u_i)_{i\in N}\rangle$ be any stage game and the action profile a^* be a Nash equilibrium of G. Let $\omega^* = u(a^*)$. Suppose there exists a payoff profile λ in $CP(G)$ such that $\lambda_i > \omega_i^*$ for all $i \in N$. Pick such a payoff profile λ. Then, there exists some $\delta^* \in (0, 1)$ such that for any $\delta \in [\delta^*, 1)$ the payoff profile λ can be obtained as the limit average payoff of a subgame-perfect Nash equilibrium strategy profile of the repeated game $G^\delta[\infty]$.*

Proof See the appendix at the end of this chapter. ■

The following example will help us better understand the above proposition.

Example 12.3 Reconsider the infinitely repeated Prisoner's Dilemma Game in Example 12.2, where we proved that if $\delta \in [1/3, 1)$, the payoff profile $(3, 3)$ can be obtained as the limit average payoff of some subgame-perfect Nash equilibrium strategy profile of the repeated game $G^{\delta}[T]$. Proposition 12.3 shows that we can extend this result for any payoff profile that Pareto dominates a Nash equilibrium payoff profile of the stage game G. Recall that the unique Nash equilibrium action profile of the Prisoner's Dilemma Game in Example 12.2 is (B, B) yielding the payoff profile $(1, 1)$. In Fig. 12.2, we plot the set of payoffs that Pareto dominate the payoff profile $(1, 1)$ (the dark blue colored region).

Proposition 12.3 ensures that we can obtain any payoff profile in the dark blue colored region in $CP(G)$ that Pareto dominates the Nash equilibrium payoff profile $(1, 1)$ as the limit average payoff of some subgame-perfect Nash equilibria of $G^{\delta}[\infty]$ if δ is sufficiently close to 1. So, let us consider the payoff profile $(1.5, 3.5)$. This profile Pareto dominates the Nash equilibrium payoff profile $(1, 1)$ of the stage game G. Also, the payoff profile $(1.5, 3.5)$ is the midpoint of the line segment connecting the payoff profiles $(0, 4)$ and $(3, 3)$ in Fig. 12.2. Therefore, we can easily see that the sequence of action profiles $(a^1, a^2, \ldots, a^{\infty})$, where $a^t = (A, B)$ if $t = 1, 3, 5, \ldots,$ and $a^t = (A, A)$ if $t = 2, 4, 6, \ldots,$, generates the limit average payoff $(1.5, 3.5)$.

Now, pick a $\delta \in (0, 1)$ and consider a strategy profile s in the repeated game $G^{\delta}[\infty]$ such that $s_i^1(\emptyset) = a_i^1$ and for each $t = 2, 3, \ldots, \infty$

$$s_i^t(h^t) = \begin{cases} a_i^t \text{ if } h^t = (a^1, \ldots, a^{t-1}) \\ B \text{ if } h^t \neq (a^1, \ldots, a^{t-1}). \end{cases}$$

The described strategy profile s induces the outcome profile $y(s) = (y^1(s), y^2(s), \ldots, y^{\infty}(s))$, where, for each $t = 1, 2, \ldots, \infty$, we have $y^t(s) = a^t$.

If the strategy profile s is a subgame-perfect Nash equilibrium of $G^{\delta}[\infty]$, it must satisfy the one-period deviation principle. First, consider player 1. If she were not sufficiently punished for deviation, she would like to deviate in any period τ from s_i to $\hat{s}_i$ such that $\hat{s}_i^{\tau} = B$. So, under the deviation strategy $\hat{s}_i$ in any period τ, player 1's average discounted sum of the payoffs from period τ to period ∞ becomes

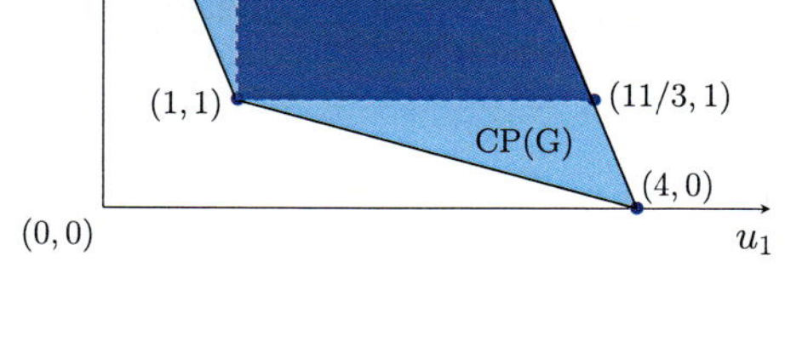

Fig. 12.2 Payoff profiles (with dark blue color) that can be obtained in Example 12.2 as the limit average payoff at some SPNE of $\Gamma^{\delta}[\infty]$ for sufficiently high values of $\delta \in (0, 1)$

$$V_1^{\delta,[\tau,\infty]}(\hat{s}_1, s_{-1}) = (1-\delta)\left(\delta^{\tau-1}u(B, s_{-1}^{\tau}) + \sum_{t=\tau+1}^{\infty} \delta^{t-1}u_1((B,B))\right)$$
$$= (1-\delta)\left(\delta^{\tau-1}u_1(B, s_{-1}^{\tau}) + \sum_{t=\tau+1}^{\infty} \delta^{t-1}\right)$$
$$= (1-\delta)\delta^{\tau-1}\left(u_1(B, s_{-1}^{\tau}) + \frac{\delta}{1-\delta}\right).$$

If player 1 sticks to the strategy s_1, her average discounted sum of the payoffs from period τ to period ∞ becomes

$$V_i^{\delta,[\tau,\infty]}(s) = (1-\delta)\sum_{t=\tau}^{\infty}\delta^{t-1}u_i(a^t) = (1-\delta)\delta^{\tau-1}(3+\delta^2(3)+\delta^4(3)+\ldots)$$
$$= (1-\delta)\frac{3\delta^{\tau-1}}{1-\delta^2}.$$

So, one-period deviation would not profitable for player 2 in period τ if and only if

$$V_1^{\delta,[\tau,\infty]}(s_1, s_{-1}) \geq V_1^{\delta,[\tau,\infty]}(\hat{s}_1, s_{-1})$$

or equivalently

$$\frac{3}{1-\delta^2} \geq u_1(B, s_{-1}^{\tau}) + \frac{\delta}{1-\delta}$$

or

$$0 \geq -3 + u_1(B, s_{-1}^{\tau})(1-\delta^2) + \delta(1+\delta). \quad (12.1)$$

Note that $u_1(B, s_{-1}^{\tau}) = 1$ if τ is odd and the inequality in (12.1) reduces to

$$0 \geq -2 + \delta,$$

which holds for all values of $\delta \in (0, 1)$. Also, note that $u_1(B, s_{-1}^{\tau}) = 4$ if τ is even and the inequality in (12.1) reduces to

$$0 \geq 1 + \delta - 3\delta^2,$$

which holds if $\delta \geq (1+\sqrt{13})/6 \simeq 0.768$. Thus, we have established that player 1 will never deviate from s_1 if and only if $\delta \in ((1+\sqrt{13})/6, 1)$.

Now, consider player 2. She has no incentive to deviate in any odd period (as she would be playing her dominant action B). On the other hand, if τ is even, she may want to deviate from s_2 (where she is playing the action A) to $\hat{s}_2$ such that $\hat{s}_2^{\tau} = B$.

So, under the deviation strategy $\hat{s}_i$ in any even period τ, player 2's average discounted sum of the payoffs from period τ to period ∞ becomes

$$\begin{aligned} V_2^{\delta,[\tau,\infty]}(\hat{s}_2, s_{-2}) &= (1-\delta)\left(\delta^{\tau-1}u_2(s_{-2}^{\tau}, B) + \sum_{t=\tau+1}^{\infty} \delta^{t-1}u_i((B, B))\right) \\ &= (1-\delta)\left(\delta^{\tau-1}4 + \sum_{t=\tau+1}^{\infty} \delta^{t-1}\right) \\ &= (1-\delta)\delta^{\tau-1}\left(4 + \frac{\delta}{1-\delta}\right). \end{aligned}$$

If player 2 sticks to the strategy s_2, her average discounted sum of the payoffs from period τ to period ∞ is

$$\begin{aligned} V_2^{\delta,[\tau,\infty]}(s) &= (1-\delta)\sum_{t=\tau}^{\infty} \delta^{t-1}u_2(a^t) \\ &= (1-\delta)\delta^{\tau-1}(4 + \delta^1(3) + \delta^2(4) + \delta^3(3) + \ldots) \\ &= (1-\delta)\delta^{\tau-1}\left(\frac{4+3\delta}{1-\delta}\right). \end{aligned}$$

So, one-period deviation would not be profitable for player 1 in period τ if and only if

$$V_2^{\delta,[\tau,\infty]}(s_2, s_{-2}) \geq V_2^{\delta,[\tau,\infty]}(\hat{s}_2, s_{-2})$$

or equivalently

$$\frac{4+3\delta}{1-\delta} \geq 4 + \frac{\delta}{1-\delta} = \frac{4-3\delta}{1-\delta},$$

which holds for all values of $\delta \in (0, 1)$. Thus, we have established that player 2 will never deviate from s_2, irrespective of the value $\delta \in (0, 1)$.

So, the strategy profile s is a subgame-perfect Nash equilibrium of the repeated game $G^{\delta}[\infty]$ if and only if $\delta \in ((1+\sqrt{13})/6, 1) \simeq (0.768, 1)$. ■

Example 12.4 (*Infinitely-Lived Cournot Duopoly*) Assume that the Cournot duopoly in Example 4.10 is infinitely lived. We denote by G the (stage) game in Example 4.10 and denote by $G^{\delta}[\infty]$ the infinitely repeated game where the firms discount future profits by the discount factor $\delta \in (0, 1)$. Recall from Example 4.10 that the payoff (profit) function of the player (firm) $i = 1, 2$ in the stage game G is

$$u_i(q) = \pi_i(q_i, q_{-i}) = P(q_i + q_{-i})q_i - cq_i = \min\{0, (a - (q_i + q_{-i}))\}q_i - cq_i,$$

where $a > c > 0$. We found that if the firms compete in quantities, the stage game has a unique Nash equilibrium, called Cournot-Nash equilibrium, in which the firms choose the quantity profile $q^{CN} = ((a-c)/3, (a-c)/3))$ and enjoy the profile of stage game profits $\pi^{CN} = ((a-c)^2/9, (a-c)^2/9)$.

Note that if there were a single firm (a monopoly) in this industry, its profit function would be $u(Q) = \pi(Q) = P(Q)qQ - cQ = (a-c-Q)Q$ which would be maximized at the quantity $Q^m = (a-c)/2$ and price $P(Q^m) = (a+c)/2$, yielding the monopoly profit $\pi(Q^m) = (a-c)^2/4$. If the two firms can make a written contract to act as a single firm, then they can produce the total output Q^m and share the monopoly profit $\pi(Q^m)$. This is what we call collusion. Formally, the firms in the duopolistic industry make a collusive agreement in output if they choose q_1 and q_2 such that they maximize the total profit $u_1(q_1, q_2) + u_2(q_2, q_1) = P(q_1+q_2)(q_1+q_2) - c(q_1+q_2)$ and share it in a way that is agreeable for both of them. Notice that the total profit function of the colluding duopoly is equivalent to the profit function of a monopoly. Thus, a profile of quantities (q_1^{col}, q_2^{col}) maximizes the collusive profit if and only if $q_1^{col} + q_2^{col} = Q^m = (a-c)/2$. Since the stage game is symmetric with respect to two players (firms), we may naturally assume that the collusive agreement is also symmetric; i.e., each firm $i = 1, 2$ produces $q_i^{col} = Q^M/2 = (a-c)/4$ and enjoys one half of the profit $\pi^M(Q^M)$. Thus, the profile of collusive profits is $\pi^{col} = ((a-c)^2/8, (a-c)^2/8))$. Notice that each firm's net benefit from collusion, with respect to Cournot competition, is $(a-c)^2/8 - (a-c)^2/9 = (a-c)^2/72 > 0$.

While collusion is beneficial for both firms, it is not self-enforcing; i.e. it does not arise as a noncooperative equilibrium. Therefore, unless a cooperative agreement for collusion exists, the collusive outcome can never arise in a finitely-lived duopoly, $G^\delta[T]$ for any finite integer $T \geq 1$, as implied by Proposition 12.2 and the fact that the stage game G has a unique Nash equilibrium (which differs from the collusive outcome). On the other hand, if the duopoly is infinitely-lived, the collusive outcome may arise as a stable equilibrium, a SPNE of $G^\delta[\infty]$, as we will illustrate below.

In Fig. 12.3, we portray $CP(G)$, the convex polytope of G, as the light blue colored triangle. Notice that the two firms would earn nothing, i.e., enjoy the payoff profile $(0, 0)$, if neither of them produced any output. Notice also that the payoff profile would be $((a-c)^2/4, 0)$ if firm 1 produced as a monopolist, while firm 2 produced nothing. Likewise, the payoff profile would be $(0, (a-c)^2/4)$ if firm 1 produced nothing, while firm 2 produced as a monopolist. Since the total profits in the industry cannot exceed (under linear prices) the monopoly profit $(a-c)^2/4$, it follows that the convex polytope of the game, $CP(G)$, is the convex hull of the payoff profiles $(0, 0)$, $((a-c)^2/4, 0)$, and $(0, (a-c)^2/4)$. The dark blue colored small triangle in Fig. 12.3 is the set of payoff profiles each of which can be obtained as the limit average payoff of a SPNE strategy profile of the repeated game $G^\delta[\infty]$. Notice that this triangle is the set of all profiles in $CP(G)$ that are in the northeast of the Cournot-Nash equilibrium payoff profile (π_1^{CN}, π_2^{CN}).

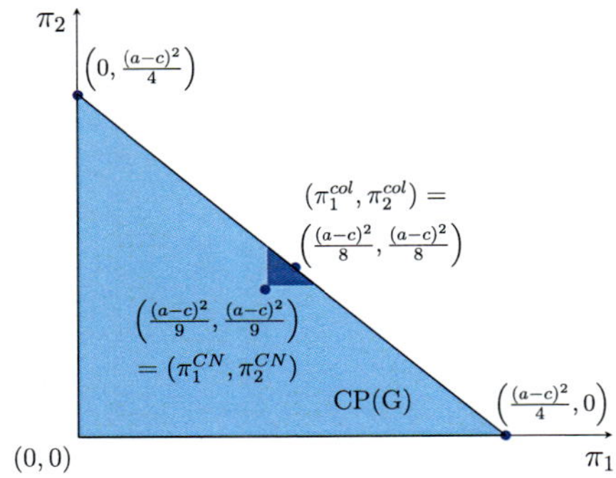

Fig. 12.3 Payoff profiles (with dark blue color) that can be obtained in an infinitely-lived Cournot duopoly as the limit average payoff at some SPNE of $\Gamma^\delta[\infty]$ for sufficiently high values of $\delta \in (0, 1)$

Given $\delta \in (0, 1)$, consider a strategy profile s in the repeated game $G^\delta[\infty]$ such that for each $i = 1, 2$ we have $s_i^1(\emptyset) = q_i^{col}$ and for each $t = 2, 3, \ldots, \infty$

$$s_i^t(h^t) = \begin{cases} q_i^{col} & \text{if } h^t = sec(q^{col}, t-1) \\ q_i^{CN} & \text{if } h^t \neq sec(q^{col}, t-1). \end{cases}$$

The described strategy profile s induces the outcome profile $y(s) = (y^1(s), y^2(s), \ldots, y^\infty(s))$, where, for each $t = 1, 2, \ldots, \infty$, we have $y^t(s) = q^{col}$.

If the strategy profile s is a subgame-perfect Nash equilibrium of $G^\delta[\infty]$, it must satisfy the one-period deviation principle. Pick any firm $i \in \{1, 2\}$. If it were not sufficiently punished for deviation, it would like to deviate in any period τ from s_i to $\hat{s}_i$ such that $\hat{s}_i^\tau = \text{argmax}_{q_i \in [0,a]} u_i(q_i, q_{-i}^{col})$. Notice that

$$u_i(q_i, q_{-i}^{col}) = P(q_i + q_{-i}^{col})q_i - cq_i = (a - c - q_i - q_{-i}^{col})q_i.$$

The first-order condition for maximizing this payoff function implies

$$\hat{s}_i^\tau = \frac{a - c - q_{-i}^{col}}{2} = \frac{a - c - q_{-i}^{col}}{2} = \frac{3(a-c)}{8}.$$

Thus, we can calculate the profit of firm i in the period it deviates from s_i as

$$\pi_i(\hat{s}_i^\tau, q_{-i}^{col}) = \left[P(\hat{s}_i^\tau + q_{-i}^{col}) - c\right]\hat{s}_i^\tau$$

$$= \left(a - c - \frac{3(a-c)}{8} - \frac{(a-c)}{4}\right)\frac{3(a-c)}{8} = \frac{9(a-c)^2}{64}.$$

So, under the deviation strategy $\hat{s}_i$ in any period τ, the average discounted sum of the payoffs of firm i from period τ to period ∞ becomes

$$V_i^{\delta,[\tau,\infty]}(\hat{s}_i, s_{-i}) = (1-\delta)\left(\delta^{\tau-1}\pi_i(\hat{s}_i^{\tau}, q_{-i}^{col}) + \sum_{t=\tau+1}^{\infty} \delta^{t-1}\pi_i(q_i^{CN}, q_{-i}^{CN})\right)$$

$$= (1-\delta)\left(\delta^{\tau-1}\frac{9(a-c)^2}{64} + \sum_{t=\tau+1}^{\infty} \delta^{t-1}\frac{(a-c)^2}{9}\right)$$

$$= (1-\delta)\delta^{\tau-1}\left(\frac{9(a-c)^2}{64} + \frac{\delta}{1-\delta^2}\frac{(a-c)^2}{9}\right).$$

If firm i sticks to the strategy s_i, its average discounted sum of profits from period τ to period ∞ is

$$V_i^{\delta,[\tau,\infty]}(s) = (1-\delta)\sum_{t=\tau}^{\infty} \delta^{t-1}\pi_i(q_i^{CN}, q_{-i}^{CN})$$

$$= (1-\delta)\frac{\delta^{\tau-1}}{1-\delta^2}\frac{(a-c)^2}{8}.$$

So, one-period deviation would not be profitable for player i in period τ if and only if

$$V_i^{\delta,[\tau,\infty]}(s_i, s_{-i}) \geq V_i^{\delta,[\tau,\infty]}(\hat{s}_i, s_{-i})$$

or equivalently

$$(1-\delta)\frac{\delta^{\tau-1}}{1-\delta^2}\frac{(a-c)^2}{8} \geq (1-\delta)\delta^{\tau-1}\left(\frac{9(a-c)^2}{64} + \frac{\delta}{1-\delta^2}\frac{(a-c)^2}{9}\right)$$

which holds if and only if

$$81\delta^2 - 64\delta - 9 \geq 0$$

or if and only if $\delta \in (\bar{\delta}, 1)$ where $\bar{\delta} \simeq 0.912$. Thus, we have established that the strategy profile s is a subgame-perfect Nash equilibrium of $G^{\delta}[\infty]$ for all $\delta \in (\bar{\delta}, 1)$. ■

12.3 Chapter Appendix

Proof of Proposition 12.3 (Folk Theorem) Since $\lambda \in CP(G)$, there exists a sequence of action profiles $(a^1, a^2, \ldots, a^{\infty})$ with $a^t \in A$ for each $t \in \{1, 2, \ldots, \infty\}$ such that $\lambda = \lim_{T\to\infty} \frac{1}{T}\sum_{t=1}^{T} u_i(a^t)$. Pick such a sequence $(a^1, a^2, \ldots, a^{\infty})$.

Also, pick any $\delta \in (0, 1)$ and consider a strategy profile s in the repeated game $G^\delta[\infty]$ such that $s_i^1(\emptyset) = a_i^1$ and for each $t = 2, 3, \ldots, \infty$

$$s_i^t(h^t) = \begin{cases} a_i^t \text{ if } h^t = (a^1, \ldots, a^{t-1}) \\ a_i^* \text{ if } h^t \neq (a^1, \ldots, a^{t-1}). \end{cases}$$

The described strategy profile s induces the outcome profile $y(s) = (y^1(s), y^2(s), \ldots, y^\infty(s))$, where, for each $t = 1, 2, \ldots, \infty$, we have $y^t(s) = a^t$.

If the strategy profile s is a subgame-perfect Nash equilibrium of $G^\delta[\infty]$, it must satisfy the one-period deviation principle. So, pick any $i \in N$ and period $\tau \geq 1$. If player i sticks to the strategy s_i, her average discounted sum of the payoffs from period τ to period ∞ is

$$\tilde{V}_i^{\delta,[\tau,\infty]}(s) = (1-\delta)\sum_{t=\tau}^{\infty}\delta^{t-1}u_i(a^t).$$

If player i makes a one-period deviation in period τ from the strategy s_i to another strategy s_i', then she should be deviating to play some action in period τ that yields a payoff v_i'. However, she would be punished starting from period $\tau + 1$ since she would need to play a_i^* as all other players would revert to play their parts in the Nash equilibrium profile of actions a^* as prescribed by the strategy profile s. Upon this one-period deviation, her average discounted sum of the payoffs from period τ to period ∞ becomes

$$\tilde{V}_i^{\delta,[\tau,\infty]}(s_i', s_{-i}) = (1-\delta)\left(\delta^{\tau-1}v_i' + \sum_{t=\tau+1}^{\infty}\delta^{t-1}u_i(a^*)\right).$$

So, one-period deviation would not be profitable for player i in period τ if and only if

$$\tilde{V}_i^{\delta,[\tau,\infty]}(s_i, s_{-i}) \geq \tilde{V}_i^{\delta,[\tau,\infty]}(s_i', s_{-i})$$

or equivalently

$$(1-\delta)\sum_{t=\tau}^{\infty}\delta^{t-1}\left(\frac{\tilde{V}_i^{\delta,[\tau,\infty]}(s_i, s_{-i})}{\delta^{\tau-1}}\right) \geq (1-\delta)\left(\delta^{\tau-1}v_i' + \sum_{t=\tau+1}^{\infty}\delta^{t-1}u_i(a^*)\right)$$

implying

$$\left(\delta^{\tau-1} + \frac{\delta^\tau}{1-\delta}\right)\left(\frac{\tilde{V}_i^{\delta,[\tau,\infty]}(s_i, s_{-i})}{\delta^{\tau-1}}\right) \geq \delta^{\tau-1}v_i' + \frac{\delta^\tau}{1-\delta}u_i(a^*)$$

or

$$\left(1 + \frac{\delta}{1-\delta}\right)\left(\frac{\tilde{V}_i^{\delta,[\tau,\infty]}(s_i, s_{-i})}{\delta^{\tau-1}}\right) \geq v_i' + \frac{\delta}{1-\delta}\omega_i^*.$$

The above inequality can be rewritten as

$$\frac{\nu_i' - \delta^{1-\tau}\tilde{V}_i^{\delta,[\tau,\infty]}}{\nu_i' - \omega_i^*} \leq \delta.$$

As δ approaches 1, the average discounted payoff $\tilde{V}_i^{\delta,[\tau,\infty]}$ approaches λ_i. Since $\lambda_i > \omega_i^*$, the left-hand side of the above inequality becomes less than 1 as δ approaches 1, whereas the right-hand side becomes 1. Hence, there exists some $\delta^* \in (0, 1)$ such that for any $\delta \in [\delta^*, 1)$, the above inequality holds, implying that player i has no incentive to use one-shot deviation in period τ. Since this is true for each value of τ, the strategy profile s, which produces the limit average payoff profile λ, is a subgame-perfect Nash equilibrium of the repeated game $G^\delta[\infty]$ for any $\delta \in [\delta^*, 1)$. ■

References

1. Nash JF (1950) Equilibrium points in n-person games. PNAS 36(1):48–49
2. Blackwell D (1965) Discounted dynamic programming. Ann Math Stat 36:226–235

Part IV
Implementation Theory

This part contains a single chapter, Chap. 13, which studies the implementation of social choice rules. In previous parts of this book, we defined static and dynamic noncooperative games and introduced several solution concepts for them. In those parts, we always assumed that the noncooperative games we dealt with were given to us. In this part, we will be designing noncooperative games to carry out (or implement) the objective of a benevolent authority in a social environment.

We assume that the social environment is, in general, such that the individuals inside are known to have unaligned (ordinal) preferences over the set of possible outcomes and a certain aspect of the environment (the state of the world) affecting the preferences of the individuals is known to them, but not to any outsider. Now, suppose that a social authority, as an outsider to this environment, can select for these individuals a subset of the outcomes available to them, by using a choice rule that depends on the preferences of the individuals. Since this authority does not know the actual state of the world, she does not know the actual preferences of the individuals in the environment. Thus, she does not know, hence cannot enforce, the outcomes that should be picked by the social choice rule for these individuals. To overcome this problem caused by asymmetric information, the social authority designs a noncooperative game (along with an outcome function mapping each strategy profile of this game to an outcome in the environment) such that whenever the individuals in the environment are enforced to play this game according to their actual preferences (unknown to the social authority) and using the solution (equilibrium) concept that the social authority expects them to use, their equilibrium profiles of strategies induce the exact outcomes that the social choice rule would pick for these individuals. If such a design is possible, we say that the social choice rule at hand is implementable.

After defining the problem of implementation formally and giving an example, Chap. 13 will introduce "the theory of implementation in Nash equilibrium" (Nash implementation) and present several examples.

Implementation of Social Choice Rules 13

This chapter will introduce "the theory of implementation in Nash equilibrium" (Nash implementation) and present several theoretical results along with their proofs. The chapter will then apply these results to several examples involving voting rules. The final section of this chapter will show how a historical problem of choice, King Solomon's Problem, which is known to be not Nash implementable, can be solved in several other ways using the results in the implementation theory.

In this chapter, we will study the topic of implementation theory, which deals with organizing activities in a social environment by designing some noncooperative game, or formally an institution called "mechanism". The pioneering works in the implementation theory with a focus on mechanism design are by Leonid Hurwicz, written in the 1960s and the 1970s. (See, for example, Hurwicz [1–3]).

For a mechanism design problem, we first define a (social) environment as a quadruple $\mathcal{E} = (N, A, \Theta, \{u_i\}_{i \in N})$, where $N = \{1, 2, \ldots, n\}$ is a finite set that denotes the society of individuals, A is the set of alternatives (outcomes) available for this society, Θ is the set of possible states of the world, and $u_i : A \times \Theta \to \mathbb{R}$ is the payoff function of player $i \in N$. So, if the outcome is $a \in A$ and the state of the world is $\theta \in \Theta$, then the payoff of individual i is $u_i(a, \theta)$. The payoff functions define preference relations. The (weak) preference relation of individual $i \in N$ in state $\theta \in \Theta$ is denoted by $R_i(\theta)$, whereas his strict preference and indifference relations are denoted by $P_i(\theta)$ and $I_i(\theta)$, respectively. These relations are generated by the payoff function $u_i(., \theta)$. That is, for any $a, b \in A$, we have

$$a R_i(\theta) b \text{ if and only if } u_i(a, \theta) \geq u_i(b, \theta),$$

$$a P_i(\theta) b \text{ if and only if } u_i(a, \theta) > u_i(b, \theta),$$

$$a I_i(\theta) b \text{ if and only if } u_i(a, \theta) = u_i(b, \theta).$$

I. Saglam, *Mastering Game Theory*, Classroom Companion: Economics,
https://doi.org/10.1007/978-3-031-82094-6_13

The preference profile of the society in state θ is denoted by $R(\theta) = (R_1(\theta), \ldots, R_n(\theta))$. Some preference profiles may be inconsistent with any states of the world in the environment, and we will be interested only with those that are consistent. Hence, we will define the preference domain for the society as the set

$$\mathcal{R}(\Theta) = \{R : \text{ there exists } \theta \in \Theta \text{ such that } R = R(\theta)\}$$

and the preference domain for individual i as the set

$$\mathcal{R}_i(\Theta) = \{R_i : \text{ there exists } R_{-i} \text{ such that } (R_i, R_{-i}) \in \mathcal{R}(\theta).\}$$

Given the above descriptions, we let $\mathcal{P}_A$ denote the set of all profiles of linear orderings (strict preferences) on A.

A social choice rule is a non-empty valued correspondence $F : \Theta \to 2^A \setminus \emptyset$ that selects at each preference profile a nonempty subset of A; i.e. $F(\theta) \subseteq A$ and $F(\theta) \neq \emptyset$ for all $\theta \in \Theta$. The set $F(\theta)$ is called the socially optimal alternatives in state θ. The image (range) of F is the set

$$F(\Theta) = \{a \in A : a \in F(\theta) \text{ for some } \theta \in \Theta\}.$$

We assume that the social choice rule F is chosen, and made public, by a benevolent social authority (the government or a regulator) whose sole aim is to enforce the outcome of this rule in any given state of the world. However, this authority is assumed to have complete knowledge only about the set of alternatives A, the preference domain $\mathcal{R}(\Theta)$ of the society, and the preference domain $\mathcal{R}_i(\Theta)$ of each individual $i \in N$. The social authority does not know the actual state of the world, θ, and therefore she does not know the socially optimal outcomes $F(\theta)$ that the social choice rule F would choose from the set A at the state of the world θ. In this chapter, we will show that under some conditions the social authority may ensure that the individuals in any state of the world θ will non-cooperatively select (when they are enforced to play a strategic form or extensive form game designed by the social authority) some strategies that will lead to the exact set of outcomes in A that the social choice rule F would choose at θ.

Before, we proceed further, we will consider a well-known social choice problem that will illustrate the meaning of some abstract definitions above.

13.1 King Solomon's Problem

The story containing the problem is mentioned in the Hebrew Bible (1 Kings, chapter 3, verses 16–28). According to the story, two women claimed, in the presence of King Solomon, to be the mother of a baby that is living in the same house with these two women. King's problem was to identify the true mother. After openly hearing and repeating the conflicting claims of the two women as to their mothership, King Solomon asked for a sword to be brought and ordered, in his wisdom, the baby to be cut in halves so that each woman could get one half. One of the women begged the King to spare the life of the baby by giving it to her rival, whereas the other women accepted the killing of the baby to be fair. The King, being convinced that only the first of these women (who opposed to the killing of the baby) could be the true mother, said "don't kill the baby" and ordered the baby to be given to this woman.

We will discuss King's solution later. Here, we will focus on the problem itself and express it as an environment in the following example.

Example 13.1 The environment in King Solomon's problem is denoted by $\mathcal{E} = (N, A, \Theta, \{u_i\}_{i \in N})$. We call the two women as Aliza ($\mathcal{A}$) and Baila ($\mathcal{B}$). Hence, $N = \{\mathcal{A}, \mathcal{B}\}$.

The set of alternatives is $A = \{a, b, c\}$, where a represents the event "Aliza gets the baby alive", b represents the event "Baila gets the baby alive", and c represents the event "The baby is cut into halves and each women gets one half". Here, we should note that the two women were facing the set of alternatives $\{a, b\}$ before they came to the presence of the King. The alternative c became a part of the set A just after the King attempted to solve women's conflict by introducing this alternative. We usually assume that the set of alternatives A is unaffected by the solution proposed by the social authority and exogenously given to both the society and the social authority before they make any interaction.

The set of the possible states of the world is $\Theta = \{\alpha, \beta\}$, where α is the state of the world where Aliza is the true mother and β is the state of the world where Baila is the true mother.

In the original statement of King Solomon's problem, there is no detail about the payoff functions of the two women. For our purpose, we will assume that these functions are such that they produce the payoffs in Table 13.1.

The above payoffs generate the following preferences for the two women:

$$R_{\mathcal{A}}(\alpha) = a, b, c$$

$$R_{\mathcal{B}}(\alpha) = b, c, a$$

$$R_{\mathcal{A}}(\beta) = a, c, b$$

$$R_{\mathcal{B}}(\beta) = b, a, c$$

Thus, we know that the preference domain is $R(\Theta) = \{R(\alpha), R(\beta)\}$ where $R(\alpha) = (R_{\mathcal{A}}(\alpha), R_{\mathcal{B}}(\alpha))$ and $R(\beta) = (R_{\mathcal{A}}(\beta), R_{\mathcal{B}}(\beta))$. Note that there are uncountably many payoff functions that can generate the preference domain $R(\Theta)$ derived above. (For example, note that changing $u_{\mathcal{A}}(\alpha, a) = 10$ to $u_{\mathcal{A}}(\alpha, a) = z$ for any $z \in (0, 10)$ would not change the ordinal preferences of Aliza in state α.) To solve some mechanism design problems, the specification of ordinal preference relations will be sufficient. For such problems, the designer need not know the payoff functions of the

Table 13.1 The assumed payoffs for the King Solomon's Problem in Example 13.1

	State of the World			
	α		β	
Outcome (x)	$u_{\mathcal{A}}(\alpha, x)$	$u_{\mathcal{B}}(\alpha, x)$	$u_{\mathcal{A}}(\beta, x)$	$u_{\mathcal{B}}(\beta, x)$
a	20	0	10	0
b	0	10	0	20
c	-100	1	1	-100

individuals. Also note that the preference domain does not involve all possible preference orderings on A. For example, the orderings c, a, b and c, b, a are in no individual's preference domain, since they are not consistent with any individual's payoff function in any state.

Finally, we will specify the social choice rule for the described problem. The social justice requires that this rule is the singleton-valued correspondence (function) F such that it gives the baby to Aliza in state α (when Aliza is the true mother) and to Baila in state β (when Baila is the true mother); i.e., $F(\alpha) = a$ and $F(\beta) = b$.

King Solomon's raising his sword and announcing his intention to cut the baby claimed by the two women into halves and to give each of them one half worked as a mechanism that induced the two women reveal their preferences over the well-being of the baby and thus allowed the King to infer which of them was the true mother. In the next section, we will formally define a mechanism and implementation in Nash equilibrium.

13.2 Nash Implementation

Formally, we define a mechanism (or game form) by a pair $\Gamma = (M, h)$, where $M = \bigtimes_i M_i$ involves a message space M_i for each player $i \in N$ and $h : M \to A$ is an outcome function that maps each message profile in M to an alternative in A. We assume that players will submit their messages simultaneously, hence we will call the mechanism Γ a strategic form mechanism. Later, we will see an example of an extensive form mechanism that will require the specifications about the game tree in addition to the message spaces of the individuals.

Let σ be a non-cooperative solution concept according to which the individuals are expected to play the strategic form game defined by a mechanism Γ. A social planner designs the mechanism Γ in such a way that in each unknown state of the world $\theta \in \Theta$, the individuals with a preference profile $R(\theta)$ will select under the perceived strategic form mechanism $\Gamma[R(\theta)]$ a set of equilibrium message profiles $\sigma(\Gamma[R(\theta)])$ that will induce a set of outcomes $h(\sigma(\Gamma[R(\theta)])) \subseteq A$. We say that a strategic form mechanism Γ implements the social choice rule F in the solution concept σ if and only if $h(\sigma(\Gamma[R(\theta)])) = F(\theta)$ for all $\theta \in \Theta$. If such a mechanism

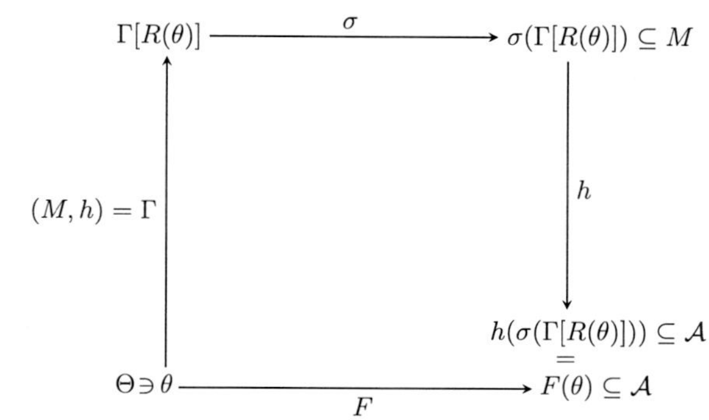

Fig. 13.1 Implementation of F in Solution σ via Mechanism Γ

exists, then we say F is σ implementable. In Fig. 13.1, we should see that F is σ implementable by the mechanism Γ if and only if the rectangular diagram commutes.

Note that the σ-implementation of a social choice rule F via a mechanism $\Gamma = (M, h)$ requires two conditions to hold: (i) $F(\theta) \subseteq h(\sigma(\Gamma[R(\theta)]))$ for all $\theta \in \Theta$ and (ii) $h(\sigma(\Gamma[R(\theta)])) \subseteq F(\theta)$ for all $\theta \in \Theta$. The first condition requires that for each outcome $a \in A$ selected by F in any state of the world θ, the mechanism Γ generates an equilibrium message that will induce the desired outcome a, whereas the second condition requires that the mechanism Γ does not generate any unwanted equilibrium: in every possible state of the world, each equilibrium message profile of the mechanism must induce an outcome that is chosen by F. To satisfy this second condition is usually the main difficulty faced by a mechanism designer at implementing a social choice rule in a given solution concept. We will illustrate this using the Nash equilibrium (NE) concept.

Let us reconsider Example 13.1 and try to find a strategic form mechanism that Nash-implements the social choice rule F that selects the outcome a at state α and b at state β. To strengthen our result, let us assume that the mechanism we are searching for contains infinitely many strategies for both Aliza and Baila as in Table 13.2.

Let $\Gamma = (M, h)$ denote such a strategic form mechanism. This mechanism must satisfy condition (i) of Nash-implementation; i.e. $F(\theta) \subseteq h(NE(\Gamma[R(\theta)]))$ must hold for all $\theta \in \Theta = \{\alpha, \beta\}$. Pick $\theta = \alpha$. The strategic form game must then involve a Nash equilibrium outcome that induces $F(\alpha) = a$. So, we have to first locate a in some cell in Table 13.2, which involves infinitely many row and column strategies

Table 13.2 A strategic form mechanism for King Solomon's Problem

		Baila			
		c_1	c_2	c_3	c_4
Aliza	r_1	**a**	y	y	y
	r_2	x			
	r_4	x			
	r_4	x			

for Aliza and Baila, respectively. Without loss of generality, let us suppose this is the cell at the upperleft corner.

Given the preferences $R_{\mathcal{A}}(\alpha) = a\,b\,c$ and $R_{\mathcal{B}}(\alpha) = b\,c\,a$ of Aliza and Baila, respectively, (r_1, c_1) becomes a Nash equilibrium message of the above game only if the social planner chooses x and y such that $x \in \{a, b, c\}$ and $y = a$. With such a choice, we have $(r_1, c_1) \in NE(\Gamma[R(\alpha)])$ and $a \in h(NE(\Gamma[R(\alpha)]))$. Note that this result is necessary to satisfy condition (i) of implementation. However, this result also violates condition (ii). To see this, consider the state of the world β and the preferences $R(\beta)$ where $R_{\mathcal{A}}(\beta) = a\,c\,b$ and $R_{\mathcal{B}}(\beta) = b\,a\,c$. In this state, the game perceived by the individuals is $\Gamma[R(\beta)]$ and the strategy profile (r_1, c_1) is a Nash equilibrium of this game, as well. However, (r_1, c_1) is an unwanted equilibrium in state β. This strategy profile induces the outcome a by design, whereas a is not chosen by F at β; i.e., $a \in h(NE(\Gamma[R(\beta)]))$, whereas $a \notin F(\beta) = \{b\}$, violating $h(NE(\Gamma[R(\beta)])) \subseteq F(\beta)$.

Above, we have established that we cannot implement King Solomon's choice rule in the Nash solution concept using any strategic form mechanism. This negative result is caused by the fact that this social choice rule does not satisfy a condition (or an axiom) called (Maskin's) Monotonicity. To present this condition, we first define individual i's lower contour set at $(a, \theta) \in A \times \Theta$, given by

$$L_i(a, \theta) = \{x \in A : \; u_i(a, \theta) \geq u_i(x, \theta)\}.$$

Any outcome $x \in A$ is contained by $L_i(a, \theta)$ if and only if individual i weakly prefers a to x in state θ.

Monotonicity. Given any environment $(N, A, \Theta, \{u_i\}_{i \in N})$, a social choice rule F satisfies monotonicity and is called monotonic if for any $\theta, \theta' \in \Theta$ and any $a \in F(\theta)$, it is true that $L_i(a, \theta) \subseteq L_i(a, \theta')$ for all $i \in N$ only if $a \in F(\theta')$.

Monotonicity requires that if an outcome $a \in A$ is chosen in some state $\theta \in \Theta$ by a social choice rule F and its relative ranking did not become worse for any individual in another state θ', then it should be chosen by F in θ', as well. In other saying, a social choice rule F is monotonic if for any $\theta, \theta' \in \Theta$ and any $a \in F(\theta)$, we have $a \notin F(\theta')$ only if there exists $i \in N$ and $b \in A$ such that $a R_i(\theta) b$ and $b P_i(\theta') a$. Note that whether a social choice rule F is monotonic may not be independent of the preference domain $R(\Theta)$.

For the problem presented in Example 13.1, we have already established that the monotonicity condition is not satisfied. The outcome a (where Aliza gets the baby alive) is chosen by the social choice rule F in state α and relative ranking of a is not worse for any woman in state β (it is still the top-ranked outcome for Aliza, while it is no longer the worst outcome for Baila). However, a is not chosen by F in state β, a violation of the monotonicity condition. If a social choice rule violates monotonicity in a given environment, then it cannot be Nash-implemented (via any strategic form mechanism). We can state this result of Maskin as follows.

Proposition 13.1 ([4]) *If a social choice rule is Nash-implementable, then it is monotonic.*

Proof Suppose the mechanism $\Gamma = (M, h)$ Nash-implements a social choice rule F in an environment $(N, A, \Theta, \{u_i\}_{i \in N})$. Then, take any $\theta \in \Theta$. If $a \in F(\theta)$, then there exists a profile $m \in NE(\Gamma[R(\theta)])$ such that $h(m) = a$. Suppose $\theta' \in \Theta$ is such that $L_i(a, \theta) \subseteq L_i(a, \theta')$ for all $i \in N$. Since m is a Nash equilibrium profile, for any $i \in N$ and any $m_i' \in M_i$ we must have $h(m_i', m_{-i}) \subseteq L_i(a, \theta)$, implying $h(m_i', m_{-i}) \subseteq L_i(a, \theta')$. Therefore, $m \in NE(\Gamma[R(\theta')])$. Since F is Nash implementable by assumption, we must then have $h(m) = a \in F(\theta')$, proving that F is monotonic. ■

The above proposition is a necessity result stating that a social choice rule is Nash implementable only if it is monotonic. Maskin [4,5] shows that it is not sufficient: there are monotonic social choice rules that are not Nash implementable. Maskin [4] shows that monotonic social choice rules are Nash implementable if they also satisfy a condition called no veto power.

No Veto Power. Given any environment $(N, A, \Theta, \{u_i\}_{i \in N})$, a social choice rule F satisfies no veto power if for any $\theta \in \Theta$ and $a \in A$ it is true that $a \in F(\theta)$ whenever there exists $j \in N$ such that $L_i(a, \theta) = A$ for all $i \in N - j$.

The no veto power condition states that an alternative $a \in A$ is chosen by a social choice rule F in a state $\theta \in \Theta$ whenever it is top ranked by $n - 1$ individuals in the society. If this condition holds, no individual has the power to veto an alternative that is top ranked by the rest of the society from being chosen by a given social choice rule. For $n = 2$, a social choice rule F satisfies this condition only if in each state of the world each individual's top-ranked alternative is among the alternatives chosen by F. The condition puts much less restriction on a social choice rule if $n \geq 3$ as there are environments where no alternative is top ranked by even more than one individual. Think, for example, an economic environment where the set of alternatives is all possible divisions of a single endowment say a cake of unit size. If each individual's utility is monotonically increasing in the size of the cake it receives, each individual's top-ranked alternative becomes the division at which she is entitled to the whole cake. Therefore, in such an environment there can be no preference profile where $n - 1$ of n individuals would agree upon that an alternative is top ranked for each of them. Hence, the no veto power condition would vacuously hold, without any need to check what alternatives a given social choice rule would choose at any state of the world.

Using monotonicity and no veto power together, Maskin [4] was able to propose the following.

Proposition 13.2 ([4]) *Let $n \geq 3$. If a social choice rule satisfies monotonicity and no veto power, then it is Nash-implementable.*

Proof See the appendix at the end of this chapter. ■

We should note that the above result is only a sufficiency result. Monotonicity and no veto power conditions together are not necessary for Nash implementation. No veto power enables the particular mechanism, known as the canonical mechanism, in the proof of Proposition 13.2 to implement a given monotonic social choice rule when the number of individuals is at least three. The next example will reveal that there are monotonic social choice rules that are Nash implementable even though they do not satisfy the no veto power condition.

Example 13.2 A social choice rule $F : \Theta \to A$ is *dictatorial* if there exists an agent $i \in N$ such that for all $\theta \in \Theta$ and all $a \in A$, $L_i(a, \theta) = A$ if $a \in F(\theta)$. So, when F is dictatorial, there exists an individual i such that every alternative $a \in A$ chosen by F in any state $\theta \in \Theta$ is the top-ranked alternative of i in that state.

Since the social choice rule F is fully aligned with the preference of a single individual, F does not satisfy the no veto power condition. But, this does not imply that F is not Nash implementable, since no veto power is only sufficient together with monotonicity.

Now, let us check whether F satisfies monotonicity when the preference domain $\mathcal{R}(\Theta)$ contains only strict preferences (no indifferences); i.e. $\mathcal{R}(\Theta) \subseteq \mathcal{P}_A$. Pick any $\theta' \in \Theta$ and any $a' \in A$ such that $a' \in F(\theta')$. Since preferences are strict, $F(\theta') = \{a'\}$. Since F is dictatorial, there exists an individual $k \in N$ such that for all $\theta \in \Theta$ and all $a \in A$, $L_k(a, \theta) = A$ if $a \in F(\theta)$. Pick this individual. Then, $L_k(a', \theta') = A$. Suppose $\theta'' \in \Theta$ is such that $L_i(a', \theta') \subseteq L_i(a', \theta'')$ for all $i \in N$. Thus, $L_k(a', \theta') \subseteq L_k(a', \theta'')$, implying $L_k(a', \theta'') = A$. Suppose $a' \notin F(\theta'')$. Then, there exists $b \in A$ such that $b \neq a'$ and $b \in F(\theta'')$. This contradicts the fact that F is dictatorial since the fact that $L_k(a, \theta'') = A$ and the assumption that preferences are strict imply that $a P_k(\theta'') b$. So, $a' \in F(\theta'')$ holds and F is thus monotonic.

Since F does not satisfy the no veto power condition, we are not able to use Maskin's Theorem. However, we can simply ascertain that F is Nash implementable. If k is the individual that is the dictator according to F, just consider a game form $\Gamma = (M, h)$ that makes him the dictator. Formally, for each $i \in N$, let $M_i = A$ be the strategy space of player i and $M = \times_{i \in N} M_i$. Given a strategy profile $m = (m_1, \ldots, m_n) \in M$, let the outcome function h select the alternative $a \in A$ only if $a = m_k$; i.e., $h(m) = m_k$ for all $m \in M$. Clearly, for all $\theta \in \Theta$, $a \in F(\theta)$ if and only if $a \in h(NE(\Gamma[R(\theta)]))$. (Note that if $L_k(a, \theta) = A$ in a given state $\theta \in \Theta$, then a strategy profile $m \in M$ is a Nash equilibrium of the game $\Gamma[R(\theta)]$ if and only if $m_k = a$.) ■

Below, we will present a social choice rule that satisfies no veto power but not monotonicity. Let $N^T(a, \theta)$ denote the set of individuals that top rank a in state θ; i.e., $N^T(a, \theta) = \{i \in N : L_i(a, \theta) = A\}$.

Example 13.3 A social choice rule $F : \Theta \to A$ is called *the pluralitarian top correspondence* if it selects at each preference profile all alternatives that are ranked as the top by the most number of agents; i.e. $F(\theta) = \{a \in A : \ |N^T(a, \theta)| \geq |N^T(b, \theta)| \text{ for all } b \in A\}$.

Let the environment involve at least three individuals; i.e., $n \geq 3$. Note that F satisfies no veto power since for any $\theta \in \Theta$ and any $a \in A$, if a is top ranked by $n - 1$ individuals in N, then a is top ranked by the highest number of agents in N and $a \in F(\theta)$ becomes true. On the other hand, F does not satisfy monotonicity. To see this, we will consider the following counter-example. Let $N = \{1, 2, 3\}$, $A = \{a, b, c\}$, and $\theta, \theta' \in \Theta$ are such that the induced preference profiles are given by

$$R(\theta) = \begin{pmatrix} R_1(\theta) & R_2(\theta) & R_3(\theta) \\ a & b & c \\ b & c & a \\ c & a & b \end{pmatrix} \quad \text{and} \quad R(\theta') = \begin{pmatrix} R_1(\theta') & R_2(\theta') & R_3(\theta') \\ a & c & c \\ b & b & a \\ c & a & b \end{pmatrix}.$$

Notice that $a \in F(\theta) = \{a, b, c\}$, and the inclusion $L_i(a, \theta) \subseteq L_i(a, \theta')$ holds (with equality) for each $i \in N$. But, $a \notin F(\theta')$ since $F(\theta') = \{c\}$. Therefore, F does not satisfy monotonicity. Since monotonicity is necessary for Nash implementation, F is not Nash implementable. ■

Below, we present a social choice rule that satisfies both no veto power and monotonicity.

Example 13.4 A social choice rule is called *the weak Pareto correspondence* if it selects at each state of the world all alternatives that are not weakly Pareto dominated by any other alternative; i.e.,

$$F(\theta) = \{a \in A : \text{ there exists no } b \in A \text{ such that } bP_i(\theta)a \text{ for all } i \in N\}.$$

Let the environment involve at least three individuals; i.e., $n \geq 3$. For any $\theta \in \Theta$ and any $a \in A$, suppose $N^T(a, \theta) = n - 1$, i.e., a is top ranked by $n - 1$ individuals in N. Since any alternative that is top ranked by any individual is not Pareto dominated by any $b \in A$, the alternative a is not weakly Pareto dominated, implying that $a \in F(\theta)$. Therefore, F satisfies no veto power.

To check whether F satisfies monotonicity as well, pick any $\theta, \theta' \in \Theta$ and $a \in A$ such that $a \in F(\theta)$ and $L_i(a, \theta) \subseteq L_i(a, \theta')$ for all $i \in N$. Suppose towards a contradiction that $a \notin F(\theta')$. Then, there exists $b \in A$ that weakly Pareto dominates a in state θ', implying that $bP_i(\theta')a$ for all $i \in N$. The supposition that $a \in F(\theta)$ implies that $aR_k(\theta)b$ for some $k \in N$. Moreover, the supposition $L_k(a, \theta) \subseteq L_k(a, \theta')$ implies $aR_k(\theta')b$, in contradiction with $bP_i(\theta')a$ for all $i \in N$. Therefore, $a \in F(\theta')$, implying that F is monotonic.

Since F satisfies both no veto power and monotonicity, we can use the sufficiency result of Maskin (Proposition 13.2) to conclude that F is Nash implementable (using the strategic form mechanism presented in the proof of that result).

Finally, we should note that whenever the preference domain involves only strict preferences, all of the above conclusions would remain to hold if the social choice rule were changed to the Pareto correspondence that selects at each state of the world all alternatives that are not Pareto dominated; i.e.,

$$F(\theta) = \{a \in A : \text{ there exists no } b \in A \text{ such that } bR_i(\theta)a \text{ for all } i \in N$$

$$\text{and } bP_i(\theta)a \text{ for some } i \in N\}$$

for all $\theta \in \Theta$. ■

Below, we present a social choice rule that does not satisfy monotonicity in any environment with at least three individuals, while it can satisfy no veto power if and only if the number of individuals in a given environment is sufficiently large.

Example 13.5 For any individual $i \in N$, let the Borda score $B_i(a, \theta)$ of an alternative $a \in A$ in state $\theta \in \Theta$ be defined as $B_i(a, \theta) = |L_i(a, \theta)|$. According to this scoring measure, for any individual an alternative receives, in any state, one point for being ranked last, two points for being ranked next to last, ..., and $|A|$ points for being ranked first. Let $B(a, \theta) = \sum_{i \in N} B_i(a, \theta)$ denote the total Borda score for each $a \in A$ and $\theta \in \Theta$. A social choice rule F is called *the Borda rule* if it selects in each state of the world the alternatives with the highest total Borda score; i.e.,

$$F(\theta) = \{a \in A : B(a, \theta) \geq B(b, \theta) \text{ for all } b \in A\}.$$

Suppose $n - 1$ individuals in N top rank an alternative a in A. Let us find a condition under which a is chosen by the Borda rule. To find the least restrictive condition, assume the remaining individual ranks the alternative a at the bottom. Then the Borda score of a is $(n - 1)|A| + 1$. Let us calculate the best score another alternative b in A can get. First, b must get the second best score from all individuals who top rank a. So, the score obtained from these individuals is $(|A| - 1)(n - 1)$. Second, b must get the top score from the remaining individual, which score is $|A|$. Thus the score of b can be at most $|A| + (n - 1)(|A| - 1)$. For the alternative a to be chosen by the Borda rule, we must have $(n - 1)|A| + 1 \geq |A| + (n - 1)(|A| - 1)$ or $n \geq |A|$. So, the Borda rule satisfies the no veto power condition if and only if the number of individuals is not less than the number of alternatives.

On the other hand, F does not satisfy monotonicity. To see this, we will consider the following counter-example. Let $N = \{1, 2, 3\}$, $A = \{a, b, c, d\}$, and $\Theta = \{\theta, \theta'\}$. Let the preference domain $R(\Theta)$ be such that the possible preference profiles are given by

$$R(\theta) = \begin{pmatrix} R_1(\theta) & R_2(\theta) & R_3(\theta) \\ a & c & b \\ b & d & c \\ c & a & d \\ d & b & a \end{pmatrix} \quad \text{and} \quad R(\theta') = \begin{pmatrix} R_1(\theta') & R_2(\theta') & R_3(\theta') \\ b & c & b \\ a & b & c \\ c & d & d \\ d & a & a \end{pmatrix}.$$

Notice that the Borda scores at θ and θ' are given by

$$\begin{pmatrix} x & B_1(x,\theta) & B_2(x,\theta) & B_3(x,\theta) & B(x,\theta) \\ a & 4 & 2 & 1 & 7 \\ b & 3 & 1 & 4 & 8 \\ c & 2 & 4 & 3 & 9 \\ d & 1 & 3 & 2 & 6 \end{pmatrix}$$

and

$$\begin{pmatrix} x & B_1(x,\theta') & B_2(x,\theta') & B_3(x,\theta') & B(x,\theta') \\ a & 3 & 1 & 1 & 5 \\ b & 4 & 3 & 4 & 11 \\ c & 2 & 4 & 3 & 9 \\ d & 1 & 2 & 2 & 5 \end{pmatrix},$$

respectively. Thus, $c = argmax_{x \in A} B(x, \theta)$ and therefore $F(\theta) = \{c\}$, whereas $b = argmax_{x \in A} B(x, \theta')$ and therefore $F(\theta') = \{b\}$. Also note that $L_i(c, \theta) \subseteq L_i(c, \theta')$ for all $i \in N$, while $c \notin F(\theta')$. Therefore, F does not satisfy monotonicity. Since monotonicity is necessary for Nash implementation, F is not Nash implementable. ■

13.3 Various Solutions to King Solomon's Problem

Recall from Sect. 13.2 that King Solomon's problem is not Nash implementable, since the preference domain involving the preference profiles of the two women who claim the baby in dispute are not monotonic. This negative result led the researchers to consider other solutions to the problem of King Solomon.

Solution 1. Implementation in Weakly Undominated Nash Equilibrium [6].

Consider the environment in Example 13.1 where we have introduced King Solomon's problem. The solution that we will present requires a new alternative, say d, to be added to the original set of alternatives $\{a, b, c\}$ in the problem, leading to a modified set of alternatives $A = \{a, b, c, d\}$. The alternative d represents the outcome where everybody (the two women and the child they are claiming) will be slayed (simply, death to all). Recall that $N = \{\mathcal{A}, \mathcal{B}\}$. We assume that this alternative will be the worst alternative for each woman $w \in N$ in each state $\theta \in \Theta = \{\alpha, \beta\}$, thus their preferences will be:

$$R_{\mathcal{A}}(\alpha) = a, b, c, d$$

$$R_{\mathcal{B}}(\alpha) = b, c, a, d$$

$$R_{\mathcal{A}}(\beta) = a, c, b, d$$

$$R_{\mathcal{B}}(\beta) = b, a, c, d$$

It should be clear that the social choice rule F that selects a in state α and b in state β remains to violate the monotonicity condition, since d was added to the bottom of the preference orderings of both women and therefore its addition does no affect arguments used to check monotonicity. Thus, given A and $R(\Theta)$, we can still say that the social choice rule F is not Nash implementable. As illustrated by Moore [7] and Moore and Repullo [8], even if one were to use an extensive form game by adding stages and consider subgame perfection, one would still be unable to implement King Solomon's social choice rule under the assumption that A and $R(\Theta)$ are just as described above.

Now, we will show that King Solomon's problem can be implemented in a strategic form mechanism if the Nash equilibrium concept is replaced with the weakly undominated Nash equilibrium, as proposed by Palfrey and Srivastava [6]. Note that a weakly undominated Nash equilibrium is a Nash equilibrium in which no player plays a weakly undominated strategy. Consider the strategic form mechanism $\Gamma = (M, h)$ where $M_i = \Theta \times \{1, 2, \ldots, n\}$ for each $w \in N = \{\mathcal{A}, \mathcal{B}\}$. So, each individual is asked to announce a message $m_i = (\theta_i, z_i) \in M_i$. The outcome function h is defined as follows: If $m \in M$ is such that $\theta_{\mathcal{A}} \neq \theta_{\mathcal{B}}$, then $h(m) = d$. If $m \in M$ is such that $\theta_{\mathcal{A}} = \theta_{\mathcal{B}}$, then $h(m)$ is determined according to the outcome tables below.

Now, we are ready to prove that F can be implemented in weakly undominated Nash equilibrium (WUNE) through the mechanism Γ described above. Formally, we have to show that for each $\theta \in \Theta$, $F(\theta)$ and $h(WUNE(\Gamma[R(\theta)]))$ will be equal. In any state $\theta \in \Theta$, any message $m \in M$ can be an equilibrium, i.e., $m \in WUNE(\Gamma[R(\theta)])$, only if $\theta_{\mathcal{A}} = \theta_{\mathcal{B}}$; for otherwise $h(m)$ will be equal to d, which is the worst alternative for each woman in each state. Then, let $\theta_{\mathcal{A}} = \theta_{\mathcal{B}}$. In state α, Aliza has no weakly undominated strategy if $\theta_{\mathcal{A}} = \theta_{\mathcal{B}} = \beta$ since she prefers b to c in state α (see the right panel in Table 13.3). The only undominated equilibrium, m, of the game $\Gamma[R(\alpha)]$ occurs at the top left-hand corner of the left-hand panel, where $m_{\mathcal{A}} = m_{\mathcal{B}} = (\alpha, 1)$, since Baila prefers c to a in state α. This equilibrium ensures $F(\alpha) = \{a\} = h(m)$.

Table 13.3 The strategic form mechanism in solution 1

$\theta_{\mathcal{A}} = \theta_{\mathcal{B}} = \alpha$

Baila

Aliza	1	2	3	4	
1	a	a	a	a	.
2	c	a	a	a	.
4	c	c	a	a	.
4	c	c	c	a	.
.	.	.	.	.	

$\theta_{\mathcal{A}} = \theta_{\mathcal{B}} = \beta$

Baila

Aliza	1	2	3	4	
1	b	c	c	c	.
2	b	b	c	c	.
4	b	b	b	c	.
4	b	b	b	b	.
.	.	.	.	.	

Similarly, in state β, Baila has no weakly undominated strategy if $\theta_{\mathcal{A}} = \theta_{\mathcal{B}} = \alpha$ since she prefers a to c in state β (see the left-hand panel in Table 13.3). The only weakly undominated equilibrium, m, of the game $\Gamma[R(\beta)]$ occurs at the top left-hand corner of the right-hand panel above, where $m_{\mathcal{A}} = m_{\mathcal{B}} = (\beta, 1)$, since Aliza prefers c to b in state β. This equilibrium ensures $F(\beta) = \{b\} = h(m)$.

Thus, we have established that $F(\theta) = h(WUNE(\Gamma[R(\theta)]))$ for each $\theta \in \Theta$, showing that F is implementable in weakly undominated Nash equilibrium.

Solution 2. Implementation in Nash Equilibrium under the Threat of Fines [9].

Let us modify King Solomon's problem by introducing monetary fines. Recall that $N = \{\mathcal{A}, \mathcal{B}\}$ and $\Theta = \{\alpha, \beta\}$. We assume that the true mother values the child at 20, the impostor values the child at 10 and that the King can either fine both women 15 or fine none of them. Thus, the set of alternatives is $A = \{(\mathcal{A}, 0), (\mathcal{B}, 0), (\mathcal{A}, 15), (\mathcal{A}, 15)\}$, where the first component in each alternative denotes the woman who gets the baby alive, and the second component denotes the amount all women are fined. We assume that the payoff of each women on $(x, t) \in A$ is her utility derived from x minus t. We further assume that the utility of each woman w derived from w' for any $w, w' \in N$ is equal to her valuation on the child if she gets the baby alive (i.e., $w = w'$) and equal to 0 if the other woman gets the baby alive (i.e., $w \neq w'$). Finally, we assume that for any $\theta \in \Theta$ and $w \in N$, a social choice rule F selects $(w, 0)$ if and only if w is the true mother.

Let us first calculate the (numerical) payoffs of Aliza and Baila, then derive their ordinal preferences, and finally check whether the social choice rule F is monotonic. In each state $\theta \in \Theta$, the utility $u_w((x, t), \theta)$ of any women $w \in N$ at any alternative $(x, t) \in A$ can be calculated as follows (Table 13.4):

Thus, the ordinal preferences of $\mathcal{A}$ and $\mathcal{B}$ are given by:

$$R_{\mathcal{A}}(\alpha) = (\mathcal{A}, 0), (\mathcal{A}, 15), (\mathcal{B}, 0), (\mathcal{B}, 15)$$

$$R_{\mathcal{B}}(\alpha) = (\mathcal{B}, 0), (\mathcal{A}, 0), (\mathcal{B}, 15), (\mathcal{A}, 15)$$

$$R_{\mathcal{A}}(\beta) = (\mathcal{A}, 0), (\mathcal{B}, 0), (\mathcal{A}, 15), (\mathcal{B}, 15)$$

$$R_{\mathcal{B}}(\beta) = (\mathcal{B}, 0), (\mathcal{B}, 15), (\mathcal{A}, 0), (\mathcal{A}, 15)$$

Table 13.4 Possible payoffs under monetary fines in solution 2

	State α		State β	
(x, t)	$U_{\mathcal{A}}((x, t), \alpha)$	$U_{\mathcal{B}}((x, t), \alpha)$	$U_{\mathcal{A}}((x, t), \beta)$	$U_{\mathcal{B}}((x, t), \beta)$
$(\mathcal{A}, 0)$	20	0	10	0
$(\mathcal{A}, 15)$	5	−15	−5	−15
$(\mathcal{B}, 0)$	0	10	0	20
$(\mathcal{B}, 15)$	−15	−5	−15	5

Without looking for any strategic form mechanism, can we say whether the social choice rule F is Nash implementable or not by any strategic form mechanism? Since F satisfies monotonicity, we cannot automatically say that F is not Nash implementable. Also, since $|N| = 2$, the condition $|N| \geq 3$ in Maskin's Theorem is not satisfied. Thus, we cannot use this theorem to validate whether F is Nash implementable. Luckily, we will be able to find a simple strategic form mechanism via which F is Nash implementable even when players use pure strategies.

Consider the mechanism $\Gamma = (M, h)$ where $M = (M_{\mathcal{A}}, M_{\mathcal{B}})$ is the message space, with $M_{\mathcal{A}} = \{r_1, r_2\}$ and $M_{\mathcal{B}} = \{c_1, c_2, c_3\}$. The outcome function h satisfies

$$h(r_1, c_1) = (\mathcal{A}, 0) \;\; h(r_1, c_2) = (\mathcal{B}, 15) \;\; h(r_1, c_3) = (\mathcal{A}, 15),$$

$$h(r_2, c_1) = (\mathcal{B}, 15) \;\; h(r_2, c_2) = (\mathcal{A}, 15) \;\; h(r_2, c_3) = (\mathcal{B}, 0).$$

The mechanism Γ can be represented by Table 13.5.

In state α, the players perceive the game $\Gamma[R(\alpha)]$, given by Table 13.6. We have $NE(\Gamma[R(\alpha)]) = \{(r_1, c_1)\}$, and $h((r_1, c_1)) = (\mathcal{A}, 0)$. So, $h(NE(\Gamma[R(\alpha)])) = F(\alpha)$.

On the other hand, in state β, the players perceive the game $\Gamma[R(\beta)]$, given by Table 13.7.

We have $NE(\Gamma[R(\beta)]) = \{(r_2, c_3)\}$, and $h((r_2, c_3)) = (\mathcal{B}, 0)$, implying that $h(NE(\Gamma[R(\beta)])) = F(\beta)$. We have thus established that F is Nash implementable via the strategic form mechanism Γ described above.

Solution 3. Implementation in Subgame Perfect Nash Equilibrium under the Threat of Fines [10].

Below, we will show that given the environment described for Solution 2 above, the social choice rule F is subgame-perfect Nash implementable via an extensive form mechanism, say Γ, proposed by Glazer and Ma [10]. The game three associated

Table 13.5 The strategic form mechanism in solution 2

$\mathcal{A}$ \ $\mathcal{B}$	c_1	c_2	c_3
r_1	$(\mathcal{A}, 0)$	$(\mathcal{B}, 15)$	$(\mathcal{A}, 15)$
r_2	$(\mathcal{B}, 15)$	$(\mathcal{A}, 15)$	$(\mathcal{B}, 0)$

Table 13.6 The strategic form game induced by the mechanism in Table 13.5 in state α

$\mathcal{A}$ \ $\mathcal{B}$	c_1	c_2	c_3
r_1	$(20, 0)$	$(-15, -5)$	$(5, -15)$
r_2	$(-15, -5)$	$(5, -15)$	$(0, 10)$

Table 13.7 The strategic form game induced by the mechanism in Table 13.5 in state β

		$\mathcal{B}$		
		c_1	c_2	c_3
$\mathcal{A}$	r_1	$(10, 0)$	$(-15, 5)$	$(-5, -15)$
	r_2	$(-15, 5)$	$(-5, -15)$	$(0, 20)$

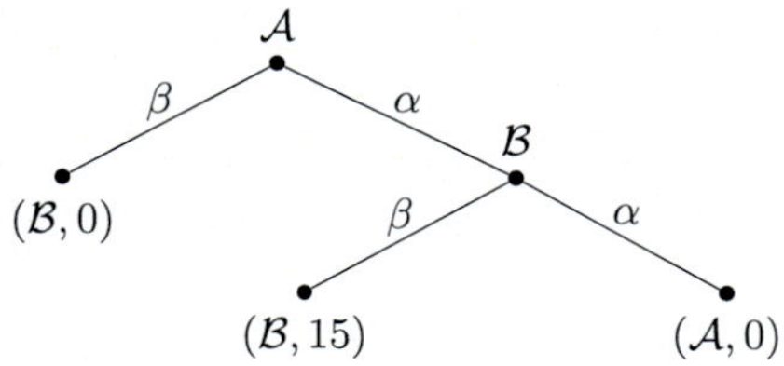

Fig. 13.2 The extensive form mechanism using the Threat of Fines to Solve King Solomon's problem

with this mechanism is portrayed in Fig. 13.2. This tree involves players' names and strategies specified at non-terminal nodes and the possible outcomes specified at the terminal nodes.

The preference domain $R(\Theta)$ is as described in Solution 2. In state α, the players perceive the game $\Gamma[R(\alpha)]$, illustrated by Fig. 13.3.
(The first and second payoffs at each terminal node belong to Aliza and Baila.) We have

$$SPNE(\Gamma[R(\alpha)]) = \{s^{\alpha}\} = \{(\mathcal{A} \text{ plays } \alpha, \mathcal{B} \text{ plays } \alpha \text{ if } \mathcal{A} \text{ plays } \alpha)\},$$

and $h(s^{\alpha}) = (\mathcal{A}, 0)$. Therefore, $h(\text{SPNE}(\Gamma[R(\alpha)])) = F(\alpha)$.

In state β, the players perceive the game $\Gamma[R(\beta)]$, illustrated by Fig. 13.4. (The first and second payoffs at each terminal node belong to Aliza and Baila.) We have

$$SPNE(\Gamma[R(\beta)]) = \{s^{\beta}\} = \{(\mathcal{A} \text{ plays } \beta, \mathcal{B} \text{ plays } \beta \text{ if } \mathcal{A} \text{ plays } \alpha)\},$$

and $h(s^{\beta}) = (\mathcal{B}, 0)$. Therefore, $h(\text{SPNE}(\Gamma[R(\beta)])) = F(\beta)$.

We have thus established that F is subgame-perfect Nash implementable via the extensive form mechanism Γ described above.

Solution 4. Implementation in Subgame Perfect Nash Equilibrium under the Possibility of Bidding [7,10].

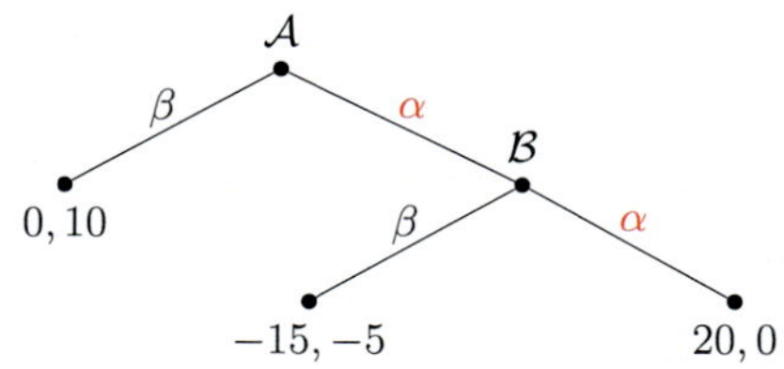

Fig. 13.3 The extensive form game induced by the mechanism in Fig. 13.2 in state α

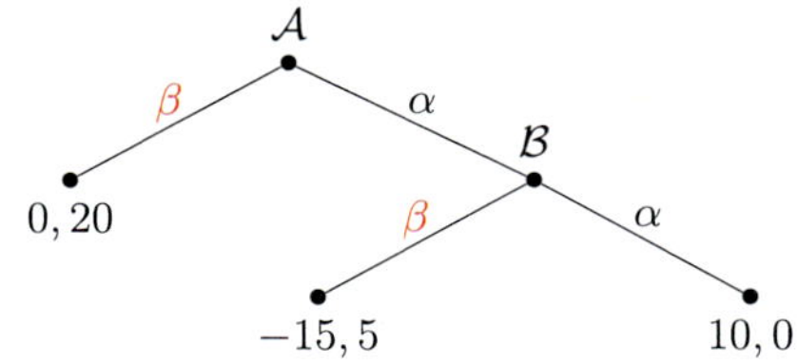

Fig. 13.4 The extensive form game induced by the mechanism in Fig. 13.2 in state β

Here, we will suppose the social authority (the King) can use, instead of fines, a bidding process as part of the extensive form mechanism. We remain to assume that the King does not know who the real mother, i.e., what the actual state of the world, is. Moreover, we relax the assumption that the King knows the valuation of each woman for the baby in any state of the world. Here, we will only require that the King knows that the valuation of the true mother for the baby is higher than the valuation of the impostor. So, if $v_{\mathcal{A}}(\theta)$ and $v_{\mathcal{B}}(\theta)$ denote the private valuations of Aliza and Baila for the child in state $\theta \in \Theta$, then the King only needs to know that $v_{\mathcal{A}}(\theta) > v_{\mathcal{B}}(\theta)$ in state $\theta = \alpha$ and $v_{\mathcal{B}}(\theta) > v_{\mathcal{A}}(\theta)$ in state $\theta = \beta$. Under this assumption, we will show that the extensive form mechanism in Fig. 13.5 will implement King's social choice rule in subgame-perfect Nash equilibrium.

According to this mechanism, Aliza has two strategies in her first node, to announce β in which case Baila gets the baby (alive) or to announce α in which case Baila will play in the game. Baila has two strategies if she plays: to announce α in which case Aliza gets the baby (alive) or to bid a price $p_{\mathcal{B}}$ for the baby to be paid to the King if she is allowed to get the baby. In case Baila bids for the baby, it is Aliza's turn to play. She will first pay a fee F to the King and then play one of following two strategies: to make a bid $p_{\mathcal{A}}$ for the baby to match the bid of Baila (not less than $p_{\mathcal{B}}$) or not to make any bid. In case Aliza makes a matching bid, Aliza will get the baby and pay $p_{\mathcal{A}}$ to the King, while Baila will pay the fee F to the King. In

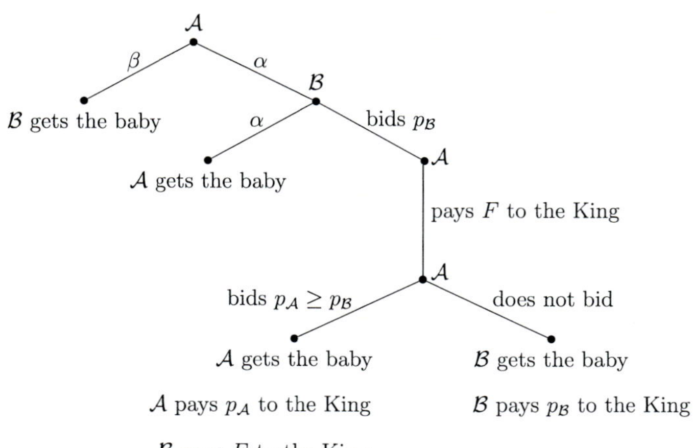

Fig. 13.5 The extensive form mechanism using the possibility of bidding to solve King Solomon's problem

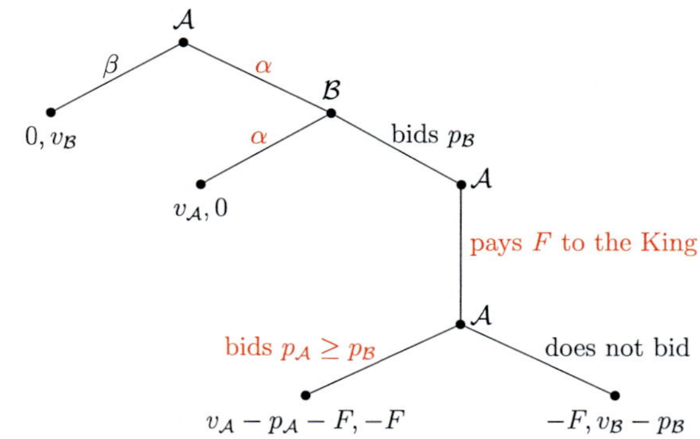

Fig. 13.6 The extensive form game induced by the mechanism in Fig. 13.5 in state α

case Aliza does not make any bid, Baila will get the baby and pay her bid $p_{\mathcal{B}}$ to the King. We assume that the fee F is sufficiently small to ensure $F < |v_{\mathcal{A}}(\theta) - v_{\mathcal{B}}(\theta)|$ for each $\theta \in \Theta$.

Given the mechanism Γ described above, the extensive form game perceived by the players in state α is portrayed in Fig. 13.6.

In the above game the highest bid Baila can make is her valuation for the baby, which is $v_{\mathcal{B}}(\alpha)$. Predicting this, Aliza should plan to choose to bid in her last node of play, since with a bid $p_{\mathcal{A}} \geq p_{\mathcal{B}}$ she would be entitled to the baby and obtain a payoff never less than $v_{\mathcal{A}}(\alpha) - v_{\mathcal{B}}(\alpha) - F$ which is higher than $-F$, the payoff she would get if she choose not to bid for the baby. Predicting this, Baila then should plan to announce α whenever she is to play which would yield to her zero payoff which is higher than $-F$, the payoff she would obtain if she chose to bid $v_{\mathcal{B}}(\alpha)$ and Aliza responded optimally by matching her bid. Predicting Baila's strategy, then Aliza should plan to choose to announce α in her first node of play as it would yield to her a payoff of $v_{\mathcal{A}}$ which is higher than zero, the payoff she would get if she were to choose to announce β. We highlight the subgame-perfect Nash equilibrium strategy profile in Fig. 13.6, by coloring the equilibrium strategies by red. This equilibrium will lead to the outcome where Aliza gets the baby, inducing the payoff profile $(v_{\mathcal{A}}, 0)$.

On the other hand, the extensive form game perceived by the players in state β is portrayed in Fig. 13.7. The highest bid Baila can make is her valuation for the baby, which is $v_{\mathcal{B}}(\beta)$. However, she knows that $v_{\mathcal{B}}(\beta) > v_{\mathcal{A}}(\beta)$ and Aliza cannot bid more than $v_{\mathcal{A}}(\beta)$. Thus, Baila should plan to choose $p_{\mathcal{B}}$ arbitrarily above $v_{\mathcal{A}}(\beta)$. Predicting this, Aliza should plan to choose not to bid in her last node of play, since with a matching bid $p_{\mathcal{A}} \geq p_{\mathcal{B}}$ she would be entitled to the baby and obtain the payoff $v_{\mathcal{A}}(\beta) - p_{\mathcal{B}}(\beta) - F$ which is (arbitrarily) lower than $-F$, the payoff she would get if she choose not to bid for the baby. Predicting this, Baila then should plan to choose to bid $p_{\mathcal{B}}$ for the baby (arbitrarily above $v_{\mathcal{A}}(\beta)$), which would yield to her a payoff of $v_{\mathcal{B}}(\beta) - p_{\mathcal{B}}$ which is higher then zero, the payoff she would obtain if she chose to announce α whenever she is to play. Predicting Baila's strategy, Aliza should then plan to choose to announce β in her first node of play as it would yield to her a payoff of zero which is higher than $-F$, the payoff she would get if she were to choose to

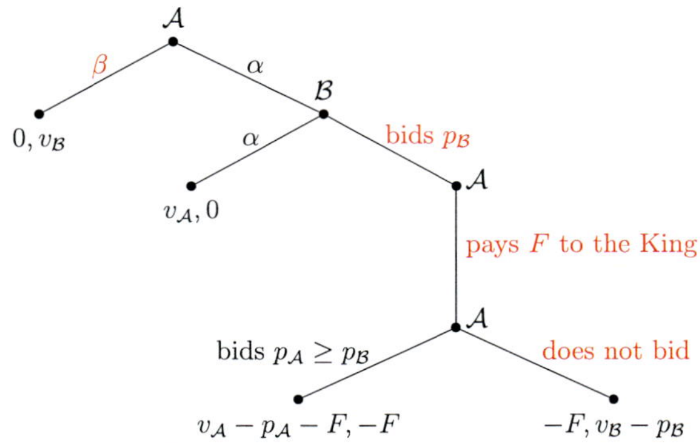

Fig. 13.7 The extensive form game induced by the mechanism in Fig. 13.5 in state β

announce α and the two women played their equilibrium strategies in the subgame starting with the node of Baila.

We highlight the subgame-perfect Nash equilibrium strategy profile in Fig. 13.7, by coloring the equilibrium strategies by red. The outcome of this equilibrium will be the outcome where Baila gets the baby, inducing the payoff profile $(0, v_{\mathcal{B}})$.

Thus, we have established that the extensive form game mechanism Γ implements King's social choice rule in subgame-perfect Nash equilibrium. A nice feature of this mechanism is that no woman actually pays any fee or any bid in any equilibrium associated with this mechanism. The fee F and the bidding process in the mechanism just act to deter the impostor, in any equilibrium, from making plans involving a bidding for the baby. Hence, the mechanism can always allocate to baby to the true mother without making her to pay any fee or bid in the realized path of any equilibrium.

13.4 Chapter Appendix

Proof of Proposition 13.2 ([4]) (The proof we state here is due to Repullo, [11]). Suppose that a social choice rule F satisfies monotonicity and no veto power and consider the mechanism $\Gamma = (M, h)$ where $M_i = A \times \Theta \times \{1, 2, \ldots, n\}$ for each $i \in N$. In this mechanism, each individual is asked to announce a message $m_i = (a^i, \theta^i, z^i) \in M_i$. The outcome function h is defined as follows:

Rule (1): If there exists (a, θ, k) with $a \in F(\theta)$ and $i \in N$ such that $m^j = (a, \theta, k)$ for all $j \neq i$, then

$$h(m) = \begin{cases} a^i \text{ if } a^i \in L_i(a, \theta) \\ a \ \text{ otherwise.} \end{cases}$$

Rule (2): If rule (1) does not apply, then $h(m) = a^i$, where $i = (\sum_{i \in N} z^i)(\bmod n)$.

We will show that the described mechanism Γ implements F. We will consider two steps.

Step 1: $F(\theta) \subseteq h(NE(\Gamma[R(\theta)]))$ for all $\theta \in \Theta$.

Let us pick any $\theta \in \Theta$ and any $a \in F(\theta)$. We have to show $a \in h(NE(\Gamma[R(\theta)]))$. Let $m \in M$ be such that $m^i = (a, \theta, 1)$ for all $i \in N$. Then, rule (1) applies and we get $h(m) = a$. If any individual i deviates from m^i, it can ensure any alternative $a^i \neq a$ to be chosen, i.e. $h(m^i, m_{-i}) = a^i$, provided that $a^i \in L_i(a, \theta)$. This implies that $h(M_i, m_{-i}) \subseteq L_i(a, \theta)$. Therefore, $m \in NE(\Gamma[R(\theta)])$. Since $h(m) = a$, it follows that $a \in h(NE(\Gamma[R(\theta)]))$.

Step 2: $h(NE(\Gamma[R(\theta)])) \subseteq F(\theta)$ for all $\theta \in \Theta$.

Let us pick any $\theta \in \Theta$ and any $m \in NE(\Gamma[R(\theta)])$. We have to show that $h(m) \in F(\theta)$. We will consider three distinct cases.

Case 1: For all $i \in N, m_i = (a, \theta', k) \in A \times \Theta \times \{1, 2, \ldots, n\}$ with $a \in F(\theta')$. Pick any $i \in N$ and $b \in L_i(a, \theta')$. Consider $m_i' = (b, \theta', k)$, which leads to $h(m_i', m_{-i}) = b$. Since $m \in NE(\Gamma[R(\theta)])$, it must be the case that $b \in L_i(a, \theta)$. Then, by monotonicity, $a \in F(\theta)$.

Case 2: For all $i \in N, m_i = (a, \theta', k) \in A \times \Theta \times \{1, 2, \ldots, n\}$ with $a \notin F(\theta')$. Pick any $i \in N$ and $b \in A$. Consider $m_i' = (b, \theta', k')$ where k' is in $\{1, 2, \ldots, n\}$ and such that $i = ((n-1)k + k')(\text{mod}\, n)$. Then, $h(m_i', m_{-i}) = b$. Since $m \in NE(\Gamma[R(\theta)])$, it must be the case that $b \in L_i(a, \theta)$. Since this is true for all $b \in A$ and $i \in N$, we have $L_i(a, \theta) = A$ for all $i \in N$. Then, by no veto power, $a \in F(\theta)$.

Case 3: There exist some $i, j \in N$ such that $m_i \neq m_j$. (Recall that $m_i = (a^i, \theta^i, z^i) \in M_i$.) Without loss of generality, assume $m_1 \neq m_2$. Then, pick any $i \in N \setminus \{1, 2\}$ and pick any $b \in A$. Consider $m_i' = (b, \theta', k')$ where k' is in $\{1, 2, \ldots, n\}$ and such that $i = (k' + \sum_{j \neq i} z^j)(\text{mod}\ n)$. Then, $h(m_i', m_{-i}) = b$. Since $m \in NE(\Gamma[R(\theta)])$, it must be the case that $b \in L_i(a, \theta)$. As this is true for all $b \in A$, we have $L_i(a, \theta) = A$. Next, observe that for any $i \in N \setminus \{1, 2\}$, we have either $m_i \neq m_1$ or $m_i \neq m_2$. Without loss of generality, assume $m_i \neq m_1$. Then, by the arguments stated above, we must have $L_2(a, \theta) = A$. Since we have established that $L_i(a, \theta) = A$ for all $i \in \{2, \ldots, n\}$, no veto power implies that $a \in F(\theta)$. ∎

References

1. Hurwicz L (1960) Optimality and informational efficiency in resource allocation processes. In: Arrow K, Karlin S, Suppes P (eds) Mathematical methods in the social sciences. Stanford University Press, Stanford
2. Hurwicz L (1972) On informationally decentralized systems. In: Radner R, McGuire CB (eds) Decision and organization: a volume in honor of Jacob Marschak. North-Holland, Amsterdam, pp 297–336
3. Hurwicz L (1973) The design of mechanisms for resource allocation. Am Econ Rev, Pap Proc 63:1–30
4. Maskin E (1999) Nash equilibrium and welfare optimality. Rev Econ Stud 66:23–38
5. Maskin E (1985) The theory of implementation in Nash equilibrium: a survey. In: Hurwicz L, Schmeidler D, Sonnenschein H (eds) Social goals and social organization. Cambridge University Press, Cambridge, pp 173–204

6. Palfrey T, Srivastava S (1991) Nash implementation using undominated strategies. Econometrica 59:479–501
7. Moore J (1992) Implementation, contracts and renegotiation in environments with complete information. In: Laffont JJ (ed) Advances in economic theory: sixth world congress, vol I. Cambridge University Press, Cambridge
8. Moore J, Repullo R (1988) Subgame perfect implementation, Econometrica 56:1191–1220
9. Reny PJ (1997) Two lectures on implementation under complete information: general results and the core. In: Hart S, Mas-Colell A (eds) Cooperation: game-theoretic approaches, NATO ASI series, vol 155. Springer, Berlin-Heidelberg
10. Glazer J, Ma CT (1989) Efficient allocation of a 'prize' - King Solomon's dilemma. Games Econ Behav 1:222–233
11. Repullo R (1987) A simple proof of Maskin's theorem on Nash implementation. Social Choice Welfare 4:39–41

Part V

Cooperative Games

This is the last part of the book, containing Chaps. 14–17. In the previous parts, we studied noncooperative games and the theory of implementation using noncooperative games. In this part, we will consider cooperative games, where individuals choose their strategic actions cooperatively. The cooperative games that we will study are static and they are with complete information. In studying these games, we will retain our earlier assumption that all players are rational and intelligent.

In Chap. 14, we study coalitional games with transferable utility. In these games, a group of individuals divides a certain amount of value (or utility) among themselves cooperatively. Each subgroup of these individuals, including the main group, is called a coalition. The utility of each coalition is assumed to be exogenously given. Also, the utility is transferable in the sense that each coalition can divide its total value among its members freely. In this framework, a plausible solution (equilibrium) concept is called the core, which is the set of utility allocations that ensures that no coalition of individuals has an incentive to leave their coalitions to form, or join into, another coalition. After formally defining a coalitional game with transferable utility, Chap. 14 defines the core, motivates it with several examples, and presents a necessary and sufficient condition under which the core is nonempty. Chapter 14 also presents a second solution concept, called the Shapley value, which always exists. This solution calculates the value of each individual in the coalitional game as her average marginal contribution over all possible orderings in which the individuals in the game may join the largest (grand) coalition. Chapter 14 ends with several examples highlighting the difference between the core and the Shapley value.

In Chap. 15, we study the problem of fair cake cutting. In this problem, the objective is to divide a cake (a metaphor for any limited and perfectly divisible resource) fairly among a group of individuals who may have different preferences over different portions of the cake. After presenting the problem formally, we present three notions of fairness, namely proportionality, envy-freeness, and equitability, and the relations between some of these three notions. Next, we present proportional cake division protocols with two players, three players, and with three or more players and also present envy-free cake division protocols with three players.

In Chap. 16, we study the problem of cooperative bargaining, which is also known as axiomatic bargaining. In a two-person cooperative bargaining problem, two individuals are faced with a set of feasible alternatives (or payoff profiles) and choose one of these alternatives using a pre-determined rule. This problem was formalized and studied by John F. Nash (while he was still an undergraduate student at Carnegie Institute of Technology). The solution Nash (1950b) proposed to the two-person bargaining problem is a rule that maximizes the product of the individuals' gains from bargaining. He showed that this rule, known as the Nash (bargaining) rule, is the only rule (among uncountable many rules) that satisfies a certain set of axioms. In Chap. 16, we present several well-known bargaining rules and prove some axiomatization results for three of them, involving the Nash rule, the Kalai-Smorodinsky rule, and the Egalitarian rule.

In the last chapter of this book, Chap. 17, we study two-sided matching between two groups of individuals or objects under nontransferable utilities. We first consider a one-to-one matching model, known as the marriage market, between a finite set of men and a finite set of women, and present several results about stable (equilibrium) matchings. Then, we extend the marriage market and some of the stability results to a many-to-one matching market, involving a set of institutions and a set of individuals, and is such that each institution can be matched to a distinct group of individuals. Finally, we study the housing market, where we show that there exists a stable matching that reallocates the houses of a finite number of individuals among them.

Coalitional Games with Transferable Utility

14

In this chapter, we will study (coalitional) games with transferable utility, also called TU games. In these games, a set of individuals will share a certain amount of value (e.g., the cost or the benefit of a project) among themselves cooperatively. Each subset of these individuals is called a coalition. The utility of each coalition is exogenously given, and the utility is transferable in the sense that each coalition can freely divide among its members the total value that this coalition can secure as a whole. In Sect. 14.1, we will provide general definitions and, in Sect. 14.2, we will define for coalitional games a solution concept, called the core, which is the set of utility allocations that ensures that no coalition of individuals has an incentive to leave their coalitions to form, or join into, another coalition. We will motivate this concept with several examples and also present a necessary and sufficient condition under which the core is nonempty. Next, in Sect. 14.3, we will present another solution concept, called the Shapley value. According to this concept, the value of each individual in the coalitional game will be equal to her average marginal contribution over all possible orderings in which the individuals in the game may join the largest coalition.

14.1 TU Games

A TU (transferable utility) game is a pair (N, v) where $N = \{1, 2, \ldots, n\}$ with $n \geq 2$ is the set of individuals (or players), also called the society, and v is a function, called the value or characteristic function, which assigns to each nonempty subset S of N a real number $v(S)$ and assigns to the empty set the value of zero, i.e., $v(\emptyset) = 0$. The subset S of N is called the coalition S. The coalition N is called the grand coalition. The set of all coalitions is denoted by 2^N.

In this chapter, we restrict our attention to superadditive games where the value of the union of any two coalitions is not less than the total value of these two coalitions. Formally, we say that a TU game (N, v) is superadditive if

$$v(S \cup S') \geq v(S) + v(S') \text{ for any } S, S' \in 2^N \text{ with } S \cap S' = \emptyset.$$

I. Saglam, *Mastering Game Theory*, Classroom Companion: Economics,
https://doi.org/10.1007/978-3-031-82094-6_14

The above assumption ensures that the players in N attain the highest total value when they form the grand coalition N.

For any nonempty coalition $S \subseteq N$, we denote a utility profile for S by a vector of real numbers $x_S = (x_i)_{i \in S}$. Given a utility profile $x \in \mathbb{R}^n$, we denote by $x(S)$ the total utility of coalition S, i.e., $x(S) = \sum_{i \in S} x_i$. Below, we have several definitions in order.

Individual rationality. A utility profile $x \in \mathbb{R}^n$ is individually rational if $x_i \geq v(\{i\})$ for each $i \in N$; i.e., each individual in the society is weakly better off under the profile x than being on its own.

Group rationality. A utility profile $x \in \mathbb{R}^n$ is groupwise rational if $x(S) \geq v(S)$ for each $S \subseteq N$; i.e., the total utility of each possible coalition S under the profile x is weakly higher than the value $v(S)$ of that coalition.

S-efficiency. Given a nonempty coalition $S \subseteq N$, a utility profile $x \in \mathbb{R}^n$ is called S-efficient if $x(S) = v(S)$.

Efficiency. A utility profile $x \in \mathbb{R}^n$ is called efficient if it is N-efficient; i.e., $x(N) = v(N)$.

Imputation. A utility profile $x \in \mathbb{R}^n$ is called an imputation if it is efficient and individually rational.

Excess. Given a coalition $S \subseteq N$ and a utility profile $x \in \mathbb{R}^n$, the excess of S under x is denoted by $\mathcal{E}(S, x)$ and is equal to $v(S) - x(S)$.

Note that for an efficient utility profile $x \in \mathbb{R}^n$, the excess of the grand coalition under x is zero; i.e., $\mathcal{E}(N, x) = v(N) - x(N) = 0$.

14.2 Core

Given the previous definitions, we are ready to define a special selection of utility profiles in a TU game.

Definition 14.1 A utility profile $x \in \mathbb{R}^n$ is in the core of a TU game (N, v) if x is an imputation and groupwise rational.

Given the above definition, we denote the core by the set $C(N, v)$, which is given by

$$C(N, v) = \left\{x \in \mathbb{R}^n \; : \; x(N) = v(N) \text{ and } x(S) \geq v(S) \text{ for each nonempty } S \subset N\right\}.$$

The above definition of the core implies that the individuals in the society form the grand coalition to maximize the total value of the society and distribute the maximized total value $v(N)$ among the individuals through a utility profile x in such a way that for any subset S of N a group of individuals that are able to form the coalition S can enjoy when they remain in the grand coalition a total value $x(S)$ which is not less than $v(S)$, which is the utility sum they would enjoy as a whole if they broke the grand coalition and form the coalition S.

Using the definition of excess, we can restate Definition 14.1 as follows.

Definition 14.2 A utility profile $x \in \mathbb{R}^n$ is in the core of a TU game (N, v) if $\mathcal{E}(N, x) = 0$ and $\mathcal{E}(S, x) \leq 0$ for each nonempty $S \subset N$.

Since a core imputation is groupwise rational, it is robust is to any possible coalitional deviation. Therefore, we have an alternative definition that is equivalent to Definitions 14.1 and 14.2.

Definition 14.3 A utility profile $x \in \mathbb{R}^n$ is in the core of a TU game (N, v) if $x(N) = v(N)$ and there exist no $S \subseteq N$ and $y_S \in \mathbb{R}^{|S|}$ such that y_S is S-efficient and $y_i > x_i$ for all $i \in S$.

The above definition says that a utility profile x is a core allocation if no coalition S of the players in N can block x by leaving the grand coalition N and redistributing the value $v(S)$ among the members of S in such a way that each member of S becomes strictly better off.

The first definition above implies that the core of a TU game is the intersection of a finite number of inequalities subject to the efficiency constraint $x(N) = v(N)$, hence the core is a convex, closed, and bounded set. In the following example, we will calculate this set.

Example 14.1 (*Hair Stylist Game*) Ariel, Blue, and Cordelia have appointments with the same hair stylist on Wednesday, Thursday, and Friday, respectively. Their dollar payoffs from having an appointment in any of these three days are shown in Table 14.1. They consider whether they can increase their payoffs by forming coalitions to reshuffle their appointment days among them.

The set of players is $N = \{1, 2, 3\}$ where $1 :=$ Ariel, $2 :=$ Blue, and $3 :=$ Cordelia. We will calculate the value function v at each nonempty coalition, considering the additional payoff available whenever a coalition is formed. Thus,

$$v(S) := u(S) - \sum_{i \in S} u(\{i\}) \text{ for each nonempty } S \subseteq N,$$

Table 14.1 Dollar payoffs of the three women from an appointment in each of the three days

	Wed	Thu	Fri
Ariel	4	5	7
Blue	8	3	4
Cordelia	8	5	5

Table 14.2 The maximal payoff and value of each possible coalition in the hair stylist game

S	$\{1\}$	$\{2\}$	$\{3\}$	$\{1, 2\}$	$\{1, 3\}$	$\{2, 3\}$	$\{1, 2, 3\}$
$u(S)$	4	3	5	13	15	9	20
$v(S)$	0	0	0	6	6	1	8

where $u(S)$ is the maximal payoff that coalition S can obtain by reshuffling the appointment days within the members of S. (We will assume that $v(\emptyset) = 0$.) In Table 14.2, we calculate $u(S)$ and $v(S)$ for each nonempty coalition S.

The core of the TU game (N, v) described above is

$$C(N, v) = \{x \in \mathbb{R}^3 : x_1 \geq 0,\ x_2 \geq 0,\ x_3 \geq 0,$$

$$x_1 + x_2 \geq 6,\ x_1 + x_3 \geq 6,\ x_2 + x_3 \geq 1,$$

$$x_1 + x_2 + x_3 = 8\}.$$

We portray the graph of $C(N, v)$ as the yellow-colored region (including the boundaries) in Fig. 14.1.

Note that any utility allocation x in the core is three dimensional. However, thanks to the efficiency constraint $x(N) = v(N)$, or $x_1 + x_2 + x_3 = 8$, which relates any

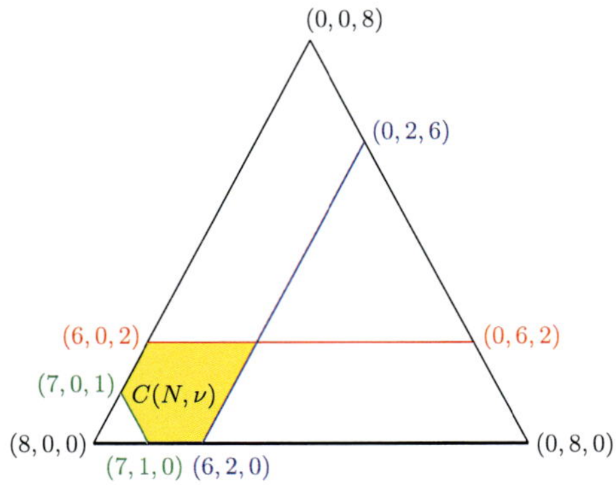

Fig. 14.1 The core, $C(N, v)$, of the hair stylist game in Table 14.2

Table 14.3 The payoffs used to calculate the value of the grand coalition in the hair stylist game

	Wed	Thu	Fri
Ariel	4	5	7
Blue	8	3	4
Cordelia	8	5	5

dimension of x to the other two, we can portray the core using a two-dimensional graph like in Fig. 14.1. The triangle in this graph contains all efficient and individually rational utility allocations ; i.e., for any allocation x in the triangle we have $x(N) = v(N) = 8$ and $x_i \geq v(\{i\}) = 0$ for each $i \in N$. For an allocation $x = (x_1, x_2, x_3)$, we should also observe that x_1 is zero on the right side of the triangle and it attains its maximum value, 8, on the bottom left corner. Similarly, x_2 is zero on the left side of the triangle and attains its maximum value, 8, on the bottom right corner, whereas x_3 is zero at the bottom side of the triangle and attains its maximum value, 8, at the top corner.

Given that the core is the yellow-colored region in Fig. 14.1, what are the final appointment days of customers supposing that they have agreed on some core allocation? The value of the grand coalition N is $v(N) = 8$, which is higher than the value of any other coalition. Recall that $v(N)$ is obtained by first summing the red-colored utilities in Table 14.3 below and then subtracting from the resulting sum all blue-valued utilities (corresponding to the initial appointments).

Since the core is nonempty, it is beneficial for the three customers to collectively agree on some allocation in the core, which necessitates the participation of each customer to the grand coalition. Subject to the formation of the grand coalition, the final appointment day of each customer can be readily found in the above table, since such a day must produce a red-colored utility in the row where the name of a customer appears. So, the final appointment days of the customers must be as follows:

$$\text{Ariel} \rightarrow \text{Fri}, \quad \text{Blue} \rightarrow \text{Wed}, \quad \text{Cordelia} \rightarrow \text{Thu}.$$

Suppose the customers agree on the core allocation $x = (6, 2, 0)$. Since $x_3 = 0$, Cordelia does not gain or lose anything from this agreement. On the other hand, $x_1 = 6$ and $x_2 = 2$, implying that Ariel gains 6 dollars and Blue gains 2 dollars by this agreement. However, Table 14.1 shows that in the absence of any utility transfer, the net increase in the payoff of Ariel is 3 ($= 7 - 4$) dollars if she moves from Wednesday to Friday, while the net increase in the payoff of Blue is 5 ($= 8 - 3$) dollars if she moves from Thursday to Wednesday. Thus, the three customers can realize the core allocation (6, 2, 0) if and only if Blue transfers 3 dollar to Ariel and the three players Ariel, Blue, and Cordelia go to the hair stylist on Friday, Wednesday, and Thursday, respectively. ■

The core in the above example is an uncountable set of allocations. The following two examples show that the core of a TU game may contain a single allocation or no allocation at all.

Table 14.4 The value of each possible coalition in Example 14.2

S	$\emptyset$	$\{1\}$	$\{2\}$	$\{3\}$	$\{1, 2\}$	$\{1, 3\}$	$\{2, 3\}$	$\{1, 2, 3\}$
$\nu(S)$	0	0	0	0	6	6	4	8

Example 14.2 Consider a TU-Game (N, ν) where $N = \{1, 2, 3\}$ and ν is given by Table 14.4.

In comparison to Example 14.1, the net worth of the coalition $\{2, 3\}$ has increased from its earlier value of 1 to its new value of 4, while everything else has remained the same. So, we can calculate the core of the above game as

$$C(N, v) = \{x \in \mathbb{R}^3 : x_1 \geq 0,\ x_2 \geq 0,\ x_3 \geq 0,$$

$$x_1 + x_2 \geq 6,\ x_1 + x_3 \geq 6,\ x_2 + x_3 \geq 4,$$

$$x_1 + x_2 + x_3 = 8\}.$$

We portray $C(N, \nu)$ in Fig. 14.2.

This figure shows that the core, $C(N, \nu)$, contains the unique allocation $(4, 2, 2)$. ■

One can see Derya [1] for a set of necessary and sufficient conditions which ensure that a TU game will have a unique core allocation. Below, we consider a TU game with an empty core.

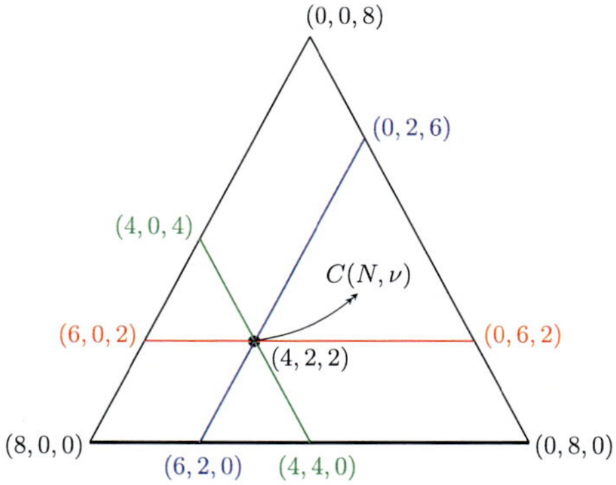

Fig. 14.2 The core, $C(N, \nu)$, of the TU game in Table 14.4

Table 14.5 The value of each possible coalition in Example 14.3

S	$\emptyset$	$\{1\}$	$\{2\}$	$\{3\}$	$\{1, 2\}$	$\{1, 3\}$	$\{2, 3\}$	$\{1, 2, 3\}$
$v(S)$	0	0	0	0	6	6	7	8

Example 14.3 Consider a TU-Game (N, v) where $N = \{1, 2, 3\}$ and v is given by Table 14.5.

In comparison to Example 14.2, the only change is in $v(\{2, 3\})$, which is increased from 4 to 7. The core of the described game is

$$C(N, v) = \{x \in \mathbb{R}^3 : x_1 \geq 0,\ x_2 \geq 0,\ x_3 \geq 0,$$

$$x_1 + x_2 \geq 6,\ x_1 + x_3 \geq 6,\ x_2 + x_3 \geq 7,$$

$$x_1 + x_2 + x_3 = 8\}.$$

Figure 14.3 shows that the core is empty.

For the given TU game, a utility allocation in the triangle could be in the core only if it were on the right of the green line segment, on the left of the blue segment, and below the red line segment. But, there is no allocation satisfying these three conditions simultaneously. ■

Now, we will present a result (independently obtained by Olga Bondareva [2] and Lloyd Shapley [3]) that characterizes a condition under which a TU game will have a nonempty core. To state this result, we have to first introduce several definitions.

Given a society of individuals $N = \{1, \ldots, n\}$, we say that a function $\lambda : 2^N \setminus \emptyset \to \mathbb{R}_+$ is balanced if

$$\sum_{S \subseteq N : i \in S} \lambda(S) = 1 \text{ for each } i \in N.$$

We can interpret the above definition as follows. Suppose each individual in the society N has 1 unit of labor to work in a plant during a year. For each nonempty

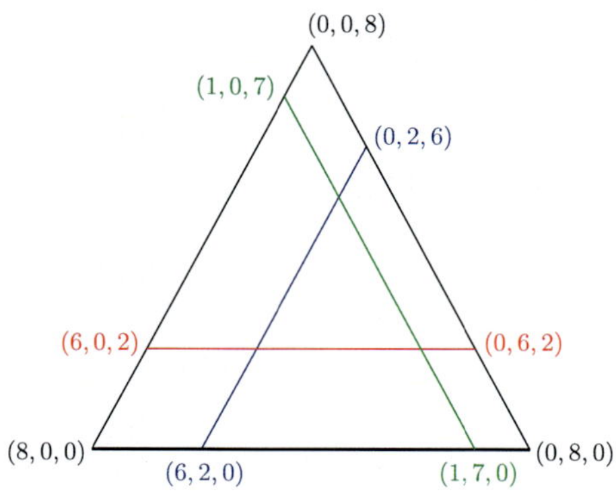

Fig. 14.3 The emptiness of core, $C(N, v)$, for the TU game in Table 14.5

coalition S in this society, suppose a social planner determines the fractional amount of labor, $\lambda(S) \in [0, 1]$, that must be spent by each member of S in such a way that each individual in the society, after spending her labor according to the prescriptions of λ on all coalitions that she is a member of, is left with zero unit of labor. In that case, we say that the function λ is balanced.

Example 14.4 For example, consider the society $N = \{1, 2, 3\}$. The set of nonempty coalitions is $\{\{1\}, \{2\}, \{3\}, \{1, 2\}, \{1, 3\}, \{2, 3\}, \{1, 2, 3\}\}$. Individual 1 is a member of coalitions in $\{\{1\}, \{1, 2\}, \{1, 3\}, \{1, 2, 3\}\}$. Then, the balancedness of λ requires

$$\lambda(\{1\}) + \lambda(\{1, 2\}) + \lambda(\{1, 3\}) + \lambda(\{1, 2, 3\}) = 1. \tag{14.1}$$

Individual 2 is a member of coalitions in $\{\{2\}, \{1, 2\}, \{2, 3\}, \{1, 2, 3\}\}$. Then, the balancedness of λ requires

$$\lambda(\{2\}) + \lambda(\{1, 2\}) + \lambda(\{2, 3\}) + \lambda(\{1, 2, 3\}) = 1. \tag{14.2}$$

Finally, individual 3 is a member of coalitions in $\{\{3\}, \{1, 3\}, \{2, 3\}, \{1, 2, 3\}\}$. Then, the balancedness of λ requires

$$\lambda(\{3\}) + \lambda(\{1, 3\}) + \lambda(\{2, 3\}) + \lambda(\{1, 2, 3\}) = 1. \tag{14.3}$$

One can find many examples of λ that solve the three equations above. One solution prescribes

$$\lambda(S) = \begin{cases} 1 \text{ if } |S| = 1 \\ 0 \text{ if } |S| \in \{2, 3\}. \end{cases}$$

Another solution prescribes

$$\lambda(S) = \begin{cases} 1/2 \text{ if } |S| = 2 \\ 0 \quad\ \text{if } |S| \in \{1, 2\}. \end{cases}$$

To show that there are uncountable many solutions, pick any $k \in [0, 1]$ and consider the associated function λ^k that satisfies

$$\lambda^k(S) = \begin{cases} k & \text{if } |S| = 1 \\ (1-k)/4 & \text{if } |S| = 2 \\ (1-k)/2 & \text{if } |S| = 3. \end{cases}$$

One can easily check that the solution family represented by $\lambda^k(S)$ is balanced. (Moreover, this family does not represent all possible balanced functions.) ■

Using the definition of balanced functions, we can define the following.

Definition 14.4 A TU game (N, v) is balanced if for each balanced function λ we have

$$\sum_{S \subset N} \lambda(S)v(S) \leq v(N). \tag{14.4}$$

The above definition implies that a TU game is balanced if the fractional membership, where each individual in the society may join to any coalition with other individuals by using a balanced fraction of her identity (or spending a balanced fraction of her labor input, according to our earlier interpretation), cannot generate a total value for the society above the value of the grand coalition where each individual spends all of her labor input.

Proposition 14.1 (Bondareva & Shapley) *A TU game is balanced if and only if its core is nonempty.*

We will skip the proof as it is involved. Below, we will illustrate how we can benefit from the above result.

Example 14.5 Consider a TU game (N, v) where $N = \{1, 2, 3\}$ and v is given by the following table.

We will assume that $\alpha \in [0, 8]$ and show whether the core of (N, v) is nonempty for any value of α. So, pick any $\alpha \in [0, 8]$ and consider the associated TU game. Proposition 14.1 tells us that to show that this game has a nonempty core, it is sufficient to show that it is balanced. Thus, we have to check whether the inequality condition in (14.4) is satisfied for each balanced function. We can explicitly write this condition as follows:

$$\lambda(\{1\})v(\{1\}) + \lambda(\{2\})v(\{2\}) + \lambda(\{3\})v(\{3\})$$
$$+\lambda(\{1, 2\})v(\{1, 2\}) + \lambda(\{2, 3\})v(\{2, 3\}) + \lambda(\{1, 3\})v(\{1, 3\})$$
$$+\lambda(\{1, 2, 3\})v(\{1, 2, 3\}) \leq v(\{1, 2, 3\}).$$

Using Table 14.6, we can reduce the above inequality to

$$6\lambda(\{1, 2\}) + \alpha\lambda(\{2, 3\}) + 6\lambda(\{1, 3\}) + 8\lambda(\{1, 2, 3\}) \leq 8. \tag{14.5}$$

For the game (N, v) to be balanced, the inequality condition in (14.5) must hold for any function λ that is balanced. So, λ must satisfy the Eqs. (14.1)–(14.3). Let us

Table 14.6 The value of each possible coalition in Example 14.5

S	$\emptyset$	$\{1\}$	$\{2\}$	$\{3\}$	$\{1, 2\}$	$\{1, 3\}$	$\{2, 3\}$	$\{1, 2, 3\}$
$v(S)$	0	0	0	0	6	6	α	8

denote the left-hand side of (14.5) by L^λ. We have to show $L^\lambda \leq 8$ for any balanced λ. Since L^λ does not contain the term $\lambda(\{i\})$ for any $i \in N$, we will consider balanced λ functions where $\lambda(\{i\}) = 0$ for each $i \in N$. If (14.5) holds for all such functions, then it will also hold for balanced λ functions where $\lambda(\{i\}) \geq 0$ for each $i \in N$.

We should immediately note that whenever $\lambda(\{i\}) = 0$ for each $i \in N$, the Eqs. (14.1)–(14.3) imply that

$$\lambda^k(S) = \begin{cases} (1-k)/2 & \text{if } |S| = 2 \\ k & \text{if } |S| = 3 \end{cases}$$

where $k \in [0, 1]$. Thus, we can calculate

$$L^\lambda = (6 + \alpha + 6)\frac{(1-k)}{2} + 8k = (12 + \alpha)\frac{(1-k)}{2} + 8k.$$

Then, the balanced condition $L^\lambda \leq 8$ can be written as

$$(12 + \alpha)\frac{(1-k)}{2} + 8k \leq 8$$

or equivalently

$$(12 + \alpha)\frac{(1-k)}{2} \leq 8(1-k)$$

implying

$$(\alpha - 4)(1-k) \leq 0.$$

Since $1 - k \geq 0$, we must have $\alpha - 4 \leq 0$, implying $\alpha \leq 4$. Thus, we have established that the core of (N, v) is nonempty if and only if $\alpha \leq 4$. Note that the game (N, v) we have studied coincides with the TU game in

Example 14.1 whenever $\alpha = 1$ and that game has a nonempty core (since $\alpha \leq 4$ holds);

Example 14.2 whenever $\alpha = 4$ and that game has a nonempty core (since $\alpha \leq 4$ holds);

Example 14.3 whenever $\alpha = 7$ and that game has an empty core (since $\alpha \leq 4$ is violated). ■

Let us consider another example below.

Example 14.6 (*The Glove Game*) Consider a TU game (N, v), a simple version of the glove game, where each individual in the society $N = \{1, 2, 3\}$ has either a left-hand glove or a right-hand glove. Let $L = \{1\}$ and $R = \{2, 3\}$ represent the set of individuals that have the left-hand glove and right-hand glove, respectively. The

Table 14.7 The value of each possible coalition in Example 14.6

S	$\emptyset$	$\{1\}$	$\{2\}$	$\{3\}$	$\{1,2\}$	$\{1,3\}$	$\{2,3\}$	$\{1,2,3\}$
$\nu(S)$	0	0	0	0	1	1	0	1

value $\nu(S)$ of a nonempty coalition $S \subseteq N$ is assumed to be equal to the number of pairs of gloves that the members of the coalition can make. Formally, the value function ν is such that $\mu(\emptyset) = 0$ and

$$\nu(S) = \min\{|S \cap L|, |S \cap R|\} \text{ for each } S \in 2^N \setminus \emptyset\}.$$

This function generates the values in Table 14.7.

Since the above game is fairly simple, we can directly attempt to calculate the core as we did in Examples 14.1–14.3. However, let us first apply Proposition 14.1 to find out whether the core of this game is nonempty or not. For this purpose, it is sufficient to check whether the game is balanced or not. Thus, we will check whether the condition in (14.4) is satisfied for each balanced function λ. Using Table 14.7, we can write this condition as

$$\lambda(\{1,2\}) + \lambda(\{1,3\}) + \lambda(\{1,2,3\}) \leq 1. \tag{14.6}$$

Since the left-hand side of (14.6) does not contain the term $\lambda(\{i\})$ for any $i \in N$, we will consider balanced λ functions where $\lambda(\{i\}) = 0$ for each $i \in N$. If (14.6) holds for all such functions, then it will also hold for balanced λ functions where $\lambda(\{i\}) \geq 0$ for each $i \in N$.

Whenever $\lambda(\{i\}) = 0$ for each $i \in N$, λ would satisfy the balancedness conditions (14.1)–(14.3) only if λ is equal to

$$\lambda^k(S) = \begin{cases} (1-k)/2 & \text{if } |S| = 2 \\ k & \text{if } |S| = 3 \end{cases}$$

where $k \in [0, 1]$. Then, given a balanced function λ^k, we can rewrite (14.6) as

$$\frac{(1-k)}{2} + \frac{(1-k)}{2} + k \leq 1$$

implying $1 \leq 1$. Thus, we have established that the core of (N, ν) is nonempty.

Simply, we can calculate the core of the above game as

$$C(N, v) = \{x \in \mathbb{R}^3 : x_1 \geq 0,\ x_2 \geq 0,\ x_3 \geq 0,$$
$$x_1 + x_2 \geq 1,\ x_1 + x_3 \geq 1,\ x_2 + x_3 \geq 0,$$
$$x_1 + x_2 + x_3 = 1\}.$$

We portray $C(N, v)$ in Fig. 14.4.

Fig. 14.4 The core, $C(N, v)$, of the glove game in Table 14.7

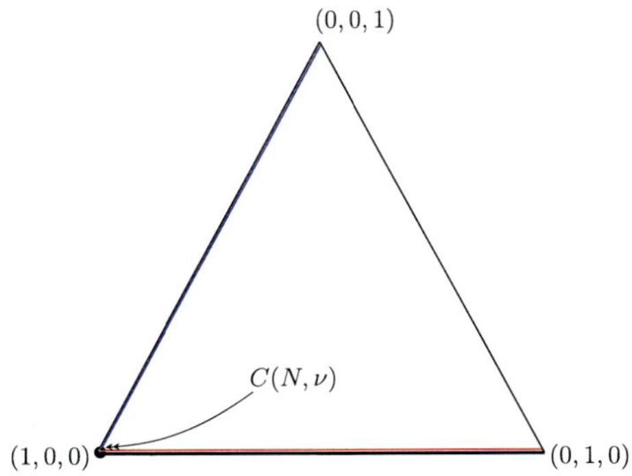

Note that a green line segment representing the core condition $x_2 + x_3 \geq 0$ is not visible in Fig. 14.4. The reason is that this line segment (with slope -1) now coincides with the allocation $(1, 0, 0)$ on the triangle representing the set of all imputations. The intersection of red and blue lines also occurs at this imputation. Thus, the core, $C(N, v)$, contains a unique imputation, which is $(1, 0, 0)$. The striking asymmetry in the utilities of individual 1 and the other two individuals is due to the asymmetry in their endowments. Since the three individuals can produce only a single pair of gloves and the right-hand gloves of individuals 2 and 3 are completely substitutable, none of these two individuals can enjoy a positive utility at a core imputation. To see why this is so, suppose towards a contradiction that an allocation $x = (1 - b - c, b, c)$, where $b, c \in (0, 1)$ and $b + c < 1$, is in the core. This allocation could only be generated by the grand coalition, which we can show would not be stable against deviations. For example, individuals 1 and 2 could leave the grand coalition and form the coalition $\{1, 2\}$ between themselves and could enjoy a mutually beneficial allocation $x = (1 - b - xc, b + xc, 0)$ for some $x \in (0, 1)$. But, this allocation would not be stable either. Individual 3 being left alone and deriving zero utility from his right-hand glove, would offer to individual 1 to leave the coalition with individual 2 to form the coalition $\{1, 3\}$ and enjoy the mutually beneficial allocation $(1 - b - yc, 0, b + yc)$ where $0 < y < x$. Thus, any coalition that generates a utility allocation x cannot be robust against coalitional deviations unless x_2 and x_3 are both zero. Finally, we should note that the core allocation $(1, 0, 0)$ can be generated by each one of the coalitions $\{1, 2\}$, $\{1, 3\}$, and $\{1, 2, 3\}$. ■

14.3 Shapley Value

In the previous section we studied the core as a solution concept in TU games. We saw that the core may not exist and it may contain multiple (even uncountably many) utility allocations whenever it exists. In this section, we will present a solution concept, called the Shapley value, that always selects a unique solution (utility allocation) for any TU game.

Definition 14.5 For a TU game (N, ν) we denote the Shapley value by the utility profile $\varphi(N, \nu)$, where

$$\varphi_i(N, \nu) = \sum_{S \subseteq N-i} \frac{|S|!\,(n-|S|-1)!}{n!}\,(\nu(S \cup \{i\}) - \nu(S)) \ \text{ for each } i \in N. \quad (14.7)$$

The Shapley value calculates the value of each individual in the grand coalition N as the average marginal contribution over all possible orderings in which individuals may join N. Note in the above definition that for any coalition S that does not involve an individual i, $\nu(S \cup \{i\}) - \nu(S)$ is the marginal value (contribution) provided by individual i if she joins to coalition S (or more precisely is she forms the coalition $S \cup \{i\}$ with the individuals in the coalition S). We know that any nonempty coalition S can be ordered in $|S|!$ ways. After agent i joins to coalition S, the remaining individuals in $N \setminus \{S \cup \{i\}\}$ can be ordered in $(n - |S| - 1)!$ ways. Thus, there are $|S|!(n - |S| - 1)!$ distinct orderings of the set N where individual i can join to the coalition S. Multiplying this number by $\nu(S \cup \{i\}) - \nu(S)$, dividing by $n!$ (the number of total orderings of N), and finally summing the resulting value over all nonempty coalitions excluding individual i, we get the average marginal value of individual i in the grand coalition.

The Shapley value can be written in a more compact way. Let $\mathcal{N}$ denote the set of all orderings of the grand coalition N. Pick any ordering $S \in \mathcal{N}$ and for any individual $i \in N$ let $S^{B,i}$ denote the set of individuals that come before individual i in the ordering S. Now suppose that individuals in N join one by one to the grand coalition N according to the order S and each of them demands her marginal contribution to the coalition of individuals preceding her. Thus, individual $i \in N$ demands the marginal contribution $\nu\left(S^{B,i} \cup \{i\}\right) - \nu\left(S^{B,i}\right)$ upon entering the coalition $S^{B,i}$ according to the order S. Summing the marginal contribution of individual i under each possible ordering $S \in \mathcal{N}$ and then dividing the calculated total sum by the number of possible orderings in $\mathcal{N}$, which is $n!$, we can obtain the Shapley value for individual i. That is,

$$\varphi_i(N, \nu) = \frac{1}{n!} \sum_{S \in \mathcal{N}} \left(\nu\left(S^{B,i} \cup \{i\}\right) - \nu\left(S^{B,i}\right)\right) \ \text{ for each } i \in N. \quad (14.8)$$

To show that (14.8) is equivalent to (14.7), note that there are $|S^{B,i}|!$ different orders in which the individuals in the set $S^{B,i}$ can enter the grand coalition before individual i. Also, the set of individuals that enter the grand coalition after individual i can make this entrance under $(n - |S^{B,i}| - 1)!$ different orders. Thus, individual i can obtain the marginal contribution $\nu\left(S^{B,i} \cup \{i\}\right) - \nu\left(S^{B,i}\right)$ in a total number of $|S^{B,i}|!(n - |S^{B,i}| - 1)!$ different orders. Note also that as the set S is varied in $\mathcal{N}$, the set $S^{B,i}$ becomes varied in $2^{N-i} \setminus \emptyset$. Thus, multiplying $\nu\left(S^{B,i} \cup \{i\}\right) - \nu\left(S^{B,i}\right)$ by $|S^{B,i}|!(n - |S^{B,i}| - 1)!$ and summing over all possible coalitions $S^{B,i} \subseteq N - i$, and then dividing by $n!$ we get the first formulation of the Shapley value, introduced by (14.7).

Using Eq. (14.8), we can easily show that the Shapley value is an efficient utility distribution.

Proposition 14.2 *For any TU game (N, ν), the Shapley value $\varphi(N\nu)$ is efficient.*

Proof Consider any TU game (N, ν). Let $\varphi(N, \nu)$ be its Shapley value. Then, using (14.8), we can calculate

$$\sum_{i\in N} \varphi_i(N, \nu) = \sum_{i\in N} \frac{1}{n!} \sum_{S\in\mathcal{N}} \left(\nu\left(S^{B,i} \cup \{i\}\right) - \nu\left(S^{B,i}\right)\right)$$

$$= \frac{1}{n!} \sum_{S\in\mathcal{N}} \sum_{i\in N} \left(\nu\left(S^{B,i} \cup \{i\}\right) - \nu\left(S^{B,i}\right)\right)$$

$$= \frac{1}{n!} \sum_{S\in\mathcal{N}} \nu(N) = \nu(N),$$

implying that the Shapley value is efficient. ■

To calculate the Shapley value of a TU game, we will use the simpler formulation in (14.8), as we illustrate next.

Example 14.7 (*Two-Player TU Games*) Consider a TU game (N, ν) containing two individuals only, i.e., $N = \{1, 2\}$. Let the value function ν be such that $\nu(\emptyset) = 0$ and ν is superadditive implying that $\nu(\{1, 2\}) \geq \nu(\{1\}) + \nu(\{2\})$.

To compute the Shapley value, we prepare the following table (Table 14.8).

Table 14.8 The marginal contributions and Shapley value in Example 14.7

$S \in \mathcal{N}$	$\nu\left(S^{B,1} \cup \{1\}\right) - \nu\left(S^{B,1}\right)$	
$\{1, 2\}$	$\nu(\emptyset \cup \{1\}) - \nu(\emptyset)$	$= \nu(\{1\}) - 0 = \nu(\{1\})$
$\{2, 1\}$	$\nu(\{2\} \cup \{1\}) - \nu(\{2\})$	$= \nu(\{1, 2\}) - \nu(\{2\})$
$\varphi_1(N, \nu)$	$\frac{1}{2}[\nu(\{1, 2\}) + \nu(\{1\}) - \nu(\{2\})]$	

Panel (i)

$S \in \mathcal{N}$	$\nu\left(S^{B,2} \cup \{2\}\right) - \nu\left(S^{B,2}\right)$	
$\{1, 2\}$	$\nu(\{1\} \cup \{2\}) - \nu(\{1\})$	
$\{2, 1\}$	$\nu(\emptyset \cup \{2\}) - \nu(\emptyset)$	$= \nu(\{2\}) - 0 = \nu(\{2\})$
$\varphi_2(N, \nu)$	$\frac{1}{2}[\nu(\{1, 2\}) - \nu(\{1\}) + \nu(\{2\})]$	

Panel (ii)

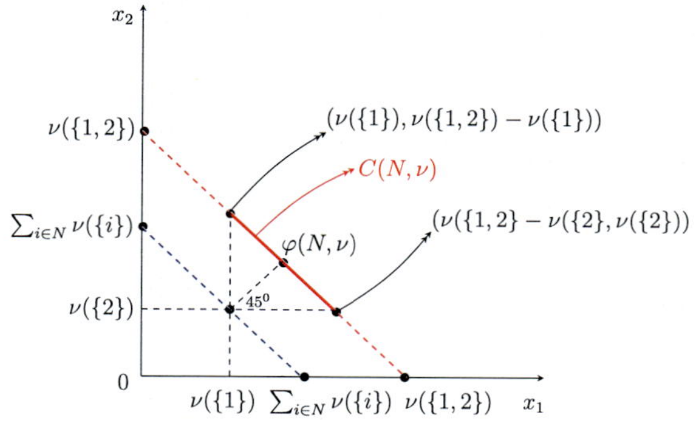

Fig. 14.5 The core, $C(N, \nu)$, and Shapley value, $\varphi(N, \nu)$, of two-player TU games

Thus, the Shapley value of the TU game (N, ν) is

$$\varphi(N,\nu)=\left(\frac{1}{2}\left[\nu(\{1,2\})+\nu(\{1\})-\nu(\{2\})\right], \frac{1}{2}\left[\nu(\{1,2\})-\nu(\{1\})+\nu(\{2\})\right]\right).$$

Note that $\sum_{i\in N}\varphi_i(N,\nu)=\nu(\{1,2\})=\nu(N)$, implying that the calculated Shapley value is efficient as predicted by Proposition 14.2.

We can calculate the core of the described two-person TU game as

$$C(N,\upsilon)=\left\{x\in\mathbb{R}^2 : x_1\geq\nu(\{1\}),\ x_2\geq\nu(\{2\}),\ x_1+x_2=\nu(\{1,2\})\right\}.$$

Note that since the TU game (N, ν) involves only two individuals, a utility profile is an imputation if and only if it is in the core.

Let us check whether the Shapley value of the game (N, ν) is contained by the core. The superadditivity of ν implies individual rationality of the Shapley Value. This is because $\nu(\{1,2\})-\nu(\{2\})\geq\nu(\{1\})$ implies $\varphi_1(N,\nu)\geq\nu(\{1\})$, and $\nu(\{1,2\})-\nu(\{1\})\geq\nu(\{2\})$ implies $\varphi_2(N,\nu)\geq\nu(\{2\})$. We also know that the Shapley value is efficient; i.e., $\varphi_1(N,\nu)+\varphi_2(N,\nu)=\nu(N)$. Therefore, the Shapley value $\varphi(N,\nu)$ is contained by the core $C(N,\nu)$.

In Fig. 14.5, we portray the core of (N, ν), involving the Shapley value. Note that the line segment between the points $(\nu(\{1\}),\nu(\{1,2\})-\nu(\{1\}))$ and $(\nu(\{1,2\})-\nu(\{2\}),\nu(\{2\}))$ is the core of the (N, ν), and the Shapley value is just at the middle of this line segment.

If the two individuals in the society cannot agree to make the grand coalition N, each individual $i\in N$ has to enjoy her value from the singleton coalition $\nu(\{i\})$. The dashed blue line passing trough the utility allocation $(\nu(\{1\}),\nu(\{2\}))$ is below the red line, since the value function ν is strictly superadditive. Thus, the distance between these two lines, $\nu(\{1,2\})-\nu(\{1\})-\nu(\{2\})$, is positive.

The Shapley value uniformly splits this excess utility value to the two individuals whenever they form the grand coalition. Thus, each individual i in the grand coalition N enjoys, in addition to her reservation value $\nu(i)$, the net individual gain of the joint coalition $\frac{1}{2}[\nu(\{1,2\})-\nu(\{1\})-\nu(\{2\})]$, thus enjoying in sum the Shapley value $\varphi_i(N,\nu)$. ■

Table 14.9 The marginal contributions and Shapley value in Example 14.8

$S \in \mathcal{N}$	$\nu\left(S^{B,1} \cup \{1\}\right) - \nu\left(S^{B,1}\right)$	
$\{1, 2, 3\}$	$\nu(\emptyset \cup \{1\}) - \nu(\emptyset)$	$= 0 - 0 = 0$
$\{1, 3, 2\}$	$\nu(\emptyset \cup \{1\}) - \nu(\emptyset)$	$= 0 - 0 = 0$
$\{2, 1, 3\}$	$\nu(\{2\} \cup \{1\}) - \nu(\{2\})$	$= 6 - 0 = 6$
$\{2, 3, 1\}$	$\nu(\{2, 3\} \cup \{1\}) - \nu(\{2, 3\})$	$= 8 - \alpha$
$\{3, 1, 2\}$	$\nu(\{3\} \cup \{1\}) - \nu(\{3\})$	$= 6 - 0 = 6$
$\{3, 2, 1\}$	$\nu(\{2, 3\} \cup \{1\}) - \nu(\{2, 3\})$	$= 8 - \alpha$
$\varphi_1(N, \nu)$	$\frac{1}{3!}(0 + 0 + 6 + 8 - \alpha + 6 + 8 - \alpha) = \frac{28-2\alpha}{6}$	

Panel (i)

$S \in \mathcal{N}$	$\nu\left(S^{B,2} \cup \{2\}\right) - \nu\left(S^{B,2}\right)$	
$\{1, 2, 3\}$	$\nu(\{1\} \cup \{2\}) - \nu(\{1\})$	$= 6 - 0 = 6$
$\{1, 3, 2\}$	$\nu(\{1, 3\} \cup \{2\}) - \nu(\{1, 3\})$	$= 8 - 6 = 2$
$\{2, 1, 3\}$	$\nu(\emptyset \cup \{2\}) - \nu(\emptyset)$	$= 0 - 0 = 0$
$\{2, 3, 1\}$	$\nu(\emptyset \cup \{2\}) - \nu(\emptyset)$	$= 0 - 0 = 0$
$\{3, 1, 2\}$	$\nu(\{1, 3\} \cup \{2\}) - \nu(\{1, 3\})$	$= 8 - 6 = 2$
$\{3, 2, 1\}$	$\nu(\{3\} \cup \{2\}) - \nu(\{3\})$	$= \alpha - 0 = \alpha$
$\varphi_2(N, \nu)$	$\frac{1}{3!}(6 + 2 + 0 + 0 + 2 + \alpha) = \frac{10+\alpha}{6}$	

Panel (ii)

$S \in \mathcal{N}$	$\nu\left(S^{B,3} \cup \{3\}\right) - \nu\left(S^{B,3}\right)$	
$\{1, 2, 3\}$	$\nu(\{1, 2\} \cup \{3\}) - \nu(\{1, 2\})$	$= 8 - 6 = 2$
$\{1, 3, 2\}$	$\nu(\{1\} \cup \{3\}) - \nu(\{1\})$	$= 6 - 0 = 6$
$\{2, 1, 3\}$	$\nu(\{1, 2\} \cup \{3\}) - \nu(\{1, 2\})$	$= 8 - 6 = 2$
$\{2, 3, 1\}$	$\nu(\{2\} \cup \{3\}) - \nu(\{2\})$	$= \alpha - 0 = \alpha$
$\{3, 1, 2\}$	$\nu(\emptyset \cup \{3\}) - \nu(\emptyset)$	$= 0 - 0 = 0$
$\{3, 2, 1\}$	$\nu(\emptyset \cup \{3\}) - \nu(\emptyset)$	$= 0 - 0 = 0$
$\varphi_3(N, \nu)$	$\frac{1}{3!}(2 + 6 + 2 + \alpha + 0 + 0) = \frac{10+\alpha}{6}$	

Panel (iii)

Example 14.8 Consider the TU game (N, ν) in Example 14.5. We use Table 14.6 to make our calculations in Table 14.9. Note that the Shapley value of the TU game (N, ν) is

$$\varphi(N, \nu) = \left(\frac{28 - 2\alpha}{6}, \frac{10 + \alpha}{6}, \frac{10 + \alpha}{6}\right).$$

We have $\sum_{i \in N} \varphi_i(N, \nu) = 8 = \nu(N)$, implying that the calculated Shapley value is efficient as predicted by Proposition 14.2.

Also, recall that the TU game we study in this example coincides with the TU games Examples 14.1–14.3 under various values of α.

Whenever $\alpha = 1$, we are in the domain of Example 14.1. The Shapley value becomes $\varphi(N, \nu) = \left(\frac{26}{6}, \frac{11}{6}, \frac{11}{6}\right)$. One can easily check that the Shapley value $\varphi(N, \nu)$ is contained by the core $C(N, \nu)$.

Whenever $\alpha = 4$, we are in the domain of Example 14.2. The Shapley value becomes $\varphi(N, \nu) = \left(\frac{20}{6}, \frac{14}{6}, \frac{14}{6}\right)$. One can easily check that the Shapley value $\varphi(N, \nu)$ is not contained by the core $C(N, \nu)$, which is the singleton set $\{(4, 2, 2)\}$.

Whenever $\alpha = 7$, we are in the domain of Example 14.3. The Shapley value becomes $\varphi(N, v) = \left(\frac{14}{6}, \frac{17}{6}, \frac{17}{6}\right)$, while the core for this example is empty; i.e., $C(N, v) = \emptyset$. ■

A TU game (N, v) is called a simple game if the value of each nonempty coalition in this game is either zero or one; i.e., $v(S) \in \{0, 1\}$ for each $S \in 2^N \setminus \emptyset$. The Glove Game we studied in Example 14.6 is a simple game. A nice feature of simple games is that a solution profile for these games can be interpreted as the power distribution of individuals in the grand coalition rather than their abstract or monetary utilities. In simple games, a coalition is called a winning coalition if its value is 1, and a losing coalition if its value is zero. If the entrance of a single individual i to a losing coalition S can transform it to a winning coalition $S \cup \{i\}$, i.e., $v(S \cup \{i\}) - v(S) = 1$, then individual i is called pivotal for the coalition S. In simple games, an individual has a higher Shapley value (power), the higher the number of coalitions in which she is pivotal.

Example 14.9 Consider the Glove Game (N, v) in Example 14.6. Using Table 14.7, we make our calculations in Table 14.10.

The Shapley value of the TU game (N, v) is $\varphi(N, v) = \left(\frac{4}{6}, \frac{1}{6}, \frac{1}{6}\right)$. We can check that $\sum_{i \in N} \varphi_i(N, v) = 1 = v(N)$, implying that the calculated Shapley value is efficient as predicted by Proposition 14.2.

Recall from Example 14.6 that the core of the Glove Game is $C(N, v) = \{(1, 0, 0)$, which does not contain the Shapley value calculated above. ■

We found that the Shapley value of the Glove Game is (4/6, 1/6, 1/6). Since individual 1 has one left-hand glove, while individuals 2 and 3 each have one right-hand gloves, individual 1 must be present in every winning coalition, and she must be pivotal for every nonempty coalition she joins. But, none of these is true for individuals 2 and 3 simply because the right-hand gloves owned by these two individuals are perfectly substitutable and one of these gloves becomes worthless when both individuals 2 and 3 are in a winning coalition. More interestingly the power (implied by the Shapley value) of individual 1 and the total power of individuals 2 and 3 are not equal despite the fact that to produce a complete pair of gloves, individuals 2 and 3 as a whole (the sole owners of a right-hand glove) are as indispensable as individual 1 (the sole owner of a left-hand glove); i.e., a winning coalition always involves individual 1 and at least one of the other two individuals. In fact, the power of individual 1 (which is 4/6) is twice as high as the total power of individuals 2 and 3 (which is 2/6). This is because the Shapley value (or the power profile it represents) is calculated for individuals in a society rather than for collections of individuals. The perfect substitutability of individuals 2 and 3 in winning coalitions reduces the powers of individuals 2 and 3 against individual 1.

Example 14.10 (*The Airport Game*) Three airline companies would like to share the cost of a runway for landing on. The companies ordered as 1,2, and 3 in the

Table 14.10 The marginal contributions and Shapley value in Example 14.9

$S \in \mathcal{N}$	$\nu\left(S^{B,1} \cup \{1\}\right) - \nu\left(S^{B,1}\right)$	
$\{1, 2, 3\}$	$\nu(\emptyset \cup \{1\}) - \nu(\emptyset)$	$= 0 - 0 = 0$
$\{1, 3, 2\}$	$\nu(\emptyset \cup \{1\}) - \nu(\emptyset)$	$= 0 - 0 = 0$
$\{2, 1, 3\}$	$\nu(\{2\} \cup \{1\}) - \nu(\{2\})$	$= 1 - 0 = 1$
$\{2, 3, 1\}$	$\nu(\{2, 3\} \cup \{1\}) - \nu(\{2, 3\})$	$= 1 - 0 = 1$
$\{3, 1, 2\}$	$\nu(\{3\} \cup \{1\}) - \nu(\{3\})$	$= 1 - 0 = 1$
$\{3, 2, 1\}$	$\nu(\{2, 3\} \cup \{1\}) - \nu(\{2, 3\})$	$= 1 - 0 = 1$
$\varphi_1(N, \nu)$	$\frac{1}{3!}(0 + 0 + 1 + 1 + 1 + 1) = \frac{4}{6}$	

Panel (i)

$S \in \mathcal{N}$	$\nu\left(S^{B,2} \cup \{2\}\right) - \nu\left(S^{B,2}\right)$	
$\{1, 2, 3\}$	$\nu(\{1\} \cup \{2\}) - \nu(\{1\})$	$= 1 - 0 = 1$
$\{1, 3, 2\}$	$\nu(\{1, 3\} \cup \{2\}) - \nu(\{1, 3\})$	$= 1 - 1 = 0$
$\{2, 1, 3\}$	$\nu(\emptyset \cup \{2\}) - \nu(\emptyset)$	$= 0 - 0 = 0$
$\{2, 3, 1\}$	$\nu(\emptyset \cup \{2\}) - \nu(\emptyset)$	$= 0 - 0 = 0$
$\{3, 1, 2\}$	$\nu(\{1, 3\} \cup \{2\}) - \nu(\{1, 3\})$	$= 1 - 1 = 0$
$\{3, 2, 1\}$	$\nu(\{3\} \cup \{2\}) - \nu(\{3\})$	$= 0 - 0 = 0$
$\varphi_2(N, \nu)$	$\frac{1}{3!}(1 + 0 + 0 + 0 + 0 + 0) = \frac{1}{6}$	

Panel (ii)

$S \in \mathcal{N}$	$\nu\left(S^{B,3} \cup \{3\}\right) - \nu\left(S^{B,3}\right)$	
$\{1, 2, 3\}$	$\nu(\{1, 2\} \cup \{3\}) - \nu(\{1, 2\})$	$= 1 - 1 = 0$
$\{1, 3, 2\}$	$\nu(\{1\} \cup \{3\}) - \nu(\{1\})$	$= 1 - 0 = 1$
$\{2, 1, 3\}$	$\nu(\{1, 2\} \cup \{3\}) - \nu(\{1, 2\})$	$= 1 - 1 = 0$
$\{2, 3, 1\}$	$\nu(\{2\} \cup \{3\}) - \nu(\{2\})$	$= 0 - 0 = 0$
$\{3, 1, 2\}$	$\nu(\emptyset \cup \{3\}) - \nu(\emptyset)$	$= 0 - 0 = 0$
$\{3, 2, 1\}$	$\nu(\emptyset \cup \{3\}) - \nu(\emptyset)$	$= 0 - 0 = 0$
$\varphi_3(N, \nu)$	$\frac{1}{3!}(0 + 1 + 0 + 0 + 0 + 0) = \frac{1}{6}$	

Panel (iii)

size of their planes, starting from the company with the smallest plane size. Let $N = \{1, 2, 3\}$.

The length of the shortest runway that can serve the planes of the company $i \in N$ is l_i kilometers. We assume that $0 < l_1 < l_2 < l_3$. To build a runway of length l_i costs c_i dollars. We assume that $0 < c_1 < c_2 < c_3$.

The airline companies can form coalitions to build a runway of any length l_i, where $i \in N$. For any nonempty coalition $S \subseteq N$, the cost incurred by coalition S to build a runway that can serve the planes of all airline companies in S is

$$C(S) = \max\{c_i : i \in S\}.$$

The value of the empty coalition is zero, i.e., $\nu(\emptyset)$, and the value of any nonempty coalition $S \subseteq N$ is the net savings from the cost to build the runway, which is given by

Table 14.11 The cost and value of each possible coalition in the airport game in Example 14.10

S	$\{1\}$	$\{2\}$	$\{3\}$	$\{1, 2\}$	$\{1, 3\}$	$\{2, 3\}$	$\{1, 2, 3\}$
$C(S)$	c_1	c_2	c_3	c_2	c_3	c_3	c_3
$\nu(S)$	0	0	0	c_1	c_1	c_2	$c_1 + c_2$

$$\nu(S) = \sum_{i \in S} c_i - C(S).$$

The set of airline companies N and the value function ν described above together define a TU game, called the Airport Game. The cost function and the value function for this game generate Table 14.11.

Using Table 14.11 we can make our calculations in Table 14.12.

Table 14.12 The marginal contributions and Shapley value in Example 14.10

$S \in \mathcal{N}$	$\nu\left(S^{B,1} \cup \{1\}\right) - \nu\left(S^{B,1}\right)$	
$\{1, 2, 3\}$	$\nu(\emptyset \cup \{1\}) - \nu(\emptyset)$	$= 0 - 0 = 0$
$\{1, 3, 2\}$	$\nu(\emptyset \cup \{1\}) - \nu(\emptyset)$	$= 0 - 0 = 0$
$\{2, 1, 3\}$	$\nu(\{2\} \cup \{1\}) - \nu(\{2\})$	$= c_1 - 0 = c_1$
$\{2, 3, 1\}$	$\nu(\{2, 3\} \cup \{1\}) - \nu(\{2, 3\})$	$= c_1 + c_2 - c_2 = c_1$
$\{3, 1, 2\}$	$\nu(\{3\} \cup \{1\}) - \nu(\{3\})$	$= c_1 - 0 = c_1$
$\{3, 2, 1\}$	$\nu(\{2, 3\} \cup \{1\}) - \nu(\{2, 3\})$	$= c_1 + c_2 - c_2 = c_1$
$\varphi_1(N, \nu)$	$\frac{1}{3!}(0 + 0 + c_1 + c_1 + c_1 + c_1) = \frac{4c_1}{6}$	

Panel (i)

$S \in \mathcal{N}$	$\nu\left(S^{B,2} \cup \{2\}\right) - \nu\left(S^{B,2}\right)$	
$\{1, 2, 3\}$	$\nu(\{1\} \cup \{2\}) - \nu(\{1\})$	$= c_1 - 0 = c_1$
$\{1, 3, 2\}$	$\nu(\{1, 3\} \cup \{2\}) - \nu(\{1, 3\})$	$= c_1 + c_2 - c_1 = c_2$
$\{2, 1, 3\}$	$\nu(\emptyset \cup \{2\}) - \nu(\emptyset)$	$= 0 - 0 = 0$
$\{2, 3, 1\}$	$\nu(\emptyset \cup \{2\}) - \nu(\emptyset)$	$= 0 - 0 = 0$
$\{3, 1, 2\}$	$\nu(\{1, 3\} \cup \{2\}) - \nu(\{1, 3\})$	$= c_1 + c_2 - c_1 = c_2$
$\{3, 2, 1\}$	$\nu(\{3\} \cup \{2\}) - \nu(\{3\})$	$= c_2 - 0 = c_2$
$\varphi_2(N, \nu)$	$\frac{1}{3!}(c_1 + c_2 + 0 + 0 + c_2 + 0) = \frac{c_1+3c_2}{6}$	

Panel (ii)

$S \in \mathcal{N}$	$\nu\left(S^{B,3} \cup \{3\}\right) - \nu\left(S^{B,3}\right)$	
$\{1, 2, 3\}$	$\nu(\{1, 2\} \cup \{3\}) - \nu(\{1, 2\})$	$= c_1 + c_2 - c_1 = c_2$
$\{1, 3, 2\}$	$\nu(\{1\} \cup \{3\}) - \nu(\{1\})$	$= c_1 - 0 = c_1$
$\{2, 1, 3\}$	$\nu(\{1, 2\} \cup \{3\}) - \nu(\{1, 2\})$	$= c_1 + c_2 - c_1 = c_2$
$\{2, 3, 1\}$	$\nu(\{2\} \cup \{3\}) - \nu(\{2\})$	$= c_2 - 0 = c_2$
$\{3, 1, 2\}$	$\nu(\emptyset \cup \{3\}) - \nu(\emptyset)$	$= 0 - 0 = 0$
$\{3, 2, 1\}$	$\nu(\emptyset \cup \{3\}) - \nu(\emptyset)$	$= 0 - 0 = 0$
$\varphi_3(N, \nu)$	$\frac{1}{3!}(c_2 + c_1 + c_2 + c_2 + 0 + 0) = \frac{c_1+3c_2}{6}$	

Panel (iii)

Thus, the Shapley value of the TU game (N, ν) is

$$\varphi(N, \nu) = \left(\frac{4c_1}{6}, \frac{c_1 + 3c_2}{6}, \frac{c_1 + 3c_2}{6} \right).$$

The intuition underlying this distribution is as follows. The value of the grand coalition is $c_1 + c_2$. Of this, the amount c_1 requires the presence of individual 1 in the coalition. Moreover, any pairwise coalition that involves individual 1 generates the value c_1. Since individuals 2 and 3 are perfectly substitutable with each other in a pairwise coalition with individual 1, their total power in the grand coalition when they distribute the value c_1 is half the power of individual 1, as we have discussed in detail in the Glove Game. Thus, four sixth of c_1 goes to individual 1, while the remaining two individuals each obtain one sixth of c_1. When we subtract the value c_1 from the value of the grand coalition, we are left with the amount c_2. The production of this value requires the presence of both individual 2 and individual 3, but not the presence of individual 1. Thus, the value c_2 should be distributed between individuals 2 and 3 only and in exactly the same way as they are in a two-player TU game, studied in Example 14.7. Thus, the Shapley value distributes the value c_2 uniformly between individuals 2 and 3, providing each of them with $\frac{c_2}{2}$. As a result, the utility ensured by the Shapley value for both individual 2 and individual 3 becomes $\frac{1}{6}c_1 + \frac{3}{6}c_2$.

Note that $\sum_{i \in N} \varphi_i(N, \nu) = c_1 + c_2 = \nu(N)$, implying that the calculated Shapley value is efficient as predicted by Proposition 14.2. We will finally check whether the Shapley value $\varphi(N, \nu)$ is contained by the core given by

$$\begin{aligned} C(N, v) = \{ x \in \mathbb{R}^3 \, : \, & x_1 \geq 0, \; x_2 \geq 0, \; x_3 \geq 0, \\ & x_1 + x_2 \geq c_1, \; x_1 + x_3 \geq c_1, \; x_2 + x_3 \geq c_2, \\ & x_1 + x_2 + x_3 = c_1 + c_2 \}. \end{aligned}$$

We portray $C(N, v)$ in Fig. 14.6.

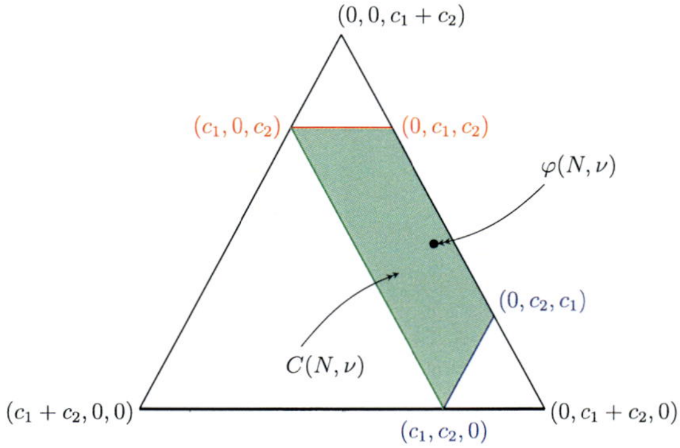

Fig. 14.6 The core, $C(N, \nu)$, and Shapley value, $\varphi(N, \nu)$, of the airport game in Table 14.11

One can easily check that the core of this game is the green-colored region, representing the convex hull of the points $(c_1, c_2, 0)$, $(0, c_2, c_1)$, $(c_1, 0, c_2)$, and $(0, c_1, c_2)$, and the Shapley value is inside this core. ■

References

1. Derya AM (2015) A decomposition of transferable utility games: structure of transferable utility games. Turkish J Math 39(2), Article 5
2. Bondareva ON (1962) Theory of the core in the n-person game (in Russian). Vestnik Leningradskii Universitet 13:141–142
3. Shapley LS (1967) On balanced sets and cores. Naval Res Logist Q 14:453–460

Cake Cutting and Fairness

15

In this chapter, we will study the problem of cutting or dividing a cake (a metaphor for any limited and perfectly divisible resource like land or time) fairly among a group of individuals who may have different preferences over different portions of the cake. In Sect. 15.1, we will first present the cake-cutting problem formally and, in Sect. 15.2, we will introduce three notions of fairness, namely proportionality, envy-freeness, and equitability, along with the relations between some of these three notions. In Sect. 15.3, we will present protocols that produce a proportional cake division among two, three, or arbitrary number of individuals and, in Sect. 15.4, we will present protocols that produce an envy-free cake division among three individuals.

15.1 Cake-Cutting Problem

Consider a cake of unit size $[0, 1]$ to be divided among $n \geq 2$ individuals. Let $N = \{1, \ldots, n\}$ denote the set of these individuals. A division, $(A_1, \ldots, A_n)$, is a partition of this cake; i.e., $A_i \subseteq [0, 1]$ for each $i \in N$ and $\cup_{i \in N} A_i = [0, 1]$. In this division, A_i is the portion of the cake received by individual $i \in N$. Note that A_i does not need to be a connected set, it may contain more than one piece of the cake. We assume that individual i has a utility function v_i that maps each portion of the cake (a subset of $[0, 1]$) to a non-negative real number and is such that

(i) $v_i(A) \geq 0$ for each $A \subseteq [0, 1]$,

(ii) $v_i(A) = 1$ if $A = [0, 1]$, and

(iii) $\sum_{j \in N} v_i(A_j) = 1$ if $\cup_{j \in N} A_j = [0, 1]$.

We will not assume that the utility function v_i is continuous because it would be too restrictive. Consider, for example, a cake half of which is made of vanilla

I. Saglam, *Mastering Game Theory*, Classroom Companion: Economics,
https://doi.org/10.1007/978-3-031-82094-6_15

and the other half is made of chocolate. We would like to allow the possibility that an individual may derive a constant positive utility from each subset of the cake involving vanilla and zero utility from any subset of the cake involving chocolate (or vice versa). Clearly, such a possibility cannot be modeled using any continuous utility function. So, we will need a weaker, but widely used, assumption, called the non-atomicity. That is for each $i \in N$ if $\nu_i(S) > 0$ for any interval subset S of $[0, 1]$, then there exists some interval subset $T \subset S$ such that $\nu_i(S) > \nu_i(T) > 0$. The above assumption implies that every single bit of the cake $[0, 1]$ is non-atomic in terms of the utility it generates for any player. We will also assume that for each $i \in N$, ν_i satisfies finite additivity. That is for any disjoint subsets S_1 and S_2 of $[0, 1]$, $\nu_i(S_1 \cup S_2) = \nu(S_1) + \nu(S_2)$.

The set of individuals N and their utility functions $(\nu_i)_{i \in N}$ together define a cake-cutting problem.

15.2 Definitions of Fairness

Given a cake-cutting problem $\langle N, (\nu_i)_{i \in N} \rangle$, we will first present a notion of efficiency.

Definition 15.1 A division $(A_1, \ldots, A_n)$ of the cake is efficient if there exists no division $(A'_1, \ldots, A'_n)$ such that for all $i \in N$, $\nu_i(A'_i) \geq \nu_i(A_i)$ and $\nu_i(A'_i) > \nu_i(A_i)$ for some $j \in N$.

The above definition is the well-known notion of Pareto efficiency. Notice that the fact that a division is a partition of the cake does not imply that every division is efficient.

Example 15.1 Let $S = [0, 1]$ be a cake such that the left half $[0, 1/2]$ is made of vanilla and the right half $(1/2, 1]$ is made of chocolate. Consider the set of players $N = \{1, 2\}$ which want to share this cake. Let the utility function ν_i of player $i \in N$ be such that for any subset T of $[0, 1]$, $u_i(T) = \int_T f_i(x)dx$ where $f_1(x) = 1$ and

$$f_2(x) = \begin{cases} 1/2 \text{ if } x \in [0, 1/2] \\ 3/2 \text{ if } x \in (1/2, 1]. \end{cases}$$

Player 1 is indifferent over any two portions of the cake as long as their sizes are equal. On the other hand, player 2 values any portion with chocolate three times as much as an equally large portion with vanilla.

First consider the division $A = (A_1, A_2)$ such that $A_1 = [1/4, 1]$ and $A_2 = [0, 1/4)$. We can calculate that $u_1(A_1) = (1)(1 - 1/4) = 3/4$ and $u_2(A_2) = (1/2)(1/4 - 0) = 1/8$. Now, consider the division $A' = (A'_1, A'_2)$ such that $A'_1 = [0, 3/4]$ and $A'_2 = (3/4, 1]$. We can calculate that $u_1(A'_1) = (1)(3/4 - 0) = 3/4$ and $u_2(A'_2) = (3/2)(1 - 3/4) = 3/8$. Note that the division A' Pareto dominates the division A since we have $u_1(A'_1) = u_1(A_1)$, while $u_2(A'_2) > u_2(A_2)$. Therefore, the division A is not efficient.

On the other, the division A' is efficient. There exists no division that Pareto dominates A'. The reason is that A'_2 is contained by the right half of the cake (with chocolate) that player 2 likes most. Moreover, player 1 only cares about the size of her share. Given the division A', one cannot make any of the two players better off without raising her share of the cake, and consequently reducing the share of the other player and making her worse off. ■

Now, we will present, in order, three notions of fairness, namely proportionality, envy-freeness, and equitability.

Definition 15.2 A division $(A_1, \ldots, A_n)$ of the cake is proportional if $\nu_i(A_i) \geq 1/n$ for each $i \in \{1, \ldots, n\}$.

Definition 15.3 A division $(A_1, \ldots, A_n)$ of the cake is envy-free if $\nu_i(A_i) \geq \nu_i(A_j)$ for each $i, j \in \{1, \ldots, n\}$ with $j \neq i$.

Definition 15.4 A division $(A_1, \ldots, A_n)$ of the cake is equitable if $\nu_i(A_i) = \nu_j(A_j)$ for each $i, j \in \{1, \ldots, n\}$ with $j \neq i$.

Proportionality requires that each player values her share from the cake at least $1/n$, the assumed relative weight of each player in the society dividing the cake, envy-freeness requires that each player values her share from the cake at least as much as the share of every other player, and equitability requires that each player values her share from the cake exactly the same as other players value their own shares. These three notions of fairness are independent of the efficiency notion, as we are going to see later through some example.

Below, we will first investigate the relationship between proportionality and envy-freeness.

Proposition 15.1 *For any cake-cutting problem with two or more individuals, a division is proportional if it is envy-free.*

Proof Consider any envy-free division $(A_1, \ldots, A_n)$ of the cake $[0, 1]$ to n individuals. Then, we must have $\nu_i(A_i) \geq \nu_i(A_j)$ for each $i, j \in N$. Now, suppose that for some player k, we have $\nu_k(A_k) < 1/n$. We know that $\sum_{j=1}^{n} \nu_k(A_j) = \nu_k([0, 1]) = 1$ by the definition of value functions. Therefore, there must exist $j \in N \setminus k$ such that $\nu_k(A_j) > 1/n$. Then, $\nu_k(A_j) > \nu_k(A_k)$, contradicting that the division $(A_1, \ldots, A_n)$ is envy-free. Therefore, we must have $\nu_i(A_i) \geq 1/n$ for all $i \in N$. ■

When $n = 2$, the converse of the above proposition is also true.

Proposition 15.2 *For any cake-cutting problem with two individuals, a division is envy-free if it is proportional.*

Table 15.1 A three-person cake division that is proportional but not envy-free

	A_1	A_2	A_3
Player 1	0.40	0.25	0.35
Player 2	0.45	0.40	0.15
Player 3	0.25	0.30	0.45

Proof Consider a proportional division (A_1, A_2) of the cake $[0, 1]$ to individuals 1 and 2. Then, we must have $v_i(A_i) \geq 1/2$ for each $i \in \{1, 2\}$, implying $v_i(A_j) \leq 1/2$ for each $i, j \in \{1, 2\}$ with $i \neq j$. Therefore, $v_i(A_i) \geq v_i(A_j)$ for each $i, j \in \{1, 2\}$ with $i \neq j$, proving that the division (A_1, A_2) is envy-free. ■

When the number of players is more than two, a proportional division does not have to be envy-free, as illustrated by the following example.

Example 15.2 Consider a set of three players, i.e., $N = \{1, 2, 3\}$. Let $A = (A_1, A_2, A_3)$ be a division at which the players have the utilities in Table 15.1.

The red-colored utilities show that for each $i \in N$, $v_i(A_i) \geq 1/|N| = 1/3$ holds. Thus, the division A is proportional. We say that the division A is envy-free for individual $i \in N$ if $v_i(A_i) \geq v_i(A_j)$ for every $j \in N - i$. We observe from the red-colored and blue-colored payoffs (if any) in each row that A is envy-free for players 1 and 3 but not so for player 2 who envies the portion of player 1. Therefore, A is not envy-free. ■

Below, we will generalize the result in the above example for any $n \geq 3$.

Proposition 15.3 *For any cake-cutting problem with $n \geq 3$ individuals, there exists a proportional division that is not envy-free.*

Proof Consider a division $(A_1, \ldots, A_n)$ of the cake $[0, 1]$ to $n \geq 3$ individuals such that for each $i \in N$ we have

$$v_i(A_i) = 1/n, \quad v_i(A_{n-i+1}) = 2/n, \quad \text{and} \quad v_i(A_k) = (n-3)/(n(n-2))$$

for all $k \in N \setminus \{i, n-i+1\}$. This division is proportional since $v_i(A_i) \geq 1/n$ holds for each $i \in N$. But, this division is not envy-free. For example, player 1 envies player n. The reason is that $v_1(A_1) = 1/n$, while $v_1(A_n) = 2/n$. (As a matter of fact, the division is constructed such that for any $i \in N$, player i envies player $n - i + 1$.) ■

The other notion of fairness, equitability, is independent of both envy-freeness and proportionality. The following example shows this fact for a society with two individuals.

Table 15.2 Axioms satisfied by several examples of two-person cake divisions

A_1	A_2	$(u_1(A_1), u_1(A_2))$	$(u_2(A_1), u_2(A_2))$	E	EQ	EF	P
$[0.00, 0.60]$	$(0.60, 1.00]$	$(0.60, 0.40)$	$(0.40, 0.60)$	✓	✓	✓	✓
$[0.00, 0.50]$	$(0.50, 1.00]$	$(0.50, 0.50)$	$(0.25, 0.75)$	✓	×	✓	✓
$[0.00, 0.40]$	$(0.40, 1.00]$	$(0.40, 0.60)$	$(0.20, 0.80)$	✓	×	×	×
$[0.60, 1.00]$	$[0.00, 0.60)$	$(0.40, 0.60)$	$(0.60, 0.40)$	×	✓	×	×
$[0.70, 1.00]$	$[0.00, 0.70)$	$(0.30, 0.70)$	$(0.45, 0.55)$	×	×	×	×
$[0.00, 0.40] \cup [0.50, 0.60]$	$(0.40, 0.50) \cup (0.60, 1]$	$(0.50, 0.50)$	$(0.35, 0.65)$	×	×	✓	✓
$[0.00, 0.40] \cup [0.50, 0.66]$	$(0.40, 0.50) \cup (0.66, 1]$	$(0.56, 0.44)$	$(0.44, 0.56)$	×	✓	✓	✓

Example 15.3 Reconsider the cake-cutting problem $\langle N, (v_i)_{i \in N} \rangle$ in Example 15.1. Table 15.2 calculates for various divisions of the cake the utilities attached by the two individuals in N and report whether these divisions are efficient (E), equitable (EQ), envy-free (EF), or proportional (P).

Recall from Propositions 15.1 and 15.2 that envy-freeness reduces to proportionality when $n = 2$. That is why the results in the last two columns of Table 15.2 always agree. Also, notice from the results presented in the rows 3, 4, and 6 of Table 15.2 that each of the axioms, involving efficiency (E), equitability (EQ), and envy-freeness (EF), are individually independent of the remaining two axioms. However, efficiency and equitability together imply envy-freeness in our example. That is why we need only seven rows in Table 15.2 to represent all possibilities in which each of the axioms E, EQ, and EF hold or fail. ■

Below, we show that the relationship between envy-freeness and the pair of axioms involving efficiency and equitability, which we observed in the above example, is actually true for all cake-cutting problems involving two individuals.

Proposition 15.4 *For any cake-cutting problem with two individuals, a division is envy-free if it is efficient and equitable.*

Proof Let $N = \{1, 2\}$. Consider a division (A_1, A_2) of the cake $[0, 1]$ such that it is efficient and equitable. Equitability implies that $v_1(A_1) = v_2(A_2)$. Suppose $v_1(A_1) < 0.5$. Then, $v_1(A_2) > 0.5$ and $v_2(A_1) > 0.5$, implying that the division (A_2, A_1) Pareto dominates the division (A_1, A_2), i.e., both individuals become better off iff they exchange their portions, in contradiction with the assumption that (A_1, A_2) is efficient. Thus, $v_1(A_1) \geq 0.5$, also implying $v_2(A_2) \geq 0.5$, by equitability. Then, individual 1 does not envy A_2 and individual 2 does not envy A_1, implying that the division (A_1, A_2) is envy-free. ■

Barbanel and Brams [1] give an algorithm that finds an efficient and equitable, hence envy-free, division in any cake-cutting problem with two players. If the number of players is more than two, the result in Proposition 15.4 fails to hold. Moreover, one can find cake-cutting problems (with more than two players) such that no division (under any finite number of cuts) is efficient, equitable, and envy-free, as demonstrated by Brams, Jones, and Klamler [2].

So far, we have defined several desirable axioms for cake-cutting, discussed the relationships between these axioms and the possibility of finding divisions that satisfy all or some of these axioms. In the next two subsections, we will present several protocols (algorithms or solutions) that produce proportional or envy-free divisions, respectively.

15.3 Proportional Cake Division Protocols

When there exist only 2 players, the following solution method, known as "the Cut and Choose Protocol", generates a proportional cake division (which is also envy-free).[1]

The Cut and Choose Protocol with 2 Players.

Player 1 cuts the cake into two equal slices, say A' and A'', according to her utility function ν_1. Then, player 2 chooses the best slice (or one of the two slices if the best slice is not unique) according to his utility function and player 1 gets the remaining slice.

Clearly, the Cut and Choose Protocol defined for 2 players yields a proportional division. Player 2 chooses a slice $A_2 \in \{A', A''\}$ such that $\nu_2(A_2) \geq 1/2$ and player 1 ends up with the slice $A_1 = \{A', A''\} \setminus \{A_2\}$ for which $\nu_1(A_1) = 1/2$, satisfying $\nu_1(A_1) \geq 1/2$. By Proposition 15.2, the proportional division generated by the Cut and Choose Protocol is also envy-free.

While the Cut and Choose Protocol is very elegant and practical, it does not necessarily produce an efficient division or an equitable division, as we illustrate below.

Example 15.4 Let $S = [0, 1]$ be a cake such that the left half $[0, 1/2]$ is made of vanilla and the right half $(1/2, 1]$ is made of chocolate. Consider the set of players $N = \{1, 2\}$ which want to share this cake. Let the utility function ν_i of player $i \in N$ be such that for any subset T of $[0, 1]$, $u_i(T) = \int_T f_i(x)dx$ where

$$f_1(x) = \begin{cases} 2 \text{ if } x \in [0, 1/2] \\ 0 \text{ if } x \in (1/2, 1] \end{cases}$$

and

$$f_2(x) = \begin{cases} 0 \text{ if } x \in [0, 1/2) \\ 2 \text{ if } x \in [1/2, 1]. \end{cases}$$

[1] This protocol is attributed to Prophet Abraham in Hebrew Bible (Genesis 13:5–13), who devised it to distribute the land where he and his family were living between his herdsmen and his nephew Lot's.

Now, suppose player 1 cuts the cake into two pieces involving [0, 0.6] and (0.6, 1]. Then, player 2 will choose the piece $(0.6, 1] \equiv A_2$ and player 1 will get the remaining piece $[0, 0.6] \equiv A_1$. Their utilities will be $(u_1(A_1), u_1(A_2)) = (1, 0)$ and $(u_2(A_1), u_2(A_2)) = (0.2, 0.8)$. First, note that $u_1(A_1) > u_2(A_2)$, implying that the division (A_1, A_2) not equitable.

Now, consider the division $(A'_1, A'_2) = ([0, 0.5], (0.5, 1])$. Note that $u_1(A'_1) = 1$ and $u_2(A'_2) = 1$, implying that (A'_1, A'_2) Pareto dominates (A_1, A_2). Thus, (A_1, A_2) is not efficient. ■

The chooser (player 2) in the Cut and Choose Protocol always gets a piece that has a utility no less than 0.5, while the cutter (player 1) always gets a piece that has a utility of 0.5. Therefore, the Cut and Choose Protocol is not equitable.

A distinction about a cake-cutting protocol is the (maximal) number of cuts for partitioning the cake. For a problem with n players, the number of cuts is said to be minimal if it is $n - 1$. Apparently, the number of cuts is minimal (one) in the Cut and Choose Protocol with two players. While this seems to be computationally attractive, it also leads to the problem that the resulting division may not be efficient.

A direct extension of the Cut and Choose Protocol to three or more players is not fair; it does not even satisfy the proportionality. To see this, suppose that player 1 cuts the cake into three equal slices, say A', A'', A''', according to her valuation v_1. Then players in the order of 2, 3, 1 consecutively choose their most favorite slices according to their utility functions. In the above protocol, player 1 ensures that $v_1(A) = 1/3$ for each $A \in \{A', A'', A'''\}$. Therefore, even though she gets the slice unchosen by the other two players, the slice she gets, say A_1, satisfies $v_1(A_1) \geq 1/3$. Since player 2 chooses his best slice, say A_2, among three slices, it can ensure that $v_2(A_2) \geq 1/3$. But, if this slice is worth more than 2/3 for player 3, i.e., $v_3(A_2) > 2/3$, then player 3 as the second chooser may end up with a slice, say A_3, giving a utility less than 1/3, i.e., $v_3(A_3) < 1/3$. Thus, the division produced by the above protocol is not fair; it does not necessarily produce a proportional division.

However, the following extension of the Cut and Choose Protocol, devised by Hugo Steinhaus in 1943 and published in 1948, always produces a proportional cake division among three players.

The Lone Divider Protocol With Three Players.

Consider a cake to be divided among three persons Amy, Brooke, and Chloe according to the following steps:

Step 1 (Cutting): Amy is the divider. She cuts the cake into three slices of equal value based on her utility function.

Step 2 (Stating): Brooke and Chloe state all slices that are acceptable (having a value no less than 1/3 based on their utility functions).

Step 3 (Choosing): This last step consists of three stages:

(i) If Brooke stated that two or more slices are acceptable for her, then the three Players choose the slices in the order of Chloe, Brooke, and Amy.

(ii) If Brooke stated that only one slice is acceptable for her, while Chloe stated that at least two slices are acceptable for her, then the three players choose their slices in the order of Brooke, Chloe, and Amy.

(iii) If both Brooke and Chloe stated that only one slice is acceptable for them, then there must be at least one slice that is unacceptable for both of them. Give such a slice to Amy. Then, mix the remaining two slices to form a new cake and let Brooke and Chloe divide it according to the Cut and Choose Protocol defined for 2 players.

A division generated by the above protocol satisfies proportionality. To show this, we will consider each player separately.

Consider first Amy. Since she cuts the cake, each slice (hence the slice she will get) must be worth $1/3$ according to her utility function.

Now, consider Brooke. If she is in step 3-i, she will get one of the two slices she finds acceptable; so her slice will be worth at least 1/3 according to her utility function. If she is in step 3-ii, she will get the best slice for her and that will be worth at least 1/3 according to her utility function. Finally, if Brooke is in step 3-iii, then the slice Amy gets is unacceptable (worth less than 1/3) for Brooke. The remaining part of the cake must be worth more than 2/3 according to Brooke's utility function. Since Brooke and Chloe will share this cake part using a two-player Cut and Choose Protocol, the share of Brooke from the total cake will be more than 1/3. So, in all possible stages of step 3, the share of Brooke will be worth at least $1/3$ according to her utility function.

Finally, consider Chloe. If she is in step 3-i, she will get the best slice for her and that will be worth at least 1/3 according to her utility function. If she is in step 3-ii, she will get one of the two slices she finds acceptable; so her slice will be worth at least 1/3 according to her utility function. Finally, if Chloe is in step 3-iii, then the slice Amy gets is unacceptable (worth less than 1/3) for Chloe. The remaining part of the cake must be worth more than 2/3 according to the utility function of Chloe. Since Brooke and Chloe will share this using the two-player Cut and Choose Protocol, the share of Chloe from the total cake will be more than 1/3. So, in all possible stages of step 3, the share of Chloe will be worth at least $1/3$ according to her utility function.

Thus, we have established that any division generated by the above protocol is proportional. Also, this protocol is computationally desirable: the number of cuts is two, which is minimal. But, this protocol does not satisfy envy-freeness: it can produce a division that is not envy-free. To show this, note that in step 3-i, Chloe and Amy do not envy any other player. However, Brooke may envy Chloe if the two acceptable slices are not of the same worth for Brooke and Chloe chooses the slice that is the more preferred one according to the utility function of Brooke. Similarly, in step 3-ii, Brooke and Amy do not envy any other player. However, Chloe may envy Brooke if the two acceptable slices are not of the same worth for Chloe and Brooke chooses the slice that is the more preferred one according to the utility function of Chloe. In step 3-iii, Brooke and Chloe do not envy one another. But, Amy may envy either Brooke or Chloe if the Cut and Choose Protocol between them generates a slice that is worth more than 1/3 according to Amy's utility function. One can also check that the division produced by the Lone Divider Protocol does not have to be equitable or efficient.

Harold Kuhn (1967) proposed a protocol that extends the Lone Divider Protocol to cake division problems with more than three players. We will not present this extension as it involves advanced mathematics. Below, we will present a simpler solution, called the Last Diminisher Protocol. This protocol, attributed by Hugo Steinhaus [3] to his students Borislav Knaster and Stephan Banach, generates a proportional division among three or more players. Below, we present a slightly formalized version of their protocol.

The Last Diminisher Protocol with Three or More Players

The set of individuals is $N = \{1, 2, \ldots, n\}$. The protocol consists of $n - 1$ steps. Let N^k denote the set of individuals who play in step $k \in \{1, 2, \ldots, n-1\}$, with N^1 being set to N. In any step k, an individual $i \in N^k$ plays before an individual $j \in N^k$ if $i < j$. Also, let S^k denote the total portion of the cake that is available in step k, with S^1 being set to $[0, 1]$.

In step 1, player 1 (the individual with the smallest integer value in N^1) cuts from the cake S^1 a slice C_1^1 that she considers to be fair (worth $1/n$ according to her valuation). Player 2 has the right to diminish (trim) this slice. Next, player 3 has the right to trim the already trimmed (or not trimmed) slice, and so on and so forth up to player n (the individual with the highest integer value in N^1). The rule requires the last player who cut or trimmed a slice has to get the slice obtained after cutting/trimming. This player is removed from the set N^1, resulting in the set N_2. Also, all trimmings in step 1 are blended together with the portion $S^1 \setminus C_1^1$, resulting in S^2, the total portion of the cake available for step 2.

In step 2, the remaining $n - 1$ individuals (the members of N^2) continue, in the same way as described in step 1, with the available portion of the cake S^2.

The protocol lasts in this way for $n - 1$ steps; a distinct player becomes the last cutter or diminisher in each step and gets a slice of the cake. Finally, after the last cutter/diminisher in step $n - 1$ leaves the protocol with her slice, the remaining player in N^{n-1} gets the remainder of the portion S^{n-1}.

Given the above protocol, suppose that each player following the first mover (the cutter) in any step trims the slice she inspects if and only if she finds it unfair in the sense that it is worth more than $1/n$ according to her valuation. Under this assumption, let us show that the protocol always yields a proportional division.

Let C_k^1 denote the slice of the cake after player $k \in N_1 - n$ completes her move in step 1. In step 1, player 1 ensures that the slice C_1^1 she cuts from the cake is worth $1/n$ according to her valuation; i.e., $v_1(C_1^1) = 1/n$. If player 2 does not trim the slice C_1^1, then $C_2^1 = C_1^1$ and $v_2(C_2^1) \leq 1/n$. If player 2 trims the slice C_1^1, then $C_2^1 \subsetneq C_1^1$ and $v_2(C_2^1) = 1/n$. Similarly, for all $k \in \{3, \ldots, n-1\}$, we have $v_k(C_k^1) \leq 1/n$ if player k does not trim the slice C_{k-1}^1 in step k and $v_k(C_k^1) = 1/n$ otherwise. The rule requires that in each step the last player who cut or trimmed a slice has to get that slice. Denoting this player by $k^*(s)$ in step $s = \{1, 2, \ldots, n-1\}$ and the slice she gets by $C_{k^*(s)}^s$, we can write that $v_{k^*(s)}(C_{k^*(s)}^s) = 1/n$. Now consider the player, say l, who was neither a cutter nor a trimmer in any step of the protocol. Then, it must be true that $\nu_l(C^s_{k^*(s)}) \leq 1/n$ for all $s \in \{1, 2, \ldots, n-1\}$, implying $\nu_l(\cup_{s=1}^{n-1} C^s_{k^*(s)}) \leq (n-1)/n$. Let C_l^{n-1} denote the slice player l gets at the end of the protocol. Note that $C_l^{n-1} = [0, 1] \setminus \cup_{s=1}^{n-1} C^s_{k^*(s)}$. Since $\nu_l([0, 1]) = 1$, it must be true that $\nu_l(C_l^{n-1}) \geq 1/n$.

Thus, we have established that each individual in N gets a slice that is worth at least $1/n$ according to her valuation, implying that the Last Diminisher Protocol satisfies proportionality. However, this protocol does not satisfy envy-freeness. The reason is simply that an individual who gets her slice in any step $k < n-1$ may envy the slices of some individuals who get their slices in later steps. It should also be obvious that the division produced by the Lone Divider Protocol does not have to be equitable or efficient.

Below, we will consider another protocol, or more precisely a moving knife solution, for cutting a cake fairly among three or more individuals. This solution was proposed by [4].

Dubins-Spanier (Moving Knife) Solution With Three or More Players

The cake to be divided in step 1 is $S^1 = [0, 1]$.

Step $k \in \{1, 2, \ldots, n-1\}$: The cake to be divided is S^k. A referee moves a knife continuously over the cake from left to right. Let $\underline{s}(S^k) = \min\{s : s \in S^k\}$. When the knife reaches a point x^k at which the piece of cake to the left, $[\underline{s}(S^k), x^k]$, is worth $1/n$ to one of the players, this player yells "stop" and receives this piece. (Ties are broken arbitrarily.) The remainder of the cake constitutes the cake to be divided (or allocated) at the next step; i.e., $S^{k+1} = S^k \setminus [\underline{s}(S^k), x^k]$.

Step n: The player that has not received any cake in the previous steps receives the final piece S^n.

To show that the division produced by the Dubins-Spanier Solution is proportional, notice that if a player receives a piece of cake in any step $k < n$, then it must have yelled "stop" in that step, implying that the piece of cake she receives is worth $1/n$ according to her valuation. If a player, say $i(n)$, receives a piece of cake in step $k = n$, then there are two cases to consider. The first case is that she yelled "stop" in every previous step together with someone but was never able to get the piece of cake she yelled for (as she always lost while ties were broken arbitrarily). In this case, the total value of the cakes allocated in the first $n-1$ steps is $(n-1)/n$ for player $i(n)$, implying that the piece of cake she received in step n is worth $1/n$ according to her valuation. The second case is that player $i(n)$ did not yell "stop" in one of the steps between step 1 and step $n-1$. Then the total value of the cakes allocated in the first $n-1$ steps is less than $(n-1)/n$ for player $i(n)$, implying that the piece of cake she received in step n is worth at least $1/n$ according to her valuation.

Thus, each player receives a cake that is worth at least $1/n$ according to her valuation, proving that the division produced by the Dubins-Spanier Solution is proportional. On the other hand, our reasoning above shows that the division produced by the Dubins-Spanier Solution is not equitable (the utility of player receiving the last piece of cake is at least as high as $1/n$, whereas the utility of every other player is $1/n$). Also, this division is not necessarily envy-free since any player who receives her piece of cake in any step $k = 1, 2, \dots, n-2$ may envy the piece received by some player in some step $l \in \{k+1, \dots, n\}$. Moreover, it should be obvious that the division produced by the Dubins-Spanier Solution does not have to be efficient. However, unlike the Last Diminisher Protocol, the Dubins-Spanier Solution is computationally desirable: the number of cuts is minimal; i.e., it requires $n - 1$ cuts in a problem with n players.

15.4 Envy-Free Cake Division Protocols

Below, we present a protocol—defined for the case of three players— that always produces an envy-free division. This protocol was found in 1960 by John Selfridge and John Horton Conway independently (neither of them published it).

Selfridge-Conway Protocol with Three Players

Step 1 (Cutting the Cake): Player 1 divides the cake into 3 equal pieces according to her utility function.

Step 2 (Trimming the Best Piece): Player 2 trims the best piece of cake such that trimming creates a tie between the two best pieces according to his utility function. After trimming, the cake involves four pieces: two untrimmed pieces C_1^u and C_2^u, one trimmed piece C^T, and the trimming T. So, $C_1^u \cup C_2^u \cup C^T \cup T = [0, 1]$. In step 3, the trimmed and untrimmed pieces of cakes will be allocated and in step 4 the trimming will be allocated.

Step 3 (Choosing Between Trimmed and Untrimmed Pieces): Player 3 chooses one of the three pieces in $\{C_1^u, C_2^u, C^T\}$. If player 3 does not choose the trimmed piece C^T, then player 2 gets it. Otherwise, player 2 chooses either C_1^u or C_2^u and player 1 receives the remaining piece. In this step, the trimmed piece C^T is chosen by either player 2 or player 3. We denote this player by P^T and the other player P^U.

Step 4 (Cutting the Trimming): Player P^U cuts the Trimming into 3 equal pieces according to his utility function.

Step 5 (Choosing Between Pieces of the Trimming): Player P^T chooses one of the three pieces of the Trimming. Player 1 chooses one of the remaining two pieces and the unchosen piece of the Trimming is received by player P^U.

Let us show that the Selfridge-Conway Protocol always produces an envy-free cake division among 3 players. Note that the cake is shared in steps 3 and 5.

First, consider step 3 and check whether any player would envy the piece of cake received by another player in this step. Player 3 is the first to choose; therefore, player 3 does not envy the pieces received by other players in step 3. Player 2 receives either C^T or the best of C_1^u and C_2^u according to his utility function. Since Player 2 is the one who trims in step 2 one of the cake pieces to make the trimmed piece C^T equivalent to the best of C_1^u and C_2^u according to his utility function, Player 2 does not envy another player in step 3. Finally, player 1 does not envy another player in step 3 since she cuts the cake into 3 equal pieces in step 1 and she receives one of the untrimmed pieces in step 3.

Now, let us consider step 5 where the trimming is allocated. Since player P^T is the first to choose, he does not envy another player in step 5. Player P^U is the last to choose; but, he is the player that cuts in step 4 the trimming into 3 equal pieces according to his utility function. Therefore, P^U does not envy another player in step 5. Since $\{P^U, P^T\} = \{2, 3\}$, we have so far established that neither player 2 nor player 3 envies another player in either step 3 or step 5. Finally, we will consider player 1. She may envy the trimming chosen by player P^T. Therefore, the division protocol in step 5 is not envy-free for player 1. However, if we consider the sum of pieces received by players 1 and P^T in steps 3 and 5, we should observe that there is no reason for why player 1 should envy player P^T. Player 1 divides, in step 1, the cake into 3 equal pieces according to her utility function and receives, in step 3, an untrimmed piece, while player P^T receives the trimmed piece. Therefore, even if player 1 were to receive no part of the trimming in step 5, she would still end up, according to her utility function, with a piece of cake better than the total portion received by player p^T in steps 3 and 5.

Therefore, the division produced by the Selfridge-Conway Protocol among three players is always envy-free. The number of cuts in the Selfridge-Conway Protocol is either two (the case of no trimming) or five (the case of trimming). By Proposition 15.1, the division produced by the Selfridge-Conway Protocol is also proportional. Sadly, the Selfridge-Conway Protocol does not generalize to more than three players.

There is another elegant method that produces an envy-free division of cake for three players. This method was discovered by Walter Stromquist in [5]. We present this method using our notations as follows.

Stromquist (Moving Knife) Solution with Three Players

A referee moves a sword continuously and slowly from left to right over the cake $[0, 1]$. The position of the sword is denoted by x_0, which starts to increase from 0 to higher values in $[0, 1]$. While the sword reaches a position x_0, it hypothetically divides the cake to a left piece $[0, x_0)$ and a right piece $[x_0, 1]$. While x_0 changes, each player $i \in \{1, 2, 3\}$ adjusts a knife in her hand over the right piece $[x_0, 1]$ of the cake at a point x_i such that $v_i([x_0, x_i)) = v_i([x_i, 1])$. The points x_1, x_2, x_3 chosen by the three players are ordered from the smallest one to the largest one and renamed as x_L, x_M, x_R. Thus, $x_L \leq x_M \leq x_R$. Accordingly, the players choosing the points x_L, x_M, x_R are renamed as L, M, R, respectively. When any player shouts "cut", the cake is cut by the sword and by the knife at the position x_M. The player who shouted "cut" receives the left piece $[0, x_0)$. The player L whose knife was at the position x_L, if he did not shout "cut", takes the centerpiece $[x_0, x_M)$; and the player R whose knife was at the position x_R, if he did not shout "cut", takes the right piece $[x_M, 1]$. The player M whose knife was at the position x_M, if he did not shout "cut", takes the piece that is left. If more than one player shouts simultaneously or if at least two of the points x_L, x_M, x_R coincide, then ties are broken arbitrarily.

Under the assumptions, that all three players know the rules of the above solution method and that they can always observe the positions of the sword and all knives while they move over the cake, we can show that this method produces an envy-free cake division. The player who shouted "cut" and received the piece $[0, x_0)$ must not envy the pieces of the other two players, because she knew all pieces and the rules of the solution method when she shouted and she chose to shout. The player L whose knife was at the position x_L and received, if she did not shout "cut", the piece $[x_0, x_M)$ must not envy the piece $[0, x_0)$ for otherwise she would choose to shout. Moreover, she must not envy the piece $[x_M, 1]$ since

$$\nu_L([x_0, x_M)) \geq \nu_L([x_0, x_L)) = \nu_L([x_L, 1]) \geq [x_M, 1]$$

due to the facts that $\nu_L(.)$ is continuous and $x_M \geq x_L$. The player R whose knife was at the position x_R and received, if she did not shout "cut", the piece $[x_M, 1]$ must not envy the piece $[0, x_0)$ for otherwise she would choose to shout. Moreover, she must not envy the piece $[x_0, x_M)$ since

$$\nu_R([x_M, 1]) \geq \nu_R([x_R, 1]) = \nu_R([x_0, x_R)) \geq \nu_R([x_0, x_M))$$

due to the facts that $\nu_R(.)$ is continuous and $x_R \geq x_M$. Finally, the player M whose knife was at the position x_M and received, if she did not shout "cut", either the piece $[x_0, x_M)$ or the piece $[x_M, 1]$ must not envy the piece $[0, x_0)$ for otherwise she would choose to shout. Moreover, the player M is indifferent between the pieces $[x_0, x_M)$ and $[x_M, x_1]$ since x_M is such that $\nu_M([x_0, x_M)) = \nu_M([x_M, 1])$. Thus, player M does not envy the piece received by any other player.

Thus, the division produced by the Stromquist Solution among three players is always envy-free. By Proposition 15.1, this division is also proportional. However, one can easily show that neither the Selfridge-Conway Protocol nor the Stromquist Solution need to produce a division that is equitable or efficient. Unlike the Selfridge-Conway Protocol, the Stromquist Solution requires a minimal number of cuts, which is two. However, like the Selfridge-Conway Protocol, this solution has a limited use since it does not generalize to more than three players.

The year 1995 marked a breakthrough in the cake-cutting literature when an algorithm that produces envy-free division among any number of players exceeding three was proposed by Steven J. Brams and Alan D. Taylor. However, their algorithm suffered from a shortcoming: the number of cuts to compute the outcome of the algorithm is unbounded, though it is finite. That is, for any $n \geq 4$ and any number of cuts T, one can always find a profile of utility functions for n players such that the algorithm of Brams and Taylor [6] requires more than T cuts to terminate. This problem with unbounded computations was ingeniously solved in 2016 by Haris Aziz and Simon Mackenzie, who proposed an algorithm which can calculate an envy-free division at most in k steps where

$$k = n^{n^{n^{n^{n^{n}}}}}.$$

(We should however note that this is a huge number. For example, for $n = 4$, k is equal to 3.2318×10^{616}.)

References

1. Barbanel JB, Brams SJ (2014) Two-person cake cutting: the optimal number of cuts. Math Intell 36(3):23–35
2. Brams SJ, Jones MA, Klamler C (2013) N-player cake-cutting: there may be no perfect division. Am Math Mon 120(1):35–47
3. Steinhaus H (1949) Sur la division pragmatique. Econometrica (supplement) 17:315–319
4. Dubins LE, Spanier EH (1961) How to cut a cake fairly. Am Math Mon 68(1):1–17
5. Stromquist W (1980) How to cut a cake fairly. Am Math Mon 88:640–644
6. Brams J, Taylor AD (1995) An envy-free cake division protocol. Am Math Mon 102(1):9–18

Cooperative Bargaining 16

In this chapter, we will study two-person cooperative bargaining problem, in which two individuals (players) are faced with an uncountable number of alternatives and choose one of them using a pre-determined rule. A typical example for cooperative bargaining is the negotiation process between workers' unions and organizations of employers about workplace issues and/or wages. The cooperative problem was theoretically formalized and studied first by John F. Nash [1]. The solution he proposed is a rule that maximizes the product of two individuals' gains from bargaining. He showed that this bargaining rule, which we call the Nash rule, is the only bargaining rule that satisfies a certain set of axioms. In Sect. 16.1, we will present the cooperative bargaining problem and, in Sects. 16.2 and 16.3, we will present some well-known bargaining rules and some well-known axioms, respectively. Finally, in Sect. 16.4, we will provide characterization results for three well-known bargaining rules, including the Nash rule, the Kalai and Smorodinsky rule, and the Egalitarian rule.

16.1 Bargaining Problem

Consider a two-person bargaining problem denoted by a nonempty subset S of $\mathbb{R}^2_+$. This set denotes the set of von Neumann-Morgenstern utility allocations which are attainable through the cooperation of two individuals.[1] If the individuals fail to agree on any allocation in S, they can only attain the disagreement allocation of $d = (0, 0)$ at which each should enjoy zero utility. We assume that the set S is compact and

[1] Von Neumann and Morgenstern [2] showed that under certain axioms the ordinal preferences of individuals over a set of alternatives can be represented by a cardinal expected utility function, called von Neumann-Morgenstern utility function. In this chapter, we will not deal with ordinal preferences.

I. Saglam, *Mastering Game Theory*, Classroom Companion: Economics,
https://doi.org/10.1007/978-3-031-82094-6_16

convex. Also, it is non-singleton; i.e., there exists a point s in S such that $s > d$.[2] Moreover, for all $s, s' \in \mathbb{R}^2_+$ if $s \in S$ and $d \leq s' \leq s$, then $s' \in S$. This is called comprehensiveness of S or the possibility of free disposal of utilities.

We denote by Σ_0^2 the set of all two-person bargaining problems, with the disagreement point $d = (0, 0)$, satisfying the above assumptions. Finally, we define a bargaining rule F on Σ_0^2 as a mapping from Σ_0^2 to $\mathbb{R}^2_+$ such that $F(S) \in S$ for each $S \in \Sigma_0^2$.

16.2 Some Well-Known Bargaining Rules

Here, we present some of the well-known bargaining rules in the literature. (We illustrate them in Fig. 16.1.)

The Nash Rule: This rule, denoted by N, was proposed by Nash [1]. It maps each problem $S \in \Sigma_0^2$ to the allocation

$$N(S) = \text{argmax}_{s \in S}\ s_1 s_2$$

at which the product of players' utility gains from agreement is maximized. Note that the boundary of S has the slope of -1 at $N(S)$.

The Generalized Nash Rule: This rule, denoted by N^α, was proposed by Harsanyi and Selten [3]. It extends the Nash rule to cases where players 1 and 2 may have asymmetric bargaining powers denoted by α and $1 - \alpha$, respectively. Formally,

$$N^\alpha(S) = \text{argmax}_{s \in S}\ (s_1)^\alpha (s_2)^{1-\alpha}.$$

For $\alpha = 0.5$, this rule reduces to the Nash rule.

The Utilitarian Rule: This rule, which we denote by U, is anonymous. For any problem $S \in \Sigma_0^2$, it selects the allocation

$$U(S) = \text{argmax}_{s \in S}\ s_1 + s_2$$

at which the sum of players' utility gains from agreement is maximized. The solution $U(S)$ is not necessarily unique unless the problem S is strictly convex.

The Egalitarian Rule: This rule was recommended by Rawls [4] and first axiomatized by Kalai [5]. Formally, this rule, denoted by E, maps each problem $S \in \Sigma_0^2$ to the allocation

$$E(S) = \max\{s \in S :\ s_1 = s_2\}$$

[2] Given any two allocations x and y in $\mathbb{R}^2_+$, we assume that $x > y$ means $x_i > y_i$ for each $i = 1, 2$ and $x \geq y$ means $x_i \geq y_i$ for each $i = 1, 2$.

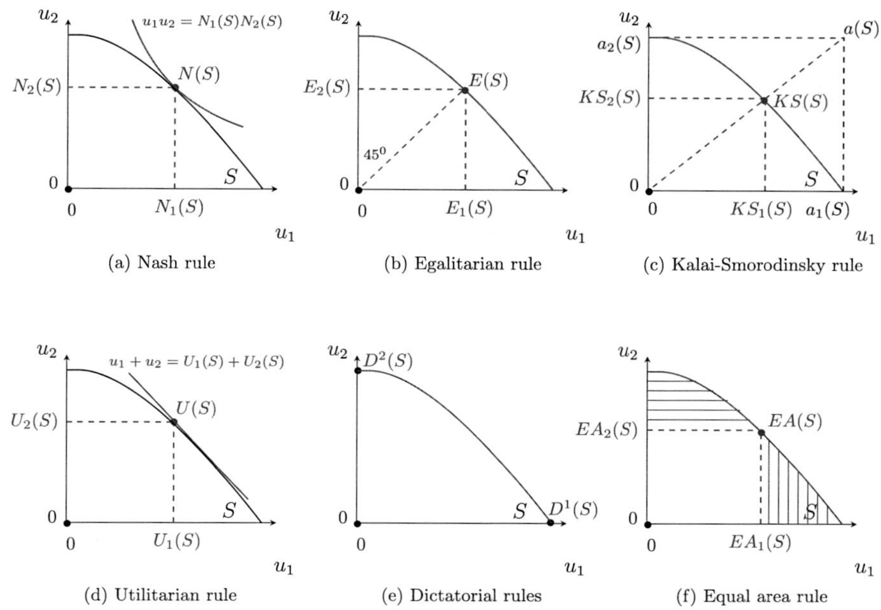

Fig. 16.1 Some well-known bargaining rules

which assigns to the players the maximum utility profile at which their utility gains are equal. Thus, this rule maximizes the utility of the worse-off player in S.

The r-Proportional Rule: This rule was proposed by Kalai [5]. Given any $r \in \mathbb{R}_{++}$, a bargaining rule is called r-Proportional and denoted by P^r if for each bargaining problem $S \in \Sigma_0^2$

$$P^r(S) = \lambda(S)(r, 1) \text{ and } \lambda(S) = \max\{t : t(r, 1) \in S\}.$$

The r-Proportional rule selects the maximum point of S on the line passing through $d = (0, 0)$ and the point $(r, 1)$. This rule coincides with the Egalitarian rule when $r = 1$.

The Kalai-Smorodinsky Rule: This rule was introduced by Raiffa [6] and first axiomatized by Kalai and Smorodinsky [7]. This rule depends on the ideal utilities of the two players. Formally, given any $S \in \Sigma_0^2$ and $i = 1, 2$, let $a_i(S)$ denote the ideal (aspiration) utility of player i; i.e., $a_i(S) = \max\{s_i : s \in S\}$. The vector $a(S) = (a_1(S), a_2(S))$ is called the ideal (aspiration) point in S. (Note that $a(S)$ may be outside of S. It is contained by S if and only if S is of a rectangular shape.) The Kalai-Smorodinsky rule is denoted by KS and chooses in any problem S the allocation

$$KS(S) = \max\left\{s \in S : \frac{s_1}{s_2} = \frac{a_1(S)}{a_2(S)}\right\}.$$

For any problem, this allocation corresponds to the largest utility profile at which the players' utility gains are proportional to their ideal utilities.

The Dictatorial-i Rule: This rule is anonymous. For any $i = 1, 2$, the dictatorial rule in which player i is dictatorial is called Dictatorial-i and denoted by D^i. This rule selects in any problem $S \in \Sigma_0^2$ the point s at which player i enjoys her ideal utility and the other player enjoys its disagreement utility (normalized to zero); i.e., $s_i = a_i(S)$ and $s_j = 0$ for $j \neq i$. Thus, $D^1(S) = (a_1(S), 0)$ and $D^2(S) = (0, a_2(S))$.

The Equal Area Rule: This rule, denoted by EA, was introduced and axiomatized by Anbarci and Bigelov [8]. It selects in any problem $S \in \Sigma_0^2$, the maximal point s in S such that the areas to the left and right of the line connecting the origin to s are the same. The definition of the EA rule implies that its solution on a problem S is obtained where the ray passing from both the origin and the center of gravity of S intersects the boundary of S, $WPO(S)$. It is also true that $EA(S)$ is the maximal point s in S such that the areas above s_2 and to the right of s_1 are equal.

The Disagreement Rule: This rule, denoted by DIS, selects in any problem $S \in \Sigma_0^2$ the disagreement point (0, 0), i.e., $DIS(S) = (0, 0)$.

16.3 Some Well-Known Axioms

Below, we will present some well-known axioms in bargaining theory and check whether they are satisfied in particular by the Egalitarian, Nash, or Kalai-Smorodinsky rules for which we will state and prove the characterization theorems in the next section of this chapter.

Our first two axioms deal with the efficiency (optimality) of the bargaining rules on a given problem. Given any problem S in Σ_0^2, we define the set of weakly Pareto optimal allocations as

$$WPO(S) = \{s \in S : \text{there exists no } s' \in S \text{ such that } s' > s\}.$$

Likewise, we define the set of (strongly) Pareto optimal allocations as

$$PO(S) = \{s \in S : \text{there exists no } s' \in S \text{ such that } s' \geq s\}.$$

Using these sets, we define the following two axioms.

Weak Pareto Optimality (WPO): For all $S \in \Sigma_0^2$, $F(S) \in WPO(S)$.

WPO requires that in each problem all gains from cooperation should be weakly exhausted. This axiom presents a minimal condition that any desirable bargaining rule must satisfy. The bargaining rules illustrated by Fig. 16.1 all satisfy WPO. In particular, the Egalitarian, Nash, and Kalai-Smorodinsky rules satisfy WPO, as should be evident from the definitions of these rules. The following axiom is a stronger version of Pareto optimality, requiring a bargaining rule to select its solution on any problem from the set of Pareto optimal allocations.

Pareto Optimality (PO): For all $S \in \Sigma_0^2$, $F(S) \in PO(S)$.

Since for any $S \in \Sigma_0^2$, we have $PO(S) \subseteq WPO(S)$, it is clear that (PO) implies (WPO). But, the converse is not true. We can illustrate this with the help of the Egalitarian rule, E, which satisfies (WPO) but not (PO). To see this, consider the problem $S = cch\{(2, 1)\}$. (Here, "cch" means convex and comprehensive hull; thus, S is a rectangle on $\mathbb{R}_+^2$ with its northeast corner located at (2, 1)). Note that $WPO(S) = \{x \in S : x_1 = 2 \text{ or } x_2 = 1\}$ and $PO(S) = \{(2, 1)\}$. Also, note that $E(S) = (1, 1)$. We have $E(S) \in WPO(S)$ and $E(S) \notin PO(S)$. Therefore, the Egalitarian rule does not satisfy (PO). Hence, we have established that (WPO) does not imply (PO).

We can easily check that the Nash and Kalai-Smorodinsky rules satisfy (PO). Let us pick any $S \in \Sigma_0^2$. Suppose towards a contradiction that $N(S) \notin PO(S)$. Then there exists $x \in S$ such that $x \neq N(S)$ and $x \geq N(S)$, implying that $x_1x_2 > N_1(S)N_2(S)$, contradicting that $N(S) = \text{argmax}_{s \in S}\, s_1s_2$. Therefore, we must have $N(S) \in PO(S)$. Likewise, suppose towards a contradiction that $KS(S) \notin PO(S)$. We know that $PO(S) \neq \emptyset$ and KS satisfies (WPO); therefore, there must exist $x \in WPO(S) \setminus P(S)$ such that $x \neq KS(S)$ and $x \geq KS(S)$, implying that either $x_1 > KS_1(S)$ or $x_2 > KS_2(S)$ is true. Without loss of generality, assume that the former inequality is true. Then, we must have $x_2 = KS_2(S)$. The convexity of S implies that $a_2(S) = KS_2(S)$. Since $KS_1(S)/KS_2(S) = a_1(S)/a_2(S)$ must hold by the definition of KS, we must have $a_1(S) = KS_1(S)$. But, from $x \in S$ and $x_1 > KS_1(S)$, it follows that $x_1 > a_1(S)$, a contradiction. Therefore, we must have $KS(S) \in PO(S)$.

Now, we will consider an axiom defined on symmetric problems. A problem $S \in \Sigma_0^2$ is symmetric if for any $(s_1, s_2) \in S$, we have $(s_2, s_1) \in S$.

Symmetry (S): If $S \in \Sigma_0^2$ is symmetric, then $F_1(S) = F_2(S)$.

The symmetry axiom requires that the solution of a bargaining rule on a symmetric problem should be symmetric with respect to the players' identities. All bargaining rules illustrated in Fig. 16.1, except for the dictatorial rules, satisfy the symmetry axiom. In particular, we should notice that the Egalitarian rule satisfies the symmetry axiom, since it always yields a symmetric solution, irrespective of whether the problem is symmetric or not.

To show that the Nash rule satisfies the symmetry axiom, suppose towards a contradiction that there exists a symmetric problem $S \in \Sigma_0^2$ such that $N_1(S) \neq N_2(S)$. Consider the point $a = (N_2(S), N_1(S))$. Note that $a \neq N(S)$. Also, since S is symmetric, $a \in S$. Now, consider the point

$$b = \frac{a}{2} + \frac{N(S)}{2} = \left(\frac{N_1(S) + N_2(S)}{2}, \frac{N_1(S) + N_2(S)}{2}\right).$$

Since a and $N(S)$ are in S and S is convex, the point b must be in S. Note also that

$$b_1b_2 - N_1(S)N_2(S) = \left(\frac{N_1(S)+N_2(S)}{2}\right)^2 - N_1(S)N_2(S) = \left(\frac{N_1(S)-N_2(S)}{2}\right)^2 > 0.$$

Then, it cannot be true that $N(S) = \text{argmax}_{s \in S}\, s_1s_2$, a contradiction. Thus, the Nash rule satisfies the symmetry axiom.

Finally, the Kalai-Smorodinsky rule satisfies the symmetry axiom simply because of the facts that $a_1(S) = a_2(S)$ for any symmetric problem $S \in \Sigma_0^2$ and $K_1(S)/K_2(S) = a_1(S)/a_2(S)$ by definition.

The following axiom says that the solution on a bargaining problem should not change when the bargaining problem contracts as long as the original solution remains feasible.

Contraction Independence (CI): For all $S, S' \in \Sigma_0^2$ with $S \subset S'$, $F(S') = F(S)$ only if $F(S) \in S'$.

The contraction independence axiom is also known as "the independence of irrelevant alternatives". To check whether the Nash bargaining rule satisfies (CI), we consider any $S, S' \in \Sigma_0^2$ such that $S' \subset S$ and $N(S) \in S'$. We know that $N_1(S)N_2(S) \geq x_1x_2$ for all $x \in S$. Since $S' \subset S$, it is trivially true that $N_1(S)N_2(S) \geq x_1x_2$ for all $x \in S'$, implying $N(S') = N(S)$. So, the Nash rule satisfies (CI). It is easy to check that the Egalitarian rule also satisfies (CI). But, the Kalai-Smorodinsky rule violates this axiom. To show this, let us pick $S = co\{(0,0), (0,2), (2,0)\}$. Clearly, $S \in \Sigma_0^2$ and $K(S) = (1,1)$. Now, consider $S' = co\{(0,0), (0,1), (1,1), (2,0)\}$. We have $S' \in \Sigma_0^2$, $S' \subset S$, and $K(S) \in S'$. One can easily check that $K(S') = (4/3, 2/3)$, implying that $K(S') \neq K(S)$. Thus, KS violates (CI).

Below, we will define another axiom, called Scale Invariance, after the following preliminary. Let $\Lambda_0^2 : \mathbb{R}_+^2 \to \mathbb{R}_+^2$ be the class of independent person by person positive linear transformations. That is $\lambda \in \Lambda_0^2$ if there exists some $k \in \mathbb{R}_{++}^2$ such that for all $x \in \mathbb{R}_+^2$, $\lambda(x) = (k_1x_1, k_2x_2)$. Given $\lambda \in \Lambda_0^2$ and $S \subset \mathbb{R}_+^2$, we can define $\lambda(S) = \{x' \in \mathbb{R}_+^2 : \exists x \in S \text{ s.t. } x' = \lambda(x)\}$.

Scale Invariance (SI): For any $S \in \Sigma_0^2$ and $\lambda \in \Lambda_0^2$, $\lambda(F(S)) = F(\lambda(S))$.

The scale invariance is also known as "the independence of equivalence utility representations". This axiom says that if a bargaining problem is scaled under a linear transformation, then the solution must also be scaled under the same transformation. This axiom prevents the bargaining solution to be based on interpersonal comparisons of utility.

The scale invariance is satisfied by the Nash rule. To show this, we pick any $S \in \Sigma_0^2$. Also, we pick any $\lambda \in \Lambda_0^2$. Then, there exists some $(k_1, k_2) \in R_{++}^2$ such that $x = (x_1, x_2) \in S$ if and only if $(k_1x_1, k_2x_2) \in \lambda(S)$. We also know that $N_1(S)N_2(S) \geq x_1x_2$ for all $x \in S$. This implies that $(k_1N_1(S))(k_2N_2(S)) \geq (k_1x_1)(k_2x_2)$ for all $x \in S$, implying that $(k_1N_1(S))(k_2N_2(S)) \geq y_1y_2$ for all $y \in \lambda(S)$, further implying that

$(k_1N_1(S), k_2N_2(S)) = N(\lambda(S))$. Since $(k_1N_1(S), k_2N_2(S))$ is nothing but $\lambda(N(S))$, we have $\lambda(N(S)) = N(\lambda(S))$. Thus we have established that the Nash rule satisfies (SI).

Another well-known rule that satisfies (SI) is the Kalai-Smorodinsky rule. To show this let us pick any $S \in \Sigma_0^2$ and any $\lambda \in \Lambda_0^2$. Then, there exists some $(m_1, m_2) \in R_{++}^2$ such that $x = (x_1, x_2) \in S$ if and only if $(m_1x_1, m_2x_2) \in \lambda(S)$. Clearly, $\lambda(a(S)) = (m_1a_1(S), m_2a_2(S))$. We also know that

$$KS(S) = \max\left\{x \in S : \frac{x_1}{x_2} = \frac{a_1(S)}{a_2(S)}\right\}.$$

This implies that

$$KS(S) = \max\left\{x \in S : \frac{m_1x_1}{m_2x_2} = \frac{m_1a_1(S)}{m_2a_2(S)}\right\},$$

further implying $(m_1KS_1(S), m_2KS_2(S)) = KS(\lambda(S))$. Since the left-hand-side is equal to $\lambda(KS(S))$, we have $\lambda(KS(S)) = KS(\lambda(S))$. Thus, the Kalai-Smorodinsky rule satisfies (SI).

On the other hand, the Egalitarian rule does not satisfy (SI). We can prove this by the following example. Consider $S = cch\{(1, 1)\}$ (the unit square). Then $E(S) = (1, 1)$. Let $\lambda \in \Lambda_0^2$ be such that $\lambda(x) = (x_1, 2x_2)$ for any $x \in \mathbb{R}_+^2$. Then, $\lambda(S)) = cch\{(1, 2)\}$ and $E(\lambda(S))) = (1, 1)$, which is not equal to $\lambda(E(S)) = (1, 2)$. Therefore, the Egalitarian rule does not satisfy (SI).

Strong Monotonicity (SM): For any $S, S' \in \Sigma_0^2$ with $S' \supset S$, $F_i(S') \geq F_i(S)$ for each $i = 1, 2$.

The strong monotonicity requires that all players should weakly benefit from any expansion of the bargaining set. This axiom is not satisfied by the Nash rule. The following example illustrates this. Let $S = co\{(0, 0), (0, 1), (1, 1), (2, 0)\}$ and $S' = co\{(0, 0), (0, 1), (1, 1), (2, 1/2), (2, 0)\}$. We have $S' \supset S$, $N(S) = (1, 1)$ and $N(S') = (3/2, 3/4)$. Notice that for $i = 2$, the condition $N_i(S') \geq N_i(S)$ fails. Thus, (SM) is not satisfied by the Nash rule.

The strong monotonicity is not satisfied by the Kalai-Smorodinsky rule, either. The following example illustrates this. Let $S = cch\{(1, 1)\}$ and $S' = co\{(0, 0), (0, 1), (1, 1), (2, 0)\}$. We have $S' \supset S$, $KS(S) = (1, 1)$ and $KS(S') = (4/3, 2/3)$. Notice that for $i = 2$, the condition $KS_i(S') \geq KS_i(S)$ fails. Thus, (SM) is not satisfied by the Kalai-Smorodinsky rule.

Let us show that the Egalitarian rule satisfies the strong monotonicity. Pick any $S, S' \in \Sigma_0^2$ with $S' \supset S$. We know that $E(S) \in WPO(S)$. Define the set $IR(E(S), S') = \{x \in S' : x \geq E(S)\}$, representing the set of utility allocations in S that are not inferior to $E(S)$. Since $S' \supset S$, we have $WPO(S') \cap IR(E(S), S') \neq \emptyset$. If $WPO(S') \cap IR(E(S), S') = \{E(S)\}$, then $E(S') = E(S)$. On the other hand, if $WPO(S') \cap IR(E(S), S') \neq \{E(S)\}$, then the geometry of the Egalitarian rule implies that $E_i(S') > E_i(S)$ for each $i \in \{1, 2\}$. In either case, the condition $E_i(S') \geq E_i(S)$ holds for each $i \in \{1, 2\}$. Thus, (SM) is satisfied by the Egalitarian rule.

Below, we present a weaker version of monotonicity.

Individual Monotonicity (IM): For any $S, S' \in \Sigma_0^2$ with $S' \supset S$, if $a_i(S) = a_i(S')$ for some $i \in \{1, 2\}$, then $F_j(S') \geq F_i(S)$ for $j \neq i$.

The individual monotonicity requires that a player should weakly benefit from any expansion of the bargaining set that does not affect her opponents' ideal point. The Kalai-Smorodinsky rule satisfies this axiom. To show this, consider any $S, S' \in \Sigma_0^2$ such that $S' \supset S$ and $a_i(S') = a_i(S)$ for some i. Without loss of generality, let $i = 1$, i.e., $a_1(S') = a_1(S)$. We must then have $a_2(S') \geq a_2(S)$, since $S' \supset S$. By definition,

$$KS(S) = \max\left\{x \in S : \frac{x_1}{x_2} = \frac{a_1(S)}{a_2(S)}\right\}$$

and

$$KS(S') = \max\left\{x \in S' : \frac{x_1}{x_2} = \frac{a_1(S')}{a_2(S')}\right\}.$$

Using $a_1(S') = a_1(S)$ and $a_2(S') \geq a_2(S)$, we can write

$$\frac{KS_1(S')}{KS_2(S')} = \frac{a_1(S')}{a_2(S')} \leq \frac{a_1(S)}{a_2(S)} = \frac{KS_1(S)}{KS_2(S)},$$

implying

$$\frac{KS_1(S')}{KS_2(S')} \leq \frac{KS_1(S)}{KS_2(S)}.$$

Now, suppose towards a contradiction that $KS_2(S') < KS_2(S)$. Then, the above inequality can be true only if $KS_1(S') < KS_1(S)$. That means $KS(S') < KS(S)$. Then $KS(S') \notin WPO(S)$. But, we know that $S' \supset S$. Therefore, $KS(S') \notin WPO(S')$, which is in contradiction with the definition of the rule KS. Thus, $KS_2(S') \geq KS_2(S)$. (Notice that if we had assumed $a_2(S') = a_2(S)$ at the beginning, we would have obtained $KS_1(S') \geq KS_1(S)$.) So, the bargaining rule KS satisfies (IM).

We can also show that the Nash rule does not satisfy (IM). Our previous example used for the axiom of (SM) illustrates this fact. We assumed in that example, $S = co\{(0, 0), (0, 1), (1, 1), (2, 0)\}$ and $S = co\{(0, 0), (0, 1), (1, 1), (2, 1/2), (2, 0)\}$. We have $S' \supset S$, $N(S) = (1, 1)$, and $N(S') = (3/2, 3/4)$. Even though $a_1(S') = a_1(S) = 2$, the inequality $N_2(S') \geq N_2(S)$ does not hold, proving that the Nash rule violates (IM).

Notice that (SM) implies (IM). To show this, we pick any bargaining rule F that satisfies (SM). (We know that such a bargaining rule exists, since we showed that the Egalitarian rule satisfies (SM).) Also, we pick any $S, S' \in \Sigma_0^2$ such that $S' \supset S$ and $a_i(S) = a_i(S')$ for some $i \in \{1, 2\}$. Pick any such i. Since F satisfies (SM), we must have $F(S') \geq F(S)$, implying that $F_i(S') \geq F_i(S)$ Thus, F satisfies (IM). Since we already know that the Egalitarian rule satisfies (SM), the above result immediately implies that it satisfies (IM), as well.

On the other hand, (IM) does not imply (SM). We know this since there are some rules (e.g., the Kalai-Smorodinsky Rule) that satisfy (IM) but not (SM).

The following version of monotonicity weakens the individual monotonicity.

Restricted Monotonicity (RM): For any $S, S' \in \Sigma_0^2$ with $S' \supset S$, if $a(S') = a(S)$, then $F(S') \geq F(S)$.

Note that (IM) implies (RM). To show this, we pick any bargaining rule F satisfying (IM). (We know such a bargaining rule exists. For example, the Kalai-Smorodinsky Rule.) Also, we pick any $S, S' \in \Sigma_0^2$ such that $S' \supset S$ and $a_i(S) = a_i(S')$ for each $i = \{1, 2\}$. Since F satisfies (IM) and $a_1(S) = a_1(S')$, we must have $F_2(S') \geq F_2(S)$. Also, since $a_2(S) = a_2(S')$, we must have $F_1(S') \geq F_1(S)$. Thus, F satisfies (RM).

Since the Kalai-Smorodinsky rule and the Egalitarian rule satisfy (IM), the above result immediately implies that they satisfy (RM), as well. We cannot use the same result to check whether the Nash rule satisfies (RM), because we know that it does not satisfy (IM). However, the example that we used in showing that the Nash rule does not satisfy the axioms of (SM) and (IM) also illustrates that the Nash rule cannot satisfy (RM), either. In that example, we considered the bargaining problems $S = co\{(0, 0), (0, 1), (1, 1), (2, 0)\}$ and $S' = co\{(0, 0), (0, 1), (1, 1), (2, 1/2), (2, 0)\}$. Thus, we have $S' \supset S$, $a_1(S') = a_1(S) = 2$, and $a_2(S') = a_2(S) = 1$. If the Nash rule satisfies (RM), then we must have $N_i(S') \geq N_i(S)$ for each $i \in \{1, 2\}$. Recall that $N(S) = (1, 1)$ and $N(S') = (3/2, 3/4)$. Thus, for $i = 2$, $N_i(S') \geq N_i(S)$ does not hold, implying that the Nash rule violates (RM).

We will now show that (RM) does not imply (IM). Consider the following rule that we may call the "Inverse Kalai-Smorodinsky (IKS) rule".

$$IKS(S) = \max\left\{x \in S : \frac{x_1}{x_2} = \frac{a_2(S)}{a_1(S)}\right\}.$$

First, we will show that the IKS rule satisfies (RM). So, pick any $S, S' \in \Sigma_0^2$ such that $S' \supset S$ and $a_i(S') = a_i(S)$ for each $i = 1, 2$. Then,

$$IKS(S') = \max\left\{x \in S' : \frac{x_1}{x_2} = \frac{a_2(S')}{a_1(S')} = \frac{a_2(S)}{a_1(S)}\right\},$$

implying that $IKS(S') \geq IKS(S)$ since $S' \supseteq S$ and $IKS(S') \in WPO(S')$. Therefore, IKS satisfies (RM). Now, we will show that the IKS rule does not satisfy (IM). Let $S = cch\{(1, 1)\}$ (the unit square) and $S' = cch\{(2, 1)\}$ (the rectangle with the northeast corner located at (2, 1)). Notice that $S, S' \in \Sigma_0^2$. We can easily calculate that $IKS(S) = (1, 1)$ and $IKS(S') = (0.5, 1)$. Notice that $S' \supset S$ and $a_2(S) =$

Table 16.1 Some well-known axioms satisfied by Egalitarian, Kalai-Smorodinsky, or Nash bargaining rules

Axiom	Bargaining Rule		
	Egalitarian Rule	Kalai-Smorodinsky Rule	Nash Rule
Weak Pareto Optimality	✓	✓	✓
Pareto Optimality		✓	✓
Symmetry	✓	✓	✓
Contraction Independence	✓		✓
Scale Invariance		✓	✓
Strong Monotonicity	✓		
Individual Monotonicity	✓	✓	
Restricted Monotonicity	✓	✓	

$a_2(S')$; but it is not true that $IKS_1(S') \geq IKS_1(S)$. We have $0.5 = IKS_1(S') < IKS_1(S) = 1$, implying that the IKS rule does not satisfy (IM). Thus, we established that (RM) does not imply (IM).

Since (SM) implies (IM) and (IM) implies (RM), we also know that (SM) implies (RM). Moreover, since (RM) does not imply (IM) and (IM) does not imply (SM), it is also true that (RM) does not imply (SM).

In the next section, we will see the axiomatization results for the Egalitarian, Kalai-Smorodinsky, and Nash rules. In Table 16.1, we list the set of axioms we have presented so far and illustrate which of them are satisfied by the aforementioned three bargaining rules.

16.4 Three Characterization Results

Below, we will state and prove characterization theorems for the Nash, Kalai-Smorodinsky, and Egalitarian rules.

Proposition 16.1 ([1]) *A bargaining rule satisfies (S), (WPO), (SI), and (CI) if and only if it is the Nash rule.*

Proof In Sect. 16.3, we checked that the Nash rule satisfies (S), (WPO), (SI), and (CI). Therefore, "if part" of the proposition holds. To show that "only if" part also holds, we pick any bargaining rule F that satisfies (S), (WPO), (SI), and (CI) and pick any problem $S \in \Sigma_0^2$. We have to show that $F(S) = N(S)$.

Let $\lambda \in \Lambda_0^2$ be such that $\lambda(N(S)) = (1, 1)$. Define $S' = \lambda(S)$. Since F and N both satisfy (SI), $F(S') = \lambda(F(S))$ and $N(S') = \lambda(N(S)) = (1, 1)$. Hence, to prove $F(S) = N(S)$, it suffices to show that $F(S') = N(S') = (1, 1)$.

So, consider the set $T = \{t \in \mathbb{R}_+^2 : t_1 + t_2 \leq 2\}$. Note that T is symmetric. Since F satisfies (S) and (WPO), we have $F(T) = (1, 1)$, which is in S'. Also, note that $PO(S')$ has the slope of -1 at $N(S') = (1, 1)$. Moreover, S' is convex. Therefore, $S' \subseteq T$. Because F satisfies (CI), we have $F(S') = F(T)$. Also, since $F(T) = N(S')$, we further have $F(S') = N(S')$, implying $F(S) = N(S)$. Since S was picked arbitrarily, we conclude that F coincides with the Nash rule on Σ_0^2. ■

Proposition 16.1 implies that any bargaining rule, other than the Nash rule, must fail to satisfy at least one of the four axioms in that result. For example, the generalized Nash rule N^α with $\alpha \neq 0.5$ satisfies (WPO), (SI), and (CI) but not (S). The Disagreement rule satisfies (S), (SI), and (CI) but not (WPO). The Egalitarian rule satisfies (S), (WPO), and (CI) but not (SI). Finally, the Kalai-Smorodinsky rule satisfies (S), (WPO), and (S) but not (CI).

Now, we present our second characterization result.

Proposition 16.2 ([7]) *A bargaining rule satisfies (S), (WPO), (SI), and (IM) if and only if it is the Kalai-Smorodinsky rule.*

Proof In Sect. 16.3, we checked that the Kalai-Smorodinsky rule satisfies (S), (WPO), (SI), and (IM). Therefore, "if part" of the proposition holds. To show "only if" part also holds, we pick any bargaining rule F that satisfies (S), (WPO), (SI), and (IM) and pick any problem $S \in \Sigma_0^2$. We have to show that $F(S) = KS(S)$.

Let $\lambda \in \Lambda_0^2$ be such that $\lambda(KS(S)) = (1, 1)$. Define $S' = \lambda(S)$. Since F and KS both satisfy (SI), $F(S') = \lambda(F(S))$ and $KS(S') = \lambda(KS(S)) = (1, 1)$. Hence, to prove that $F(S) = KS(S)$, it suffices to show that $F(S') = KS(S') = (1, 1)$.

Note that we must have $a_1(S') = a_2(S')$ since $KS(S') = (1, 1)$. Consider the set $T = co\{(0, 0), (0, a_2(S')), (1, 1), (a_1(S'), 0)\}$. Clearly, T is symmetric. Since F satisfies (S) and (WPO), we have $F(T) = (1, 1)$. Also, note that $T \subseteq S'$ and $a_1(S') = a_1(T)$. Since F satisfies (IM), we must have $F_2(S') \geq F_2(T) = 1$. Likewise, since $a_2(S') = a_2(T)$, we must have $F_1(S') \geq F_1(T) = 1$. Thus, $F(S') \geq F(T) = (1, 1)$. We also know that $F(T) \in PO(S')$, since $F(T) = KS(S')$ and KS satisfies (PO). Therefore, we must have $F(S') = F(T) = KS(S')$, implying that $F(S) = KS(S)$. Since S was picked arbitrarily, we conclude that F coincides with the Kalai-Smorodinsky rule on Σ_0^2. ■

Proposition 16.2 implies that any bargaining rule, other than the Kalai-Smorodinsky rule, must fail to satisfy at least one of the four axioms in that result. For example, for any $i \in \{1, 2\}$, the Dictatorial rule D^i satisfies (WPO), (SI), and (IM) but not (S). The Disagreement rule satisfies (S), (SI), and (IM) but not (WPO). The Egalitarian rule satisfies (S), (WPO), and (IM) but not (SI). Finally, the Nash rule satisfies (S), (WPO), and (SI) but not (IM).

Now, we give the characterization result for the Egalitarian rule.

Proposition 16.3 ([5]) *A bargaining rule satisfies (S), (WPO), and (SM) if and only if it is the Egalitarian rule.*

Proof In Sect. 16.3, we checked that the Egalitarian rule satisfies (S), (WPO), and (SM). Therefore, "if part" of the proposition holds. To show "only if" part also holds, let us pick any bargaining rule F that satisfies (S), (WPO), and (SM) and pick any problem $S \in \Sigma_0^2$. We have to show that $F(S) = E(S)$.

Let $x = E(S)$ and $T = cch\{x\}$. Clearly, $S \supseteq T$ and T is symmetric. Thus, by (S) and (WPO), we have $F(T) = x$. Moreover, since $S \supseteq T$, (SM) implies that $F(S) \geq x = F(T)$. However, $F(S) \neq x$ cannot be true. To prove this, consider two cases.

Case (i) $x \in PO(S)$. Then $F(S) \geq x$ is true only if $F(S) = x$.

Case (ii) $x \in WPO(S) \setminus PO(S)$. Suppose there exists $i \in \{1, 2\}$ such that $F_i(S) > x_i$. Without loss of generality, assume $i = 1$; i.e., $F_1(S) > x_1$. Pick $\epsilon > 0$ such that $\epsilon < F_1(S) - x_1$ and consider the points $x^a = x + (\epsilon, \epsilon)$ and $x^b = (0, x_2)$ and a new problem $V = co\{x^a, x^b, S\}$. Note that $E(V) = x^a$. Since $x^a \in PO(V)$, we have $F(V) = x^a$ by the result in Case (i). Also, since $V \supseteq S$ and F satisfies (SM), we must have $F(V) \geq F(S)$, implying $x^a \geq F(S)$, further implying $x_1^a = x_1 + \epsilon \geq F_1(S)$, a contradiction. Therefore, $F(S) = x$. ■

Proposition 16.3 shows that any bargaining rule, other than the Egalitarian rule, must fail to satisfy at least one of the three axioms in that result. For example, any Proportional rule P^r with $r \neq 1$ satisfies (WPO) and (SM) but not (S). The Disagreement rule satisfies (S) and (SM) but not (WPO). Finally, the Nash and Kalai-Smorodinsky rules satisfy (S) and (WPO) but not (SM).

We should note that there may exist more than one set of axioms that can characterize a given bargaining rule. Economists who believe that a particular set of axioms that characterizes a bargaining rule is problematic or controversial either search for alternative characterizations involving more defendable axioms or recommend bargaining rules that have more attractive characterizations.

Many axiomatic rules in cooperative bargaining can be obtained as the outcomes of noncooperative negotiation processes, in line with the Nash Program, a research agenda introduced by John F. Nash [9] to bridge cooperative and noncooperative elements of game theory. (For more on this issue, see a survey by Serrano [10]). For example, the solution implemented by the two-person Nash (bargaining) rule can be obtained as the equilibrium of a risk-involving variation of Rubinstein's [11] infinite horizon bilateral bargaining game (which we studied in Example 10.10). This variation, considered by Binmore, Rubinstein, and Wolinsky [12], assumes that players face a probability p that the negotiations will break down, leading to the payoffs b_1 and b_2 for players 1 and 2, respectively. As the probability p goes down to zero, Binmore [13] shows that the subgame-perfect Nash equilibrium payoff profile in the noncooperative bargaining game considered by Binmore, Rubinstein, and Wolinsky [12] approaches in the limit to the bargaining solution selected by the Nash rule in a bargaining problem (S, d) where the bargaining set S contains all possible distributions of the total surplus in the associated noncooperative game and the disagreement point d is equal to b (the breakdown payoff profile in the associated noncooperative game).

References

1. Nash JF (1950) The bargaining problem. Econometrica 18:155–162
2. von Neumann J, Morgenstern O (1944) Theory of games and economic behavior. Princeton University Press, Princeton
3. Harsanyi JC, Selten R (1972) A generalized Nash solution for two-person bargaining games with incomplete information. Manag Sci 18:80–106
4. Rawls J (1972) A theory of justice. Oxford University Press, Oxford
5. Kalai E (1977) Proportional solutions to bargaining situations: interpersonal comparisons of utility. Econometrica 45:1623–1630
6. Raiffa H (1953) Arbitration schemes for generalized two-person games. In: Kuhn HW, Tucker AW (eds) Contributions to the theory of games, vol II. Princeton University Press, Princeton, pp 361–387
7. Kalai E, Smorodinsky M (1975) Other solutions to Nash's bargaining problem. Econometrica 43:513–518
8. Anbarcı N, Bigelow JP (1994) The area monotonic solution to the cooperative bargaining problem. Math Soc Sci 28(2):133–142
9. Nash JF (1953) Two-person cooperative games. Econometrica 21:128–140
10. Serrano R (2021) Sixty-seven years of the Nash program: time for retirement? SERIEs 12:35–48
11. Rubinstein A (1982) Perfect equilibrium in a bargaining model. Econometrica 50:97–109
12. Binmore KG, Rubinstein A, Wolinsky A (1986) The Nash bargaining solution in economic modelling. Rand J Econ 17:176–188
13. Binmore KG (1987) Nash bargaining theory I, II, III. In: Binmore KG, Dasgupta P (eds) The economics of bargaining. Blackwell, Oxford

Matching Markets

17

In this chapter, we will study two-sided matching between two groups of individuals or objects under nontransferable utilities. Section 17.1 will introduce a one-to-one matching environment, known as the marriage market involving a set of men and a set of women, and present several results about stable matchings in this environment. Section 17.2 will extend some of these results to a many-to-one matching environment involving a set of institutions and a set of individuals. In this environment, a many-to-one matching assigns each institution to a distinct group of individuals. Finally, Sect. 17.3 will study a particular one-to-one matching problem, known as the housing market, where the aim is to find a stable matching that reallocates the houses of a finite number of individuals among them.

17.1 Marriage Market

Consider a set of men and a set of women denoted by $M = \{m_1, m_2, \ldots, m_p\}$ and $W = \{w_1, w_2, \ldots, w_q\}$, respectively. Each individual x in $X \in \{M, W\}$ has preferences (or preference lists) over individuals in the opposite gender set $Y = \{M, W\} \setminus X$. Formally, the preferences of each man $m \in M$ is an ordered list, $P(m)$, which is defined on the set $W \cup \{m\}$. For example, if $W = \{w_1, w_2, w_3, w_4\}$, the preferences of a man, say m_1, may be like $P(m_1) = w_4, w_3, w_1, m_1, w_2$, meaning that w_4 is the first choice of m_1, followed by his second and third choices w_3 and w_1. According to this ordered list, m_1 does not consider w_2 as an acceptable mate; he prefers being single, m_1, to w_2. This example involves strict preferences only. The preferences of individuals may contain indifferences. For example, according to the preferences $P(m_2) = [w_1, w_2, w_3], m_2, w_4$, the man m_2 is indifferent between w_1, w_2, and w_3 and finds only these three women acceptable as a mate. Similarly, we define the preferences of each woman $w \in W$ as an ordered list, $P(w)$, on the

I. Saglam, *Mastering Game Theory*, Classroom Companion: Economics,
https://doi.org/10.1007/978-3-031-82094-6_17

set $M \cup \{w\}$. We denote by $\mathcal{P} = \{P(m_1), \ldots, P(m_p), P(w_1), \ldots, P(w_q)\}$ the set of preference lists of all individuals in the society $M \cup W$. We denote a marriage market by the triple $\langle M, W, \mathcal{P}\rangle$.

We will write

$w \succ_m w'$ to say that m strictly prefers w to w',

$w \sim_m w'$ to say that m is indifferent between w and w', and

$w \succeq_m w'$ to say that m weakly prefers w to w' (that is, either $w \succ_m w'$ or $w \sim_m w'$).

Formally, we say that the woman w is acceptable to the man m if and only if $w \succeq_m m$, i.e., m finds w weakly better than being single. Similarly, the man m is acceptable to the woman w if and only if $m \succeq_w w$.

Definition 17.1 A matching is a one-to-one correspondence μ from $M \cup W$ onto itself such that

$$\mu(\mu(x)) = x, \quad \text{for all } x \in M \cup W,$$

$$\text{if } \mu(m) \neq m, \text{ then } \mu(m) \in W \text{ for all } m \in M,$$

$$\text{if } \mu(w) \neq w, \text{ then } \mu(w) \in M \text{ for all } w \in W.$$

According to the above definition, if an individual (man or woman) is not single under a matching, then he/she is matched to an individual of the opposite gender. Moreover, the matching between man m and woman w is mutual; a matching μ matches m to w if and only if it matches w to m.

For example, if $M = \{m_1, m_2, m_3, m_4\}$ and $W = \{w_1, w_2, w_3\}$, then we can represent a possible matching that matches m_1 to w_3, m_2 to w_2, and leaves m_3, m_4, and w_1 single as

$$\mu = \begin{array}{cccc} m_1 & m_2 & m_3 & m_4 \\ w_3 & w_2 & m_3 & m_4 \end{array}$$

or alternatively

$$\mu = \begin{array}{ccc} w_1 & w_2 & w_3 \\ w_1 & m_2 & m_1. \end{array}$$

We let $\mathcal{M}$ denote the set of all matchings. Given a marriage market $\langle M, W, \mathcal{P}\rangle$, we assume that we can derive the preferences of individuals in $M \cup W$ over the set of matchings $\mathcal{M}$ from their preferences $\mathcal{P}$ as follows: For any two matchings $\mu, \mu' \in \mathcal{M}$, the individual $x \in M \cup W$ strictly prefers matching μ to matching μ' if and only if x strictly prefers his/her mate at μ to his/her mate at μ'; i.e.,

$$\mu \succ_x \mu \text{ if and only if } \mu(x) \succ_x \mu'(x).$$

Similarly, we assume that

$$\mu \sim_x \mu \text{ if and only if } \mu(x) \sim_x \mu'(x),$$

and

$$\mu \succeq_x \mu \text{ if and only if } \mu(x) \succeq_x \mu'(x).$$

The above assumptions imply that matchings involve no externalities. In ordering two matchings μ and μ', the individual $x \in M \cup M$ does not care how the other individuals in the society would be matched under μ and μ'. The preference ordering of the individual x between μ and μ' relies only on his/her preference ordering between $\mu(x)$ and $\mu'(x)$ implied by his/her preferences $P(x)$. Now, we are going to define for matchings an equilibrium notion called stability. We will say that a matching is stable if no individual or no pair of individuals block a matching.

Formally, we say that a man or a woman $x \in M \cup W$ individually blocks a matching $\mu \in \mathcal{M}$ if x strictly prefers being single to his/her mate at μ; i.e.,

$$x \succ_x \mu(x).$$

(Note that x can individually block a matching μ only if $\mu(x) \neq x$.) We say that $\mu \in \mathcal{M}$ can be individually blocked if and only if there exists $x \in M \cup W$ who can individually block μ.

Also, we say that a man m and a woman w pairwise block a matching $\mu \in \mathcal{M}$ if they strictly prefer each other to their mates at μ; i.e.,

$$w \succ_m \mu(x) \text{ and } m \succ_w \mu(w).$$

(Note that m and w can pairwise block a matching μ only if $\mu(m) \neq w$.) We say that $\mu \in \mathcal{M}$ can be pairwise blocked if and only if there exists $(m, w) \in M \times W$ who can pairwise block μ.

Definition 17.2 A matching $\mu \in \mathcal{M}$ is called stable if it cannot be individually or pairwise blocked.

Example 17.1 Consider a marriage market $\langle M, W, \mathcal{P}\rangle$ with $M = \{m_1, m_2, m_3\}$, $W = \{w_1, w_2, w_3\}$, and the set of preferences $\mathcal{P}$ given by:

$$\begin{array}{ll} P(m_1) = w_1\ w_2\ w_3\ m_1 & P(w_1) = m_2\ m_1\ m_3\ w_1 \\ P(m_2) = w_2\ w_1\ w_3\ m_2 & P(w_2) = m_1\ m_3\ m_2\ w_2 \\ P(m_3) = w_1\ w_3\ w_2\ m_3 & P(w_3) = m_1\ m_2\ m_3\ w_3 \end{array}$$

Also, consider the following matching.

$$\mu = \begin{array}{ccc} m_1 & m_2 & m_3 \\ w_3 & w_2 & w_1 \end{array}$$

Note that μ cannot be individually blocked since $\mu(x) \succ_x x$ for each $x \in M \cup W$. On the other hand, μ can be pairwise blocked by (m_1, w_1) since $w_1 \succ_{m_1} \mu(m_1) = w_3$ and $m_1 \succ_{w_1} \mu(w_1) = m_3$. Therefore, μ is not stable.

Now consider the following matching.

$$\mu' = \begin{matrix} m_1 & m_2 & m_3 \\ w_1 & w_2 & w_3 \end{matrix}$$

One can easily check that μ' is stable; i.e., it cannot be individually or pairwise blocked. ■

We let $\mathcal{M}^S$ denote the set of stable matchings. Note that $\mathcal{M}^S \subseteq \mathcal{M}$. Gale and Shapley [1] proved that the set of stable matchings, $\mathcal{M}^S$, is nonempty for each marriage market $\langle M, W, \mathcal{P}\rangle$. To show this result, they introduced the Deferred Acceptance Algorithm that generates at least one stable matching for every marriage market. This algorithm has two versions depending on whether men or women are making proposals.

The Deferred Acceptance Algorithm with Men Proposing (DAA-MP).

The algorithm has a number of steps. Let M_k denote the set of men who are not engaged at the beginning of step k (where $k \geq 1$ is an integer). At the beginning of step 1 no man is engaged to any woman; i.e., $M_1 = M$.

In step $k \geq 1$ of the algorithm,

first, each man in M_k proposes to his most preferred woman in the set of women acceptable to him, excluding those who rejected him in the previous steps (if $k > 1$),

next, each woman engages with her favorite man in the set of men that involves the men who made acceptable proposals to her in step k and the man (if any) she was engaged in step $k-1$ (if $k > 1$), and rejects all the others,

finally, the set M_{k+1} is formed to include all men who are not engaged at the end of step k.

If a man becomes indifferent between two or more women to propose to (or if a woman becomes indifferent between two or more men to engage with) at any step k, then ties can be broken arbitrarily (or by any tie-breaking rule).

The algorithm stops at the end of step $K \geq 1$ such that either $M_{K+1} = \emptyset$ (all men are engaged at the end of step K) or each man in M_{K+1} has been rejected from all acceptable woman in the first K steps of the algorithm. Consequently, the engagements at the end of step K become the actual matchings, leading to the final outcome of the algorithm. ■

By changing the role of men and women in the above algorithm, we can obtain the Deferred Acceptance Algorithm with Women Proposing (DAA-WP). The reason why the word "deferred" is used in the title of Deferred Acceptance Algorithm is that when men (women) are proposing, each woman (man) can remain to be engaged to her best available man (woman) at any step without accepting him (her) outright. We should also notice that at each step of the DAA-MP (DAA-WP) the welfare of each man (woman) becomes weakly worse, whereas the welfare of each woman (man) becomes weakly better.

Example 17.2 Reconsider the marriage market $\langle M, W, \mathcal{P}\rangle$ in Example 17.1, with $M = \{m_1, m_2, m_3\}$, $W = \{w_1, w_2, w_3\}$, and the set of preferences $\mathcal{P}$ given by

$$\begin{array}{ll} P(m_1) : w_1\ w_2\ w_3\ m_1 & P(w_1) = m_2\ m_1\ m_3\ w_1 \\ P(m_2) : w_2\ w_1\ w_3\ m_2 & P(w_2) = m_1\ m_3\ m_2\ w_2 \\ P(m_3) : w_1\ w_3\ w_2\ m_3 & P(w_3) = m_1\ m_2\ m_3\ w_3. \end{array}$$

Let us first run the DAA-MP on the marriage market $\langle M, W, \mathcal{P}\rangle$. In each step, we indicate the new proposals by the red color and the new engagements by the blue color.

	w_1	w_2	w_3
Step 1 (Proposals)	$\{m_1, m_3\}$	$\{m_2\}$	$\{\}$
Step 1 (Engagements)	$\{m_1\}$	$\{m_2\}$	$\{\}$
Step 2 (Proposals)	$\{m_1\}$	$\{m_2\}$	$\{m_3\}$
Step 2 (Engagements)	$\{m_1\}$	$\{m_2\}$	$\{m_3\}$

Since no man is left to make a proposal, the algorithm stops at this point and each woman marries to the man she is engaged with. Thus, the outcome of the DAA-MP is the stable matching given by

$$\mu^{MP} = \begin{array}{l} m_1\ m_2\ m_3 \\ w_1\ w_2\ w_3. \end{array}$$

Now, let us run the DAA-WP on the marriage market $\langle M, W, \mathcal{P}\rangle$.

	m_1	m_2	m_3
Step 1 (Proposals)	$\{w_2, w_3\}$	$\{w_1\}$	$\{\}$
Step 1 (Engagements)	$\{w_2\}$	$\{w_1\}$	$\{\}$
Step 2 (Proposals)	$\{w_2\}$	$\{w_1, w_3\}$	$\{\}$
Step 2 (Engagements)	$\{w_2\}$	$\{w_1\}$	$\{\}$
Step 3 (Proposals)	$\{w_2\}$	$\{w_1\}$	$\{w_3\}$
Step 3 (Engagements)	$\{w_2\}$	$\{w_1\}$	$\{w_3\}$

Since no woman is left to make a proposal, the algorithm stops at this point and each man marries to the woman he is engaged with. Thus, the outcome of the DAA-WP is the stable matching given by

$$\mu^{WP} = \begin{matrix} m_1 & m_2 & m_3 \\ w_2 & w_1 & w_3, \end{matrix}$$

which is different from μ^{MP}. ■

For the marriage market in Example 17.2, the DAA-MP and the DAA-WP lead to different stable matchings; i.e., $\mu^{MP} \neq \mu^{WP}$. But, this is not a general result. As the next example shows, there exist marriage markets where the DAA-MP and the DAA-WP lead to the same stable matching.

Example 17.3 Consider the marriage market $\langle M, W, \mathcal{P}\rangle$ with $M = \{m_1, m_2, m_3\}$, $W = \{w_1, w_2, w_3\}$, and the set of preferences $\mathcal{P}$ given by:

$$\begin{array}{ll} P(m_1) : w_1\ w_2\ w_3\ m_1 & P(w_1) = m_1\ m_3\ m_2\ w_1 \\ P(m_2) : w_2\ w_3\ w_1\ m_2 & P(w_2) = m_2\ m_3\ m_1\ w_2 \\ P(m_3) : w_3\ w_2\ w_1\ m_3 & P(w_3) = m_3\ m_1\ m_2\ w_3 \end{array}$$

Let us first run the DAA-MP on the marriage market $\langle M, W, \mathcal{P}\rangle$. In each step, we indicate the new proposals by the red color and the new engagements by the blue color.

	w_1	w_2	w_3
Step 1 (Proposals)	$\{m_1\}$	$\{m_2\}$	$\{m_3\}$
Step 1 (Engagements)	$\{m_1\}$	$\{m_2\}$	$\{m_3\}$

Thus, the outcome of the DAA-MP is the stable matching given by

$$\mu^{MP} = \begin{matrix} m_1 & m_2 & m_3 \\ w_1 & w_2 & w_3. \end{matrix}$$

Now, let us run the DAA-WP on the marriage market $\langle M, W, \mathcal{P}\rangle$.

	m_1	m_2	m_3
Step 1 (Proposals)	$\{w_1\}$	$\{w_2\}$	$\{w_3\}$
Step 1 (Engagements)	$\{w_1\}$	$\{w_2\}$	$\{w_3\}$

Thus, the outcome of the DAA-WP is the stable matching given by

$$\mu^{WP} = \begin{matrix} m_1 & m_2 & m_3 \\ w_1 & w_2 & w_3, \end{matrix}$$

which coincides with μ^{MP}. ■

The above examples have illustrated how the algorithms DAA-MP and DAA-WP can work. Now, we are ready to state and prove the following existence result.

Proposition 17.1 ([1]) *Every marriage market has a stable matching.*

Proof Consider any version of the Deferred Acceptance Algorithm, say the DAA-MP. The algorithm must eventually stop (the total number of steps, K, must be finite) because there are only a finite number of men and women, and no man proposes more than once to any woman. Also note that the outcome of this algorithm is a matching since each individual at any step is engaged to at most one individual of the opposite gender. Let us denote this matching by μ. To show that μ is stable, we have to prove that it cannot be individually or pairwise blocked. Note that μ cannot be individually blocked since no man or woman is ever engaged under the DAA-MP to an unacceptable individual (of the opposite gender). To show that μ cannot be pairwise blocked, assume that there is some man m and woman w who are not matched under μ, while m strictly prefers w to his mate $\mu(m)$; i.e., $w \succ_m \mu(m)$. Since μ is not individually blocked, we must have $\mu(m) \succeq_m m$. It then follows that $w \succ_m m$ implying that w is acceptable to m. Since m is not matched to w, it must be that under the DAA-MP m had proposed to, and been rejected by, w before he was finally engaged with $\mu(m) \in W \cup \{m\}$. Because of the facts that w must have rejected m under the DAA-MP and that the welfare of w weakly improves at each step of this algorithm, she must be weakly strictly preferring her mate $\mu(w)$ to m; i.e., $\mu(w) \succeq_w m$. This means that m and w cannot block the matching μ. Since the matching μ cannot be individually or pairwise blocked, it is stable. ■

Any matching in the set of stable matchings, $\mathcal{M}^S$, cannot be individually or pairwise blocked. Moreover, any matching in $\mathcal{M}^S$ cannot be blocked by any group of individuals in $M \cup W$, as we will prove next.

We say that a matching μ' dominates another matching μ if there exists a coalition $T \subseteq M \cup W$, such that, for all men m and women w in T, $\mu'(m) \in T$, $\mu'(w) \in T$, $\mu'(m) \succ_m \mu(m)$, and $\mu'(w) \succ_w \mu(w)$.

Definition 17.3 Given a marriage market, the set of all matchings that cannot be dominated by any coalition of any size is called the core and denoted by C.

Proposition 17.2 ([1]) *The core of a marriage market coincides with the set of stable matching; i.e., $C = \mathcal{M}^S$.*

Proof Since the set of stable matchings contains matchings that cannot be dominated by coalitions involving one individual or two individuals only, it is clear that $C \subseteq \mathcal{M}^S$. To show that $\mathcal{M}^S \subseteq C$ is also true, let us pick a stable matching $\mu \in \mathcal{M}^S$ and suppose towards a contradiction that $\mu \notin C$, implying that there exists a coalition $T \subseteq M \cup W$, such that, for all men m and women w in T, $\mu'(m) \in T$, $\mu'(w) \in T$, $\mu'(m) \succ_m \mu(m)$, and $\mu'(w) \succ_w \mu(w)$. Now, pick any $w \in T$. Since $\mu \in \mathcal{M}^S$, we know that $\mu(w) \succeq_w w$. Also, since $\mu'(w) \succ_w \mu(w)$, we must have $\mu'(w) \in M$. Let $m = \mu'(w)$. Since $m \in T$ and $\mu'(m) \succ_m \mu(m)$ by our earlier supposition, it follows that m and w strictly prefer each other to their mates under μ, implying that μ can be blocked pairwise, which contradicts that μ is stable. Thus, $\mathcal{M}^S \subseteq C$. Together with $C \subseteq \mathcal{M}^S$, this implies that $C = \mathcal{M}^S$. ■

The above result holds as long as the preferences of individuals do not exhibit externalities; i.e., for each individual the desirability of possible mates is independent of the matchings taking place among other individuals.

One can ask here whether the existence result in Proposition 17.1 would be still valid if the marriage market did not contain two sides. Gale and Shapley [1] shows that the answer to this question is "no" if individuals are matched in a one-sided market, known as the roommate problem.

Example 17.4 There is a set of male students, $M = \{m_1, m_2, \ldots, m_n\}$, who can be matched in pairs to be roommates in a college dorm. Each student m has preferences $P(m)$ over the other students in M. The set $\langle M, (P(m))_{m \in M} \rangle$ defines a roommate problem. A matching in this problem is a partition of the students in M into pairs. A matching μ is stable if no two students who are not roommates both prefer each other to their roommates at μ.

Suppose that $n = 4$. Also, suppose that the preferences of students are as follows:

$$
\begin{aligned}
P(m_1) &= m_2, m_3, m_4, m_1 \\
P(m_2) &= m_3, m_1, m_4, m_2 \\
P(m_3) &= m_1, m_2, m_4, m_3 \\
P(m_4) &= [m_1, m_2, m_3], m_4
\end{aligned}
$$

Note that each student in $\{m_1, m_2, m_3\}$ is the first acceptable mate for some other student in $\{m_1, m_2, m_3\}$, while student m_4 is the last acceptable mate for all students in $\{m_1, m_2, m_3\}$. Therefore, no matching can be stable. This is because any matching must pair the student m_4 with some student $m \in \{m_1, m_2, m_3\}$ and that student m will be able to find a student m' in $\{m_1, m_2, m_3\}$ such that m is the first acceptable choice for m'. In that case, both m and m' will be strictly better off if they become a pair to share a room and thus to block the matching that pairs m_4 and m. ■

Example 17.4 shows that stable matchings need not exist in one-sided assignment problems, called roommate problems. Alkan [2] showed with an example that stable matchings need not exist in three-sided assignment problems either. This result was later generalized by Boros et al. [3] to s-sided matching problems for any $s \geq 3$. Below, we use a simple example provided by them for the possible non-existence of stable matchings in three-sided assignment problems.

Example 17.5 ([3], pp. 3–4) Consider a three-sided marriage market $\langle M, W, C, \mathcal{P}\rangle$ where $M = \{m_1, m_2\}$ is the set of men, $W = \{w_1, w_2\}$ is the set of women, $C = \{c_1, c_2\}$ is the set of children, and $\mathcal{P}$ denotes the sets of preferences given by

$$
\begin{aligned}
P(m_1) &= (w_2, c_1),\ (w_1, c_2),\ (w_2, c_2),\ (w_1, c_1),\\
P(m_2) &= (w_2, c_1),\ (w_1, c_1),\ (w_2, c_2),\ (w_1, c_2),\\
P(w_1) &= (c_2, m_2),\ (c_1, m_2),\ (c_1, m_1),\ (c_2, m_1),\\
P(w_2) &= (c_2, m_2),\ (c_1, m_1),\ (c_1, m_2),\ (c_2, m_1),\\
P(c_1) &= (m_1, w_2),\ (m_2, w_1),\ (m_1, w_1),\ (m_2, w_2),\\
P(c_2) &= (m_1, w_1),\ (m_1, w_2),\ (m_2, w_2),\ (m_2, w_1).
\end{aligned}
$$

A family is a triple $(m, w, c) \in M \times W \times C$. We say that a triple (m, w, c) is the family of individual $x \in M \cup W \cup C$ if $x \in \{m, w, c\}$. There are 8 possible families. A matching involves two families, since there are 6 individuals in the society $M \cup W \cup C$. Thus, 8 possible families can be grouped into four different matchings given by

$$
\begin{aligned}
\mathcal{M}_1 &= \{(m_1, w_1, c_1), (m_2, w_2, c_2)\},\\
\mathcal{M}_2 &= \{(m_1, w_2, c_2), (m_2, w_1, c_1)\},\\
\mathcal{M}_3 &= \{(m_1, w_2, c_1), (m_2, w_1, c_2)\},\\
\mathcal{M}_4 &= \{(m_1, w_1, c_2), (m_2, w_2, c_1)\}.
\end{aligned}
$$

A triplet $(m, w, c) \in M \times W \times C$ can block a matching $\mathcal{M}_i$, where $i \in \{1, \dots, 4\}$, if all individuals $x \in \{m, w, c\}$ strictly prefer the family (m, w, c) to their families in $\mathcal{M}_i$. One can easily check that each of the matchings, $\mathcal{M}_1, \mathcal{M}_2, \mathcal{M}_3, \mathcal{M}_4$ can be blocked by a triple of man, woman, and child. For example, $\mathcal{M}_1$ can be blocked by (m_2, w_1, c_1), $\mathcal{M}_2$ and $\mathcal{M}_4$ can be blocked by (m_1, w_2, c_1), and $\mathcal{M}_3$ can be blocked by (m_2, w_2, c_2). Therefore, no stable matching exists. ■

Examples 17.4 and 17.5, together with the more general results of Boros et al. [3] for multi-sided matching problems, show that the existence result of Gale and Shapley [1] for (two-sided) marriage markets cannot be extended to markets that involve less than, or more than, two sides (groups of individuals).

Now, we can return back to two-sided matchings. Since in a marriage market the set of stable matchings need not be a singleton set like in the simple case of Example 17.2, a natural question is whether the set of stable matchings, whenever it is not a singleton, must always contain some members that are optimal for all individuals or at least for one of the gender groups in the marriage market.

Definition 17.4 Given a marriage market $\langle M, W, \mathcal{P}\rangle$, a stable matching μ is *men-optimal* if $\mu \succeq_m \mu'$ for all $m \in M$ and all $\mu' \in \mathcal{M}^S$ and *women-optimal* if $\mu \succeq_w \mu'$ for all $w \in W$ and all $\mu' \in \mathcal{M}^S$.

Definition 17.5 Given a marriage market $\langle M, W, \mathcal{P}\rangle$, a pair of man and woman $(m, w) \in M \times W$ are *achievable* for each other if there exists a stable matching $\mu \in \mathcal{M}^S$ such that $\mu(m) = w$.

Now, we are ready to state and prove the following result.

Proposition 17.3 ([1]) *In any marriage market where all men and women have strict preferences, a men-optimal stable matching and a women-optimal stable matching exist (and they are unique). Moreover, the outcomes of the DAA-MP and the DAA-WP are the men-optimal stable matching and a women-optimal stable matching, respectively.*

Proof Pick any marriage market $\langle M, W, \mathcal{P}\rangle$, where all men and women have strict preferences. Consider first the DAA-MP, the deferred acceptance algorithm with men proposing. We will below prove by induction that no man is rejected at any step by an achievable woman.

Assume that the induction statement, given by "no man has yet been rejected by a woman who is achievable for him" is true up to some step $k > 1$. Let us show that this statement is true at step k, as well. (If no men is rejected by any woman at step k, then there is nothing to prove.) Suppose that some man m is rejected by some woman w at step k. There are two possibilities:

Possibility 1: Man m made a proposal to woman w at step k, but he was rejected since m is unacceptable for w; i.e., $w \succ_w m$. In that case, w is not achievable for m; therefore the induction statement is true at step k.

Possibility 2: Man m was engaged to woman w at the end of step $k - 1$ and rejected by her at step k since a man $m' \in M$ that w strictly prefers to m made a proposal to, and engaged with, w at step k. Since in the DAA-MP each man proposes according to his preference list starting from the most preferred woman, we know that m' prefers w to any woman in his preference list, except for the women who have rejected him in steps 1 through $k - 1$, and hence are unachievable for m' by the inductive assumption. Thus, any matching that matches m and w will be pairwise blocked by the pair (m', w), implying that there is no stable matching that matches m and w, further implying that w is not achievable for m. Hence, the inductive statement is true at step k, as well.

Thus, we have proved that no man is rejected at any step of the DAA-MP by an achievable woman. So, each man, who does not remain single under the DAA-MP, is matched with his most preferred achievable woman (who is unique since the preferences of men are strict). Therefore, the outcome of the DAA-MP is the (unique) men-optimal stable matching.

Finally, changing the roles of men and women above, we can show that the above proof is also true under the DAA-WP; i.e., the outcome of the DAA-WP is the (unique) women-optimal stable matching. ■

While there exists an optimal stable matching for each side of the marriage market, no stable matching can be optimal for both sides, as implied by the following result.

Proposition 17.4 ([4]) *Let $\langle M, W, \mathcal{P}\rangle$ be a marriage market where all men and women have strict preferences. Then, for any two stable matchings $\mu, \mu' \in \mathcal{M}^S$, we have $\mu \succeq_m \mu'$ for all $m \in M$ and $\mu \succ_m \mu'$ for some $m \in M$ if and only if $\mu' \succeq_w \mu$ for all $w \in W$ and $\mu' \succ_w \mu$ for some $w \in W$.*

Proof Let us prove "the only if" part of the proposition. Suppose there exist $\mu, \mu' \in \mathcal{M}^S$ such that $\mu \succeq_m \mu'$ for all $m \in M$ and $\mu \succ_m \mu'$ for some $m \in M$. Now, suppose that the following claim is wrong.

Claim (*): $\mu' \succeq_w \mu$ for all $w \in W$ and $\mu' \succ_w \mu$ for some $w \in W$.

Clearly, Claim (*) can be wrong only if one of the following possibilities is true.

Possibility 1: $\mu' \sim_w \mu$ for all $w \in W$.
This possibility cannot be true. The reason is that $\mu \succ_m \mu'$ for some $m \in M$ by supposition. So, there exists some $w \in W$ such that $\mu(m) = w$ and $\mu'(m) \neq w$, further implying that $\mu'(w) \neq m = \mu(w)$. Since the preferences of each woman is strict by assumption, $\mu'(w) \sim_w \mu(w)$ cannot be true for this particular woman, w.

Possibility 2: $\mu \succ_w \mu'$ for some $w \in W$.
Pick such a woman w. Then, $\mu(w) \succ_w \mu'(w)$. So, w must be matched to a man at μ, since μ' is stable (and hence it cannot be individually blocked by w) by assumption. Let $\mu(w) = m$. Since we know that $\mu(w) \neq \mu'(w)$, we also know that $m \neq \mu'(w)$, implying that $\mu'(m) \neq w = \mu(m)$. Since, the preferences of each man is strict by assumption, we must then have $\mu(m) \succ_m \mu'(m)$. It then follows that μ' can be blocked by the pair (m, w), contradicting that μ' is stable. Thus, Possibility 2 cannot be true, either.

Since neither Possibility 1 nor Possibility 2 can be true, Claim (*) cannot be wrong. Thus, we have proved that the "only if" part of the proposition holds. Finally, the proof of the "if part" can be symmetrically obtained from the above lines. ■

The above proposition directly implies that when all individuals in the marriage market have strict preferences, the men-optimal stable matching matches each woman with her least preferred choice in the set of achievable men and the women-optimal stable matching matches each man with his least preferred choice in the set of achievable women. Here, one can, of course, ask why we do not consider a criterion other than optimality to choose a particular matching in the set of stable matchings when it contains more than one member. An attractive criterion might be to choose the stable matching with the highest number of matched (non-single) individuals. Sadly, this appealing criterion is totally ineffective to distinguish among stable matchings. An interesting result of McVitie and Wilson [5], known as the lone wolf theorem or the rural hospital theorem, shows that the set of matched (non-single) individuals in a marriage market are the same in all stable matchings.

17.2 Many-to-One Matching

In this section, we will study many-to-one matchings in problems that involve a group of institutions and a group of individuals, such as firms and workers, hospital and interns, or colleges and students. Of these examples, let us focus on the matching problem between colleges and students, also known as college admissions market.

Formally, a college admissions market is denoted by the list $\langle C, S, q_C, \mathcal{P}\rangle$. The first two components are nonempty, finite and disjoint sets of colleges $C = \{c_1, \ldots, c_m\}$ and students $S = \{s_1, \ldots, s_n\}$. The third component is a set of positive integers $q_C = \{q_{c_1}, \ldots, q_{c_m}\}$, where q_c is the quota of college c. The last component is a set of preferences $\mathcal{P} = \{P(c_1), \ldots, P(c_m), P(s_1), \ldots, P(s_n)\}$. For any $c \in C$, $P(c)$ is an ordered list defined on the set 2^S. That is, each college orders all possible subsets of S. On the other hand, for any $s \in S$, $P(S)$ is an ordered list defined on the set $C \cup \{s\}$. That is, each student orders colleges in C and $\{s\}$, i.e., the prospect of being unmatched to any college. For simplicity, we assume that the preferences of all colleges and students are strict. We denote the binary strict preference relations of college $c \in C$ and student $s \in S$ by $\succ_c$ and $\succ_s$, respectively.

We also assume like [6] that the preferences of all colleges are responsive. That is, for any $c \in C$, the preferences $P(C)$ are such that

(i) for any $T \subset S$ with $|T| < q_c$ and $s \in S \setminus T$,

$$(T \cup \{s\}) \succ_c T \text{ if and only if } \{s\} \succ_c \emptyset, \text{ and}$$

(ii) for any $T \subset S$ with $|T| < q_c$ and $s, s' \in S \setminus T$,

$$(T \cup \{s\}) \succ_c (T \cup \{s'\}) \text{ if and only if } \{s\} \succ_c \{s'\}.$$

The above conditions require that the preferences of a college on the sets of students reflect an underlying ranking of individual students. (Thus, colleges perceive students as substitutes rather than complements.) Notice that the preferences of students over the individual colleges are trivially responsive.

Definition 17.6 A matching in the college admissions market $\langle C, S, q_C, \mathcal{P}\rangle$ is a correspondence μ from $C \cup S$ to $C \cup S$ such that

(i) $\mu(s) \subseteq C \cup \{s\}$ with $|\mu(s)| = 1$ for all $s \in S$

(ii) $\mu(c) \subseteq S$ with $|\mu(c)| \leq q_c$ for all $c \in C$, and

(iii) $\mu(s) = \{c\}$ if and only if $s \in \mu(c)$ for all $s \in S$ and $c \in C$.

Let $\mathcal{M}$ denote the set of all matchings in $\langle C, S, q_C, \mathcal{P}\rangle$. We say that a matching $\mu \in \mathcal{M}$ is

- blocked by a student $s \in S$ if $s \succ_s \mu(s)$,

- blocked by a college $c \in C$ if there exists $s \in \mu(c)$ such that $\emptyset \succ_c \{s\}$,

- blocked by a college-student pair $(c, s) \in C \times S$ if
  (i) $|\mu(c)| = q_c$, and there is $s' \in \mu(c)$ such that $\{s\} \succ_c \{s'\}$ and $c \succ_s \mu(s)$; or
  (ii) $|\mu(c)| < q_c$, $\{s\} \succ_c \emptyset$, and $c \succ_s \mu(s)$.

Definition 17.7 A matching $\mu \in \mathcal{M}$ is called stable if it cannot be blocked by any student, any college, and any college-student pair.

Example 17.6 Consider the college admissions market $\langle C, S, q_C, \mathcal{P}\rangle$ with $C = \{c_1, c_2, c_3\}$, $S = \{s_1, s_2, s_3\}$, $q_C = (q_{c_1}, q_{c_2}, q_{c_3}) = (2, 1, 2)$, and the set of preferences $\mathcal{P}$ given by:

$P(c_1)$	$P(c_2)$	$P(c_3)$	$P(s_1)$	$P(s_2)$	$P(s_3)$
$\{s_1, s_2\}$	$\{s_1\}$	$\{s_1, s_3\}$	c_1	c_3	c_1
$\{s_1, s_3\}$	$\{s_3\}$	$\{s_2, s_3\}$	c_2	c_2	c_2
$\{s_2, s_3\}$	$\{s_2\}$	$\{s_1, s_2\}$	s_1	c_1	c_3
$\{s_1\}$	$\emptyset$	$\{s_3\}$		s_2	s_3
$\{s_2\}$		$\{s_1\}$			
$\{s_3\}$		$\{s_2\}$			
$\emptyset$		$\emptyset$			

Consider first the following matching.

$$\mu = \begin{matrix} c_1 & c_2 & c_3 \\ \{s_2\} & \{s_1\} & \{s_3\} \end{matrix}$$

Note that μ cannot be individually blocked by any student since $\mu(s) \succ_s s$ for each $s \in S$. Also, μ cannot be individually blocked by any college since for each $c \in C$ we have $\mu(c) \in S$ and $\mu(c) \succ_c \emptyset$. However, μ can be blocked by the pair (c_1, s_1) since $|\mu(c_1)| = 1 < 2 = q_{c_1}$, $\{s_1\} \succ_{c_1} \emptyset$, and $c_1 \succ_{s_1} c_2 = \mu(s_1)$. Therefore, μ is not stable.

Now, consider the following matching.

$$\mu' = \begin{matrix} c_1 & c_2 & c_3 \\ \{s_1, s_3\} & \emptyset & \{s_2\} \end{matrix}$$

One can easily check that μ' is stable. It cannot be individually blocked by any student or any college, and it cannot be blocked by any college-student pair. (Note that μ' assigns s_1 and s_3 to their top-ranked college c_1 and s_2 to her top-ranked college c_3. Thus, no student would accept to be in a blocking pair.) ■

The Deferred Acceptance Algorithms in the Marriage Market can be extended to the College Admissions Problem. Again, there will be two versions of the algorithm depending upon whether students are applying to colleges or colleges are proposing to students.

The Deferred Acceptance Algorithm with Students Applying (DAA-SA). The algorithm has a number of steps. Let S_k denote the set of students who are not in the waiting list of any college at the beginning of step k (where $k \geq 1$ is an integer). At the beginning of step 1 no student is in the waiting list of any college; i.e., $S_1 = S$.

In step $k \geq 1$ of the algorithm,

first, each student in S_k applies to her most preferred college in the set of colleges acceptable to her, excluding those who rejected her in the previous steps (if $k > 1$),

next, each college c with quota q_c puts on its waiting list the most preferred q_c applicants from a pool of students including all acceptable applicants at step k and those on its waiting list, and it rejects all others,

finally, the set S_{k+1} is formed to include all students who are not in the waiting list of any college.

The algorithm stops at the end of step $K \geq 1$ such that either $S_{K+1} = \emptyset$ (all students are in the waiting list of some college at the end of step K) or each student in S_{K+1} has been rejected by every acceptable college in the first K steps of the algorithm. Consequently, all students in the waiting list of a college is accepted to that college, leading to the final outcome of the algorithm. ■

Example 17.7 Consider the college admissions market $\langle C, S, q_C, \mathcal{P} \rangle$ in Example 17.6. Let us run DAA-SA.

	c_1	c_2	c_3
Step 1 (Applications)	$\{s_1, s_3\}$	$\{\}$	$\{s_2\}$
Step 1 (Waiting Lists)	$\{s_1, s_3\}$	$\{\}$	$\{s_2\}$

The algorithm stops at this point since each student is in the waiting list of some college. Thus, the final outcome of the DAA-SA is the matching

$$\mu = \begin{matrix} c_1 & c_2 & c_3 \\ \{s_1, s_3\} & \emptyset & \{s_2\} \end{matrix}$$

at which c_1 fills its quota, c_2 is not matched with any student, and c_3 has an unfilled seat. ■

By changing the roles of students and colleges, the DAA-SA can be reformulated as an algorithm in which colleges are proposing; i.e., the Deferred Acceptance Algorithm with Colleges Proposing (DAA-CP). In this alternative algorithm, students are giving temporary acceptances to the acceptable colleges proposing at each step and a college who cannot (temporarily) fill its quota (seats) in any step must consider new proposals for the unfilled seats until this college either fills all of its seats or is rejected by all acceptable students.

Example 17.8 Consider the college admissions market $\langle C, S, q_C, \mathcal{P}\rangle$ in Example 17.6. Let us run DAA-CP on this market.

	s_1	s_2	s_3
Step 1 (Proposals)	$\{c_1, c_2, c_3\}$	$\{c_1\}$	$\{c_3\}$
Step 1 (Tentative Acceptances)	$\{c_1\}$	$\{c_1\}$	$\{c_3\}$
Step 2 (Proposals)	$\{c_1\}$	$\{c_1, c_3\}$	$\{c_2, c_3\}$
Step 2 (Tentative Acceptances)	$\{c_1\}$	$\{c_3\}$	$\{c_2\}$
Step 3 (Proposals)	$\{c_1\}$	$\{c_3\}$	$\{c_1, c_2\}$
Step 3 (Tentative Acceptances)	$\{c_1\}$	$\{c_3\}$	$\{c_1\}$
Step 4 (Proposals)	$\{c_1\}$	$\{c_2, c_3\}$	$\{c_1\}$
Step 4 (Tentative Acceptances)	$\{c_1\}$	$\{c_3\}$	$\{c_1\}$

The algorithm stops at this point. The final outcome of the DAA-CP is the matching

$$\mu = \begin{matrix} c_1 & c_2 & c_3 \\ \{s_1, s_3\} & \emptyset & \{s_2\} \end{matrix}$$

which is also the stable matching produced by the DAA-SA. ■

As informally argued by Gale and Shapley [1] and formally proposed by Gale and Sotomayor [7], there is an alternative representation of a college admissions market. In this representation, a college c with quota q_c can be represented by q_c replicas, $(c^1, c^2, \ldots, c^{q_c})$, each of which has the preferences of college c over individual students in S and aims to be matched with one acceptable student. In line with this representation, a college c can be replaced in the preferences of students, wherever it appears, by the ordered list $c^1, c^2, \ldots, c^{q_c}$, where each student strictly prefers c^k to c^l if $k < l$. Using this representation, it is possible to construct a marriage market from any college admissions market. When the preferences of colleges satisfy the responsiveness assumption of Roth [6], there is a one-to-one correspondence between the stable matchings of the college admissions market and the associated marriage market. In line with this result of Roth [6], we illustrate in the following example that to find the outcome of a deferred acceptance algorithm on a college admissions market, it is sufficient to apply the algorithm on the associated marriage market.

Example 17.9 We reconsider the college admissions market $\langle C, S, q_C, \mathcal{P}\rangle$ in Example 16.6. We denote the associate marriage market by $\langle M, W, \mathcal{P}'\rangle$ where $M = \{c_1^1, c_1^2, c_2, c_3^1, c_3^2\}$, $W = \{s_1, s_2, s_3\}$, and $\mathcal{P}'$ is the set of preferences that satisfy:

$P'(c_1^1)$	$P'(c_1^2)$	$P'(c_2)$	$P'(c_3^1)$	$P'(c_3^2)$	$P'(s_1)$	$P'(s_2)$	$P'(s_3)$
s_1	s_1	s_1	s_3	s_3	c_1^1	c_3^1	c_1^1
s_2	s_2	s_3	s_1	s_1	c_1^2	c_3^2	c_1^2
s_3	s_3	s_2	s_2	s_2	c_2	c_2	c_2
					$\emptyset$	c_1^1	c_3^1
					$\emptyset$	c_1^2	c_3^2

To find the outcome of the Deferred Acceptance Algorithm with Colleges Proposing (DAA-CP) on the college admissions market $\langle C, S, q_C, \mathcal{P}\rangle$, we can run the Deferred Acceptance Algorithm with Men Proposing (DAA-MP) on the associated marriage market $\langle M, W, \mathcal{P}'\rangle$ to obtain:

	s_1	s_2	s_3
Step 1 (Proposals)	$\{c_1^1, c_1^2, c_2\}$	$\{\}$	$\{c_1^3, c_3^2\}$
Step 1 (Matchings)	$\{c_1^1\}$	$\{\}$	$\{c_1^3\}$
Step 2 (Proposals)	$\{c_1^1, c_3^2\}$	$\{c_1^2\}$	$\{c_1^3, c_2\}$
Step 2 (Matchings)	$\{c_1^1\}$	$\{c_1^2\}$	$\{c_2\}$
Step 3 (Proposals)	$\{c_1^1, c_3^1\}$	$\{c_1^2, c_3^2\}$	$\{c_2\}$
Step 3 (Matchings)	$\{c_1^1\}$	$\{c_3^2\}$	$\{c_2\}$
Step 4 (Proposals)	$\{c_1^1\}$	$\{c_3^1, c_3^2\}$	$\{c_1^2, c_2\}$
Step 4 (Matchings)	$\{c_1^1\}$	$\{c_3^1\}$	$\{c_1^2\}$
Step 5 (Proposals)	$\{c_1^1\}$	$\{c_3^1, c_2\}$	$\{c_1^2\}$
Step 5 (Matchings)	$\{c_1^1\}$	$\{c_3^1\}$	$\{c_1^2\}$

The algorithm stops at this point. The resulting matching μ is such that $\mu(s_1) = c_1^1$, $\mu(s_2) = c_3^1$, and $\mu(s_3) = c_1^2$.

Likewise, to find the outcome of the Deferred Acceptance Algorithm with Students Applying (DAA-SA) on the college admissions market $\langle C, S, q_C, \mathcal{P}\rangle$, we can run the Deferred Acceptance Algorithm with Women Proposing (DAA-WP) on the associated marriage market $\langle M, W, \mathcal{P}'\rangle$ to obtain:

	c_1^1	c_1^2	c_2	c_3^1	c_3^2
Step 1 (Proposals)	$\{s_1, s_3\}$	$\{\}$	$\{\}$	$\{s_2\}$	$\{\}$
Step 1 (Matchings)	$\{s_1\}$	$\{\}$	$\{\}$	$\{s_2\}$	$\{\}$
Step 2 (Proposals)	$\{s_1\}$	$\{s_3\}$	$\{\}$	$\{s_2\}$	$\{\}$
Step 2 (Matchings)	$\{s_1\}$	$\{s_3\}$	$\{\}$	$\{s_2\}$	$\{\}$

The algorithm stops at this point. The resulting matching μ is such that $\mu(s_1) = c_1^1$, $\mu(s_2) = c_3^1$, and $\mu(s_3) = c_1^2$.

One can easily check that the stable matchings that we have found above by applying the DAA-MP and the DAA-WP on the marriage market $\langle M, W, \mathcal{P}' \rangle$ are the same and coincide with the stable matching calculated in Example 17.8 by applying the DAA-CP and the DAA-SA on the college admissions market $\langle C, S, q_C, \mathcal{P} \rangle$. ■

Many results obtained for a marriage market remain to hold for a college admissions market, as well. For example, if the preferences of colleges and students are strict and responsive, then a stable matching which is optimal for colleges as well as a stable matching which is optimal for students always exist. Moreover, these two stable matchings are produced by the deferred acceptance algorithms DAA-CP and DAA-SA, respectively.

The lone wolf (rural hospital) theorem of McVitie and Wilson [5] obtained for marriage markets has a stronger version for college admissions markets, as proposed by Roth [8]: Any college that does not fill all of its seats at some stable matching is matched to the same set of students at each stable matching.

17.3 Housing Market

In this section, we will study a particular one-to-one matching problem, known as the housing market, which was first studied by Shapley and Scarf [9]. This problem deals with reallocating the houses of a finite number of individuals among them in a stable way.

Formally, a housing market is denoted by the list $\langle N, H, \mathcal{P}, \omega \rangle$, where $N = \{1, \ldots, n\}$ is the set of individuals, $H = \{h_1, \ldots, h_n\}$ is the set of houses, $\mathcal{P} = \{P_1, \ldots, P_n)$ is the list of strict preferences of individuals over houses, and $\omega : N \to H$ is a bijection from N to H, prescribing the initial endowments of houses. A (house) matching is a bijection from N to H. We let $\mathcal{M}$ denote the set of all matchings and denote any matching $\mu \in \mathcal{M}$ by a permutation $\hat{h}_1, \ldots, \hat{h}_n$ of houses in H such that $\mu(i) = \hat{h}_i$ for all $i \in N$. Finally, we denote the strict preference P_i of individual $i \in N$ by a permutation $\tilde{h}_1, \ldots, \tilde{h}_n$ of houses in H such that $\tilde{h}_k \succ_i \tilde{h}_l$ (individual i strictly prefers $\tilde{h}_k$ to $\tilde{h}_l$) for any $k, l \in \{1, \ldots, n\}$ with $k > l$.

Definition 17.8 Given a housing market $\langle N, H, \mathcal{P}, \omega \rangle$, a coalition of individuals $S \subseteq N$ strongly blocks a matching μ via another matching μ' if

(i) $\{\mu'(i) : i \in S\} = \{\omega(i) : i \in S\}$,

(ii) $\mu'(i) \succ_i \mu(i)$ for each $i \in S$.

Condition (i) says that the coalition S can enforce μ' by reallocating the endowments of the individuals in S, and condition (ii) says that every member of S strictly prefers the house she is assigned at μ' to the house at μ.

The weak core of a housing market is the set of matchings that are not strongly blocked. Shapley and Scarf [9] showed that the weak core of a housing market is always nonempty and a matching in the weak core can be found by David Gale's *Top Trading Cycles* (TTC) Algorithm.

Definition 17.9 *Top Trading Cycles* (TTC) Algorithm. Each individual $i \in N$ points to her most preferred house in H and each house $h \in H$ points back to its owner, i.e., the individual $j \in N$ such that $\omega(j) = h$. This creates a directed graph. All cycles in this graph are identified. In each cycle, each individual is assigned to the house she points at and removed from the housing market with her assigned house. The algorithm is iterated until no individual (or unassigned house) remains in the market.

Example 17.10 Consider a housing market $\langle N, H, \mathcal{P}, \omega\rangle$, where $N = \{1, \ldots, 7\}$, $H = \{h_1, \ldots, h_7\}$, μ satisfy $\mu(i) = h_i$ for each $i \in N$, and $\mathcal{P} = (P_1, \ldots, P_7)$ is given by:

$$\begin{aligned}
P_1 &: \{h_3, h_5, h_2, h_4, h_1, h_6, h_7\}\\
P_2 &: \{h_7, h_5, h_3, h_1, h_4, h_2, h_6\}\\
P_3 &: \{h_2, h_7, h_4, h_6, h_3, h_5, h_1\}\\
P_4 &: \{h_5, h_2, h_1, h_4, h_7, h_3, h_6\}\\
P_5 &: \{h_2, h_6, h_3, h_7, h_1, h_4, h_5\}\\
P_6 &: \{h_1, h_4, h_7, h_3, h_5, h_6, h_2\}\\
P_7 &: \{h_3, h_4, h_2, h_6, h_7, h_1, h_5\}
\end{aligned}$$

If we run the TTC algorithm, we get:

$$\text{Cycle 1: } 3 \to h_2 \to 2 \to h_7 \to 7 \to h_3 \to 3.$$

So, 3 gets h_2, 2 gets h_7, 7 gets h_3, and they are removed from the market. Individuals 1, 4, 5, and 6 are left in the market with the updated preferences on the remaining set of houses $\{h_1, h_4, h_5, h_6\}$:

$$\begin{aligned}
P_1' &: \{h_5, h_4, h_1, h_6\}\\
P_4' &: \{h_5, h_1, h_4, h_6\}\\
P_5' &: \{h_6, h_1, h_4, h_5\}\\
P_6' &: \{h_1, h_4, h_5, h_6\}
\end{aligned}$$

These preferences yield

$$\text{Cycle 2: } 1 \to h_5 \to 5 \to h_6 \to 6 \to h_1 \to 1.$$

So, 1 gets h_5, 5 gets h_6, 6 gets h_1, and they are all removed from the market. Individual 4 is left in the market with the updated preference on the remaining house h_4:

$$P_4'' : \{h_4\}$$

This preference yields

$$\text{Cycle 3: } 4 \rightarrow h_4 \rightarrow 4.$$

So, 4 gets h_4 and is removed from the market. There is no individual (unassigned house) is left in the market. Thus, the algorithm terminates. The resulting matching is $\mu = (h_5, h_7, h_2, h_4, h_6, h_1, h_3)$. ■

References

1. Gale D, Shapley L (1962) College admissions and the stability of marriage. Am Math Mon 69:9–15
2. Alkan A (1988) Nonexistence of stable threesome matchings. Math Soc Sci 16:207–209
3. Boros E, Gurvich V, Jaslar S, Krasner D (2004) Stable matchings in three-sided systems with cyclic preferences. Discrete Math 289:1–10
4. Knuth DE (1976) Marriages stables. Les Presses de l'Université de Montreal, Montreal
5. McVitie DG, Wilson LB (1970) Stable marriage assignment for unequal sets. BIT Numer Math 10:295–309
6. Roth AE (1985) The college admissions problem is not equivalent to the marriage problem. J Econ Theory 36:277–288
7. Gale D, Sotomayor M (1983) Some remarks on the stable matching problem. Mimeo
8. Roth AE (1986) On the allocation of residents to rural hospitals: a general property of two-sided matching markets. Econometrica 54:425–427
9. Shapley L, Scarf H (1974) On cores and indivisibility. J Math Econ 1(1):23–37

Index

I. Saglam, *Mastering Game Theory*, Classroom Companion: Economics,
https://doi.org/10.1007/978-3-031-82094-6

9783031820939